Ethno-mycological Studies

No. 1

SOMA

DIVINE MUSHROOM OF IMMORTALITY

by

R. Gordon Wasson

HARCOURT BRACE JOVANOVICH, INC.

B. 3.73

ISBN 0-15-683800-1

LIBRARY OF CONGRESS CATALOG CARD NUMBER: 68-11197

PRINTED IN ITALY

TABLE OF CONTENTS

PART ONE

SOMA: DIVINE MUSHROOM OF IMMORTALITY

PART TWO

THE POST-VEDIC HISTORY OF THE SOMA PLANT

by Wendy Doniger O'Flaherty

PART THREE

NORTHERN EURASIA AND THE FLY-AGARIC

★

EXHIBITS AND INDEX

1. THE FLY-AGARIC IN SIBERIA

2. THE FLY-AGARIC IN SCANDINAVIAN WRITINGS

ILLUSTRATIONS

PLATES

ILLUSTRATIONS IN THE TEXT

ILLUSTRATIONS

MAPS AND LINGUISTIC CHART
accompanying Part III

PART ONE

SOMA

DIVINE MUSHROOM OF IMMORTALITY

I

THE PROBLEM

IN the second millennium before our Christian era a people who called themselves 'Aryans' swept down from the Northwest into what is now Afghanistan and the Valley of the Indus. They were a warrior people, fighting with horse-drawn chariots; a grain-growing people; a people for whom animal breeding, especially cattle, was of primary importance; finally, a people whose language was Indo-European, the Vedic tongue, the parent of classical Sanskrit, a collateral ancestor of our European languages. They were also heirs to a tribal religion, with an hereditary priesthood, elaborate and sometimes bizarre rituals and sacrifices, a pantheon with a full complement of gods and other supernatural spirits, and a mythology rich with the doings of these deities. Indra, mighty with his thunderbolt, was their chief god, and Agni, the god of fire, also evoked conspicuous homage. There were other gods too numerous to mention here.

Unique among these other gods was Soma. Soma was at the same time a god, a plant, and the juice of that plant. So far as we know now, Soma is the only plant that man has ever deified. (The Mexican Indians seem to regard the hallucinogenic plants, whether mushrooms, *peyotl*, or morning glories, as mediators with god, not as a god. The Nahua – Aztecs and other groups speaking the same tongue – called the mushrooms *teo-nanácatl*, 'god's flesh', but the mushrooms do not figure in their pantheon.) In the course of the Soma sacrifice the juice was pounded out with stones on resounding planks and was drunk by the officiating priests. Soma – the three Somas – inspired hymns vibrant with ecstasy, composed over centuries by priests who lived in centers remote from each other. In the end, at an early period in the first millennium before Christ, these hymns were gathered together, and the canon of that text has come down to us intact. Compared with ours, the Vedic civilization was simple, but their verses – the figures of speech with which they embellished their thoughts and feelings, their play with the meanings and sounds of words, the rules of their prosody–were sometimes subtle and sophisticated. Some of the hymns are of so exalted, even delirious, a tenor that the modern reader is led to exclaim: 'This surely was composed under the influence of a divine inebriant'. It takes little perception to sense the difference in tone between the awe-inspired hymns to Soma and the rowdy drinking songs of the West prompted by alcohol.

In the hierarchy of Vedic gods certain others took precedence over Soma, but since Soma was a tangible, visible thing, its inebriating juice to be ingested by the human organism in the course of the ritual, a god come down and manifesting himself to the Aryans, Soma played a singular rôle in the Vedic pantheon. The poets never tire of stressing Soma's sensuous appeal. In appearance it was brilliant, reminding them of the Sun, of Fire, of the rays of the Sun, of the round bowl of the heavenly firmament, 'the back of the sky'. The dried plants were first freshened with water. They were then macerated with stone pestles, and the tawny yellow juice as it came coursing through the conduits of the press was compared, in the hyperbole of the day, with thunder.[1] The priests, after imbibing the juice, seem to have known, for the nonce, the ecstasy of existence in the World of the Immortals. The Divine Element was not just a symbol of a spiritual truth as in the Christian communion: Soma was a miraculous drink that spoke for itself.

A book of 1028 hymns is our sole contemporary source of information about the period it deals with. There is agreement on the text, but many words are of doubtful meaning and doubly so in their allusions that often escape us. They have been preserved for us, both words and melodies, by oral tradition; by achievements of the human memory that have no parallel in other cultures, preserved better than our ancient learning in the manuscripts with which our Western scholars are familiar. When a Brahman schooled in the ancient tradition sings us a hymn from the ṚgVeda, with intonation precisely right, there is reason to think that it is as though we were listening to a tape recording 3,000 years old. These hymns, still elicit-

1. Parjanya, the god of thunder, was the father of Soma, according to one tradition, ṚgVeda IX 82[3]; Abel Bergaigne, *La Religion Védique*, Vol. 1, pp. 172-3.

ing the utmost reverence among countless Hindus, constitute the earliest monument in their religious and literary heritage, as well as one of the earliest cultural legacies of our own Indo-European world. As the eminent French Vedist Louis Renou has said, the whole of this immense collection, known as the ṚgVeda, is present *in nuce* in the themes that Soma presents to us.[1]

But what manner of plant was this Soma? No one knows. For twenty-five centuries and more its identity has been lost. The Hindus, probably for the reasons that will emerge as my argument progresses, allowed this authentic Soma to fall into disuse and early on began to resort to sundry substitutes, substitutes that were frankly recognized as such and that to this day are met with in India in their peculiar religious rôle. The West discovered the Sanskrit and Vedic cultures almost two centuries ago. For two centuries we have been absorbing the lessons that India has to teach us, in linguistics, in mythology and philosophy and religion, in literature, art, history, anthropology, archæology. This difficult assimilation is still going on, though possibly at a slower, more sober pace than in the enthusiasm of the initial discoveries.

But the identity of Soma is as obscure today as two centuries ago, and what is more, this mystery is compounded by what I will call the mystery of the mystery of Soma. When I first embarked on the problem in 1963, I could hardly believe what I found to be the situation. Here is a clear-cut botanical question – a psychotropic plant that calls for identification. The clues should be in the Vedic hymns. True, the poems contain no botanical description such as a scientist would ask for. Those remote singers were no modern botanists. The poet-priests of the ṚgVeda were composing for contemporaries, certainly not for posterity 3,000 years away, and their imagery and terms often elude our understanding, just as our hymns would make hard going for our collateral descendants 3,000 years hence. But the hymns are all shot through with Soma, and 120 of them are entirely devoted to the plant-god. The hymns are not make-believe about a fictitious plant: if ever there was genuine poetry about a genuine plant, here

1. *Etudes Védiques et Pāṇinéenes*, E. de Boccard, Editeur, Paris, tome IX, 1961, p. 8.

it is. Was it possible that so much could have been written by lyric poets about a plant, over centuries, in many centers of priestly activity, and its identity not be revealed? It was no secret for the poet-priests. How extraordinary it would have been if all of them, numbering scores, perhaps a hundred, had withheld from their verses the revealing descriptive terms, the tell-tale metaphors, that the trained reader today needs to spot the plant! But this did not happen. All that has happened is that no ethno-botanist with an interest in psychotropic plants has applied himself to the examination of the hymns.

The Vedic culture has been primarily a subject for Vedic scholars and secondarily for those scholars who interest themselves in comparative mythology and religion. The specialists in Vedic learning have included a handful of brilliant men, of extraordinary perception, who are assured of a lasting reputation by the works they have left behind. They have been tilling a field, a rich field it is true, in a remote frontier province of the humanities, and they must often have felt that they were leading an isolated existence. The early translations of the ṚgVeda were execrable. Only now, in the annotated translations of the late K. F. Geldner and the late Louis Renou, does a student with no Vedic possess a rendering that permits him to sense some of the nuances of the original.[1] (Certain it is that I, for one, could have done nothing without Renou's translation and commentary.[2]) Throughout the dawn and early morning of Vedic studies, in the 19th century, many cultivated people in the West, not merely Vedic scholars but educated laymen, knew at least what the Soma problem was. More recently, whenever the subject has been broached, scholars and editors have sighed and wearily turned away: the Soma question has dropped into oblivion simply because it has seemed to defy solution. One would have thought that students of religion

1. In *The Soma-Hymns of the ṚgVeda* S. S. Bhawe has begun an excellent translation into English, with a lengthy commentary. Oriental Institute, Baroda, 1957, 1960, 1962. He has published 70 of the hymns to Soma, out of 114 in Maṇḍala IX.

2. His translation, so far as it had gone before his untimely death in August 1966, had been completed before I had an opportunity to lay before him my thesis, in three long talks that I had with him in Paris and his country house in the Eure, in the spring of that year. He found my thesis *séduisante* and encouraged me to prepare as quickly as possible a full-dress presentation, but naturally he did not commit himself.

would have focussed on it, but our contemporary religious thinkers (with notable exceptions) have no sympathy with drugs as a way to religious experience, though accepting a rôle for other external aids such as music, architecture, and liturgy; and (unconsciously of course) many of them are nothing loath to gloss over and ignore the historical rôle of Soma in the higher reaches of our own religious history. Yet it cannot be gainsaid that Vedic culture with Soma unidentified is the play of Hamlet with Hamlet left out. The identification of Soma remains one of the chief desiderata of Vedic studies. That the question continues to be slighted constitutes, as I have said, the mystery of the Soma enigma.

There is no sure evidence that in Vedic times Soma was drunk by others than the priests. By a deliberate decision the priests must finally have decided to discontinue the use of the Sacred Plant. That familiarity with the genuine Soma seems to have been confined to the priesthood would make the enforcement of such a decision effective, and the very memory of the plant itself was finally lost. For more than two millennia the hymns to Soma have been sung and revered in the absence of Soma. The record of these intervening centuries is strangely, significantly, silent about the holy herb. In India discussions of such matters were confined to the priestly circles: others were uninformed and uninterested. The priestly compositions that have come down to us show acute concern over the relative merits of various substitutes, but no concern over the plant itself. Moreover, the Hindus traditionally have not been interested in the facts of history, considering them irrelevant when weighed in the balance against the future incarnation of the individual's soul. Brahmans among my friends in India have assured me that the many substitutes for Soma used over the large expanse of the sub-continent and through the ages were always known by the inner circles of the priestly caste to be substitutes. This cannot be proved but must have been a fact. The contrast between the ecstasy of Soma inebriation as sung in the hymns and the effects, often vile, of any of the many substitutes was always too glaring to be ignored.

To the Europeans knowledge about Soma came late. It was not

until 1784 that the word itself, spelled *Sōm*, made its début on the European scene, in Charles Wilkins's English translation of the *Bhagavad Gītā*. (The earliest citation for Soma that the Oxford Dictionary gave was 1827.) In the 1830's European scholars finally awoke to the mystery of Soma's identity, but their efforts to solve the problem from that day to this have been singularly futile. They have merely worked over the Brahmanic discussions of the substitutes, apparently under the misapprehension that the original Soma must have closely resembled the substitutes, if it was not in fact one of them. Or they volunteered suggestions pulled out of the air, flouting the ṚgVedic text and Brahmanic practice. In *Part Two* of this book Wendy Doniger O'Flaherty gives the revealing story of this search.

And so in the fullness of time it has fallen on me, a retired banker, to be greatly daring and to submit to the intellectual world a new theory of Soma's identity. As I entered into the extraordinary world of the ṚgVeda, a suspicion gradually came over me, a suspicion that grew into a conviction: I recognized the plant that had enraptured the poets. For this purpose the text of the hymns, the epithets and tropes pertaining to the plant, are abundantly clear. As I went on to the end, as I immersed myself ever deeper in the world of Vedic mythology, further evidence seeming to support my idea kept accumulating. By Jove, I said, this is familiar territory!

At the same time there hovered constantly in my mind's eye the admonitory finger of Tristram Shandy as he warned against the occupational hazard of those who advance hypotheses:

> It is in the nature of a hypothesis when once a man has conceived it, that it assimilates everything to itself, as proper nourishment, and from the first moment of your begetting it, it generally grows stronger by everything you see, hear or understand.

And so, while well aware of the perils that attend my enterprise, I present my case. If I am wrong, how quickly will my proposal be forgotten! If I am right, I claim the privilege of adding a new and exciting chapter to the world's knowledge of the remote past, the proto-history of our own Indo-European culture.

In a word, my belief is that Soma is the Divine Mushroom of Immortality, and that in the early days of our culture, before we made use of reading and writing, when the ṚgVeda was being composed, the prestige of this miraculous mushroom ran by word of mouth far and wide throughout Eurasia, well beyond the regions where it grew and was worshipped.

II

THE FLY-AGARIC OF EURASIA

My candidate for the identity of Soma is *Amanita muscaria* (Fr. ex L.) Quél., in English the fly-agaric, the *Fliegenpilz* of the Germans, the *mukhomor* of the Russians, the *fausse oronge* or *tue-mouche* or *crapaudin* of the French, the brilliant red mushroom with white spots familiar in forests and folklore throughout northern Eurasia.

This is the first time that a mushroom has been proposed in the Soma quest.

The fly-agaric is an inebriant but not alcoholic. As far back as our records go, it has been the Sacred Element in the shamanic rites of many tribes of northern Siberia, tribes that are concentrated in the valleys of the Ob and the Yenisei, and then, after an interruption, other tribes in the extreme northeast of Siberia. Apparently some of these tribesmen scarcely knew alcohol until the Russians introduced it in the 16th and 17th centuries, but the fly-agaric had been their precious possession long before then. The available records about its religious rôle are adequate to reveal its main properties but fall short of what we would have them be in the light of the proposal made in this paper. Our earliest eye-witness account of its use is by a Pole, Adam Kamieński, in 1658, among the Ostyak of the Irtysh River (tributary of the Ob), an Ugrian people of the Finno-Ugrian family.[1] Today we know its use is common to the Ostyak and their kin the Vogul, the Ket of the Yenisei Valley, the Samoyed peoples (who together with the Finno-Ugrians make up the Uralic group), and three sister tribes, unrelated linguistically to the others, on the north Pacific Coast, the Chukchi, the Koryak, and the Kamchadal. Responsible observers have reported that the Yukagir, who survive in Siberia in tiny communities near the Arctic Ocean, and the Inari Lapps in Finland, preserve oral traditions of having consumed the fly-agaric in times past, though they no longer do so.

1. *Vide* Exhibit [1], p. 233. Throughout this book the numbers between heavy brackets refer to the Exhibits.

The Russians began their conquest of Siberia at the end of the 16th century and our knowledge of these Siberian tribes virtually starts then. The use of the fly-agaric has been in retreat ever since the arrival of the White Man. It has ceded ground to the vodka of the Russians and, until the Russian Revolution, the fire-water of the western whalers. A more powerful influence has been the aggressive cultural thrust of the Russians, who frown on some of the practices peculiar to the natives. (In a conflict of cultures the stronger one always assumes right away, without examination, that all its ways, its vices as well as its virtues, are incomparably superior.) To the Russian conquest of Siberia, therefore, we owe our knowledge of the use of the fly-agaric and also the approaching end of that use. In the Exhibits we publish *in extenso* the accounts of travellers, linguists, and anthropologists that deal with the fly-agaric in Siberia. There is linguistic and folkloric evidence to indicate that formerly, in proto-history, it was used by some of the Indo-European peoples, as well as the Hungarians.

III

THE GROUND RULES OF THE SEARCH

My primary source, the basis of my identification of Soma, is the ṚgVeda, as made accessible to me in the recent annotated translations, to wit, those of Renou, Geldner, and Bhawe.[1] It is certain that the poets of the ṚgVeda knew the original Soma at first hand, and they never strayed from it for long. I invoke later texts and the Avesta only where they help us to know what the ṚgVeda means.

Any suggestion for the identity of Soma must meet the following criteria: 1) Is it in conflict with the ṚgVeda? 2) Does it fit comfortably into such descriptive terms as the ṚgVeda poets apply to Soma, and into the indications as to the source of supply, methods of handling, *etc.*? 3) Does the proposal happily resolve some of the many *cruces* in the ṚgVeda? On all three counts I am hopeful that my suggestion meets the requirements.

It was only after I had finished my examination of the ṚgVeda text, and after Dr. O'Flaherty had laid before me her paper written at my request on the history of Soma since Vedic times (here published as *Part Two*), that I realized how unorthodox I had been in my approach. It turns out that I was unique. Everyone else, literally everyone, had relied on later sources, sources composed at a time when the original Soma had been superseded by substitutes, and on plants called in various modern vernaculars by names derived from 'Soma'. Or they had had recourse to guesses, without supporting evidence; guesses such as rhubarb, or hashish, or wild grape wine, in disregard of their incompatibility with the ṚgVeda. At a later stage I shall have more to say on this. But here the question of substitutes calls for some discussion.

The Aryans came from the north but no one knows from where.

1. I exclude from consideration the latest hymns to have been written, the last to be included in the canon before it was closed. These hymns differ from the others considerably in tone and language, and there is reason to believe that substitutes, which I think had always been occasionally used, had now almost completely replaced Soma in the sacrifice. These hymns are mostly in Maṇḍala X from 85 through 191.

Thomas Burrow made headway on this question when in *The Sanskrit Language*[1] he marshalled the latest linguistic evidence about the pre-history of the Indo-European peoples. The Indo-Iranian branch of the race, before migrating to the areas that they were to conquer, had occupied the marches of the Indo-European domains and had lived in long and intimate contact with a race that spoke proto-Finno-Ugric, with whom they exchanged lexical elements.[2] The words that they borrowed from their Finno-Ugric neighbors do not appear in other Indo-European languages.

If as I believe the Aryans brought down with them from their homeland a cult of the sacred fly-agaric, it must needs follow that their priests had from the start wrestled with the problem of substitutes. The fly-agaric is not always available. Like most species of mushrooms, it does not lend itself to cultivation. The supply is therefore limited and varies with the season and the year. It can be dried and thus preserved: the ṚgVeda speaks on several occasions of water being added to the (presumably dry) Soma, so that it would swell up again.[3] The fly-agaric is a mycorrhizal mushroom: in Eurasia it grows only in an underground relationship with the pines, the firs, and above all the birches. Where these trees are not, neither does the fly-agaric grow. It can be transported, but channels of trade must be set up and maintained. A well organized priesthood would not allow the cult to falter for want of the Sacred Element, and in the Indus Valley settlements, far but not too far from birch and conifer, the use of substitutes, while always regrettable, must have been so common as to arouse no comment. This would make the final abandonment of the fly-agaric easier than if no substitutes had ever been used. Wendy O'Flaherty in her paper on the history of the Soma question has pointed out the consistent association, in early times, of the color red with the substitutes for Soma that were preferred – a significant pointer when the fly-agaric, flaming red, as the original Soma is under consideration. Friends in the Vaidika Saṁśodhana Maṇḍala in Poona, where such

1. Faber & Faber, London, 1955, Chap. 1.
2. *Ibid.*, pp. 23-27
3. *Vide, e.g.*, IX 74².

questions as this are studied, have told me that three criteria governed the choice of substitutes: the plant should be small, it should be leafless, and it should possess fleshy stalks. If these are ancient criteria, it is easy to see how they were suggested by the fly-agaric – small, leafless, with juicy stalk. That the criteria seem arbitrary and permitted the choice of plant substitutes bearing no outward resemblance to the fly-agaric is not incompatible with the sacerdotal habit of mind. The substitute had to be constantly available, and what plant that met this condition could resemble the fly-agaric? In the hot arid regions of northwest India, Afghanistan, and Iran the available species of three botanical genera, Ephedra, Sarcostemma, and Periploca, seemed to meet these requirements with fair success and they came to be widely used in the post-Vedic Soma sacrifice.

There has been an impression in India and elsewhere that Soma was a creeper, *vallī* in Sanskrit. No ṚgVedic authority exists for this term, in fact none in the whole corpus of Vedic literature. But the three genera that I have cited could be construed as *vallī* and therefore I refrain from contending that *vallī* was not used in ṚgVedic times. Indeed in the second half of the Maṇḍala X, in the last batch of hymns to be added before the canon was closed, there is a verse that may refer to such substitutes:

> X 85³
> One thinks one drinks Soma because a plant is crushed. The Soma that the Brahmans know – that no one drinks.
> *sómaṃ manyate papivā́n yát saṃpiṃṣánty óṣadhim*
> *sómaṃ yám brahmā́ṇo vidúr ná tásyāśnāti káś caná*

(For the convenience of scholars we shall always add the Vedic text of the verses that we quote.) This verse, and the late hymn of which it forms a part, may have been composed at the very moment when the original Soma had already fallen into disuse, when only priests still remembered what it was, and when substitutes were currently accepted as the genuine article.

IV

SOMA WAS NOT ALCOHOLIC

In the West it has been repeatedly suggested that Soma was an alcoholic drink. (The culture of the modern world has long been obsessed with alcohol as the sole inebriant.) This can be denied with assurance. The difference in tone between the bibulous verse of the West and the holy rapture of the Soma hymns will suffice for those of any literary discrimination or psychological insight. But there are criteria other than the literary and the subjective. The stalks were pressed as a liturgical act and before the liturgy was finished the juice was drunk. Three sacramental offerings could be made in one day. Even if we allow for the heat of the summer in the Indus Valley, fermentation could not have advanced far in a religious rite repeated thrice in a day. Moreover, those who know the fermenting process must find it hard, indeed impossible, to imagine anyone, no matter how far removed from our culture, waxing lyrical, even ecstatic, over a drink in active fermentation. It is not as though the Indo-Aryans were unfamiliar with fermented drinks. They had their *súrā*, mentioned several times in the ṚgVeda but without reverential pæans; quite the contrary:

> VII 86[6ab]
> Malice has not been of my own free will, O Varuṇa;
> it was *súrā*, anger, dice, a muddled head.
> *ná sá svó dákṣo varuṇa dhrútiḥ sā́ súrā manyúr*
> *vibhī́dako ácittiḥ*

A mere difference between fermented drinks would not cause the gulf that separates *súrā* from the Soma that inspired the great hymns.

Yet it has been suggested, as will be seen in *Part Two*, that recollections of the mead of their ancestors might have clothed this particular beverage for the Aryans with rich associations entitling it to a superior footing than other fermented drinks. But no one knows whether the ancestors of the Aryans had been drinking mead: many peoples knowing honey do not ferment it. In any case bees with their honey exist in the Indus Valley, and there would have been no reason to give up mead, if indeed it had been Soma, and even less to exchange it for the vile tasting substitutes that took its place. Honey, *mádhu*, is mentioned frequently in the ṚgVeda but mead never. Honey is cited for its sweetness and also is often applied as a metaphor of enhancement to Soma. There is reason to think it was used on occasions to mix with Soma, but the two were never confused. Honey offers no stalks to pound in the course of the liturgy, nor does it ferment in a twinkling.

The most astonishing of candidates for Soma was espoused by Sir Aurel Stein, the explorer-scholar, *viz.*, rhubarb. He had observed it growing wild in the mountains. According to Stein himself, no Indian in recorded history has made a fermented drink of rhubarb, though of course with the addition of sugar or honey the juice lends itself to fermentation. There is no reason to think any Indian ever did so. True, there are stalks to press, but the juice would not ferment and become alcoholic at once. All the rich Soma adjectives and metaphors of the ṚgVeda make odd reading when linked to rhubarb: resplendent, Born of Thunder, Mainstay of the Earth, Navel of the Way, Immortal Principle. Stein must have forgotten either his ṚgVeda or his sense of humour.

In 1921 an Indian advanced the notion that Soma, after all, was nothing but *bhang*, the Indian name for marijuana, *Cannabis sativa*, hemp, hashish. He conveniently ignored the fact that the ṚgVeda placed Soma only on the high mountains, whereas hemp grows everywhere; and that the virtue of Soma lay in the stalks, whereas it is the resin of the unripened pistillate buds of hashish that transport one into the beyond; or, much weaker, the leaves, which are never mentioned in the ṚgVeda. The stalks of hemp are woody.

The Indo-Iranians did not know the distillation process and therefore Soma could not have been a strong drink, *i.e.*, brandy or the distillate of grains. It should be unnecessary to argue this point, but scholars – Indian and Western alike – use the terms loosely, terms properly applicable only to distilled liquor. In IX 107[12] the poet compares Soma with the 'intoxicating one'; Geldner supplies *Branntwein* (brandy) as an example of what he might have meant, and Renou follows his lead with *alcool* (any distilled beverage). I doubt whether they would have made these suggestions had they been alert to the anachronism. V. S. Agrawala in his *India as Known to Pāṇini* seems to mean decanting when he mistakenly says distilling.[1] Dr. O'Flaherty in her account of the Soma search finds it necessary to record many such confusions and ambiguities.

The Orientalist Berthold Laufer wrote, 'Certain it is that distillation was a western invention, and was unknown to the ancient Chinese.'[2] The Dutch scholar R. J. Forbes in his *Short History of the Art of Distillation*[3] concludes that there is no evidence for distilled beverages before the distillation process was discovered and practiced around A. D. 1100 in Italy, probably at Salerno. ('Alcohol' is of Arabic origin but in Arabic it meant something other than what we mean by it.) The making of *aquavitæ* (as it was called for a long time, and still is, in parts of Europe) remained a secret of the alchemists and some monastic establishments for centuries, until the Reformation, when in the course of a decade, in the second quarter of the 16th century, thanks to the dissolution of the monasteries by the Protestants, the 'secret' became common property, and the product, after having changed hands for a king's ransom, sold suddenly dirt cheap. There was no contemporary comment on the social implications of this unanticipated fruit of religious reform.

1. 2nd edition, Varanasi, 1963, p. 121.
2. *Chinese Contributions to the History of Civilization in Ancient Iran*, Field Museum, Chicago, 1919, p. 238.
3. E. J. Brill, Leiden, 1948, pp. 32, 88-89.

V

THE ROOTS, LEAVES, BLOSSOMS, SEED OF SOMA: WHERE ARE THEY?

In the ṚgVeda (excluding the latter half of Maṇḍala X, last to be admitted to the canon) there is no reference to the root of the Soma plant, nor to its leaves, nor to its blossoms, nor to its seed. In a lengthy anthology of lyric poetry written over centuries in the far flung Valley of the Indus and its tributaries, how odd it is that no poet ever speaks of these conspicuous parts of almost all chlorophyll-bearing plants, not even casually or incidentally. There must have been a conspiracy of silence, laid down perhaps by the dictates of their very religion, but why?

Alternatively, they were speaking of a plant that had neither seed nor blossom nor leaf nor root; *viz.*, a mushroom. As for seed, there is positive evidence that Soma was thought to lack seed: Soma was procreated from on high, the Somic germ having been placed by the gods. Soma was divinely engendered:

> IX 83[3d]
> The [gods, those] fathers with a commanding glance, laid the [Somic] germ.
> *nṛcákṣasaḥ pitáro gárbham ā́ dadhuḥ*

In a world where farming was already well developed one would expect Soma to be cultivated. A plant with properties so extraordinary would elicit the utmost attention, though owing to its sacred character we would rather expect its growth to be confined to the gardens of the higher priesthood. But the fact is that there is never a mention of its cultivation. Perhaps it did not lend itself to man's efforts to make it grow. To this day the fly-agaric (like almost all other species of mushrooms) refuses to be cultivated. Even in the laboratory it will not sprout.

Note. While our argument in favour of the fly-agaric is founded squarely on the ṚgVeda, there is one verse in the Avesta that seems to speak of the 'trunks' and 'branches' of the sacred plant, and discussion of that verse is in order if only to forestall objections.

The Avesta is the Bible of the religion of Zoroaster (= 'Zarathustra'), a religion that lingers on to this day in the Parsi community centered in Bombay. Tradition has it that most of the Avesta was lost when Alexander the Great overran Iran, but the surviving fragments are still substantial. There is no consensus among scholars on the dates of Zoroaster. Some assign him to about the 10th century B.C., this early date being based on reasonable arguments derived from linguistics and comparative cultural studies. Others prefer to accept the tradition according to which he was living around B.C. 600, the proposed dates being 630-553, or 628-551, or 618-541. The *ipsissima verba* of the great Prophet are enshrined in the Avesta, but most of the text is of later date and some of its traditions stem back long before his time. Religiously and linguistically the Avesta and ṚgVeda are siblings. The text of the ṚgVeda is, however, much purer owing to its marvelous preservation through the ages by the disciplined human memory. The Avesta like the ṚgVeda knew an inebriating plant that was the object of worship, and the Avestan 'Haoma' and the Vedic Soma were certainly identical, at least at the start. Three chapters, Yasna 9, 10, and 11, consist of numinous phrases of adoration addressed to Haoma; collectively they are known as the Hōm Yašt. James Darmesteter, who gave us our standard translation of the Avesta, felt sure that the Hōm Yašt had been interpolated late into the text, between B.C. 140 and A.D. 50. One might expect its authority to be impugned by reason of this late date, but an interpolated text may incorporate ancient words and traditions, and precisely this is the situation here, the Hōm Yašt preserving for us in its words and matter some of the truly archaic elements in the Avesta, perhaps antedating Zoroaster himself. Thus it becomes necessary to examine Yasna 10.5, which I give below in the original Iranian dialect peculiar to the Avesta, in Darmesteter's French translation, and in my English rendering of Darmesteter. The speaker is addressing Haoma:

varəδayaŋuha mana vaca
vīspə̄sča paiti varəšajīš
vīspə̄sča paiti frasparəyə̄
vīspə̄sča paiti fravāxšə̄

(In the original text the last three lines are repeated.)

Grandis par ma parole
dans tous tes troncs, dans toutes tes branches, dans toutes tes tiges
dans tous tes troncs, dans toutes tes branches, dans toutes tes tiges.

Darmesteter's rendering.

Grow by my word
in all thy trunks, in all thy branches, in all thy stems
in all thy trunks, in all thy branches, in all thy stems.

English translation of Darmesteter.

I have underlined in the Avestan text the three words that Darmesteter renders by 'trunks', 'branches', and 'stems'. The question is how certain are these meanings. (We reproduce on p. 122 the picture of Haoma that Darmesteter gives us, in what he says is its natural size. It is a leafless plant without 'trunk' or 'branches'.) I shall take up these three words in the inverse order of their appearance in the verse.

3. *fravāxši* –. This word had three meanings or uses: (a) stem, (b) the *membrum virile*, and (c) antler. The prefix *fra-* conveys the idea of forward movement, of growth, thrust, erection. This is lacking in our lifeless 'stem', but how felicitous for our fast-growing mushroom tribe! In many of the world's cultures there is a semantic overlap between 'mushroom' and the *membrum virile*, for obvious reasons; *e.g.*, Greek μύκης, Japanese *matsutake*. In mycology we use 'stipe' rather than 'stem', but perhaps 'sprout' would convey better the thrust of the original.

2. *frasparəya* –. According to the lexicons this word meant 'shoot', 'sprout', 'sucker', and again the prefix *fra-* conveys the feel of forward thrust. There seems no warrant for Darmesteter's 'branches', except for the later Pahlavi and Sanskrit translations of the word, which may have been influenced by the current Haoma-substitutes. Darmesteter was presupposing that Haoma was a tree or shrub and translating the word to conform to his presupposition. The meaning and feel of the word closely resembles *fravāxši* –.

1. *varəšaji* –. This word presents the Avestan student with diffi-

culties. It occurs only three times, in Yasna 10.5, Yasna 71.9, and Yašt 8.42. The contexts are similar and do not help. The word is compound, noun + verb, *varəša*, 'tree', and *gay*, 'to live'. The grammatical relationship between these two elements is not clear. The suggestion has been made that the word's meaning was 'that which gives life to the tree', whence 'root' seems a viable guess. Bartholomæ suggested this meaning in his famous Old-Iranian Dictionary, but he was not sure as he added a question-mark to the explanation. Darmesteter was following Middle-Iranian traditions when he translated the word as 'trunk', and this meaning was confirmed by H. W. Bailey who at the same time rejected the rendering 'root'. (*Journal of the Royal Asiatic Society of Great Britain and Ireland*, 1934, pp. 507-8.) Bailey's conclusion was based on the Pahlavi translation of the word *varəšaǰi–* by *bun* = 'stalk' or 'trunk', and *advan* = 'stalk', the latter originally meaning 'the upper part of a tree', which rules out 'root'.

The Pahlavi tradition is not a completely reliable guide to the meaning of uncertain terms in the Avesta, but so long as no serious contradiction arises, it may help to make the rendering more probable. If Haoma was a mushroom, the translation must accommodate itself to that fact, and we are relying on the R̥gVeda to establish this. Now if Bailey's 'stalk' be accepted as a possible meaning of *varəšaǰi* –, this is again close to the other two words that we examined before. Here then we have perfect harmony of style. Yasna 10.5 was written in an elevated rhythmic prose by a bard who was addicted to poetic parallelism. We have seen that most of the verse was repeated word for word. We have seen that in the verse that concerns us the second and third substantives virtually duplicate each other. Does it not therefore become a poetic necessity that the first of the triad should be a close synonym? On this assumption and on the assumption that Haoma was a mushroom, I suggest the following translation:

> Swell, (then,) by my word!
> in all thy stalks, and in all thy shoots, and in all thy sprouts.
> in all thy stalks, and in all thy shoots, and in all thy sprouts.

⋆

For help in dissecting Yasna 10.5 I am indebted and grateful to Dr. Heinrich von Stietencron, of Heidelberg.

VI

SOMA GREW IN THE MOUNTAINS

Time and again the ṚgVeda speaks of Soma as hailing from the mountains, from the tops of the mountains, which in the case of the Indo-Aryans meant either the Hindu Kush or the Himalayas:

V 43[4c]

. . . plant from the mountain, . . .

mádhvo rásaṃ sugábhastir giriṣṭhā́ṃ

V 85[2d]

. . . he has placed the Soma on the mountain top.

diví sū́ryam adadhāt sómam ádrau

IX 18[1a]

. . . the Soma seated on the mountain top . . .

pári suvānó giriṣṭhā́ḥ pavítre sómo akṣāḥ

IX 46[1b]

. . . these Somas grown on the mountain top.

kṣárantaḥ parvatāvṛ́dhaḥ

IX 62[4a]

. . . the Soma stalk . . . seated on the mountain top;

ásāvy aṃśúr mádāyāpsú dákṣo giriṣṭhā́ḥ

IX 62[15a]

. . . Born on the mountain top, . . . the Soma juice is placed for Indra

girā́ jātá ihá stutá índur índrāya dhīyate

IX 71[4a]

. . . [This Soma] . . . that grows in the mountain . . .

pári dyukṣáṃ sáhasaḥ parvatāvṛ́dhaṃ

IX 82[3b]

. . . . . . at the navel of the Earth, in the mountains, [Soma] has placed his residence . . .

nā́bhā pṛthivyā́ giríṣu kṣáyaṃ dadhe

IX 85[10ab]

. . . In the firmament of heaven the Seers milk . . . the bull-Soma seated on the mountain top;

divó nā́ke mádhujihvā asaścáto venā́ duhanty ukṣáṇaṃ giriṣṭhā́m

IX 87[8a]
. . . Here is the flow of Soma that is come from within the most distant mountain . . .
eṣā́ yayau paramā́d antár ádreḥ

IX 95[4b]
. . . This [Soma], . . . [this] stalk, [this] bull seated on the mountain top . . .
aṃśúṃ duhanty ukṣáṇaṃ giriṣṭhā́m

IX 98[9c]
. . . This Soma juice, god [himself], sitting on the mountain . . .
devó devī giriṣṭhā́

The poets say that Soma grows high in the mountains. They make a point of this. They never speak of it as growing elsewhere. They must mean what they say. What a useless business it is for us to go chasing in the valleys after rhubarb, honey, hashish, wild Afghan grapes; in hot arid wastes after species of Ephedra, Sarcostemma, Periploca! For the Vedic poet this lofty birthplace was additional testimony to its divine origin, bringing it closer to the celestial sphere, to Indra, to Parjanya. It is unlikely that the poets of the ṚgVeda should have conspired together to attribute a fictitious habitat to Soma.

In Northern Eurasia the birch and conifer grow at sea level. South of the Oxus and in India they are found only at a great height in the mountains, around 8,000 to 16,000 feet. As I have mentioned already, the fly-agaric grows in mycorrhizal relationship with the birch and the conifer. (Centuries later, when the art of writing began to play a rôle in Indian culture, the bark of the Himalayan birch quickly gained renown in Northern India for writing purposes.)

The Indo-Aryans, having conquered only the valleys, did not control the source of their Soma supply. The mountains were still held by their enemies, probably the Dasyus, the hated and despised dark skinned Dasyus. Under the circumstances there could be no ceremony attending the gathering of the sacred Soma, such as had perhaps attended it in the homeland, and such as we know attended the gathering of the hallucinogenic mushrooms of Mexico. All that had been left behind, and now, in Vedic times, a ceremony attended the buying of the holy plant from the natives who came down from the heights where the birch and the conifer grow. In the *Śatapatha Brāhmaṇa* (as well as elsewhere) there is an account, absurd by our standards, of the ceremonial purchase, complete to speckled cane to beat the seller with, wherein a cow of a particular hue of skin and eye is exchanged for the Sacred Element.[1] This barter price for Soma reminds us, curiously, of the high price paid for the fly-agaric by the native tribesmen of the Maritime Provinces of Siberia: they are said to give as much as a reindeer for one fly-agaric, or even two or three reindeer.[2]

1. Eggeling translation, Part II, pp. 66 ff.
2. *Vide infra,* [11], p. 252.

VII

THE *TWO FORMS* OF SOMA

I now come to a crucial argument in my case.

The fly-agaric is unique among the psychotropic plants in one of its properties: it is an inebriant in *Two Forms.*

> *First Form:*
> Taken directly, and by 'directly' I mean by eating the raw mushroom, or by drinking its juice squeezed out and taken neat, or mixed with water, or with water and milk or curds, and perhaps barley in some form, and honey; also mixed with herbs such as Epilobium sp.
>
> *Second Form:*
> Taken in the urine of the person who has ingested the fly-agaric in the *First Form.*

The *Second Form,* as urine, was first called to the attention of the Western World by a Swedish army officer, Filip Johann von Strahlenberg, after having served 13 years as a captive of the Russians in Siberia. His book, first published in German in Stockholm, appeared in 1730;[1] and an English translation in London in 1736 and again in 1738 under a lengthy title beginning *An Historico-Geographical Description of the North and Eastern Parts of Europe and Asia.* Since then many other travellers and anthropologists have set forth the facts, usually going to extremes in characterizing the practice as revolting, disgusting, filthy, and the like. So far as our records go, none of them has tried the urine, not even the anthropologists, among whom there are usually some who pride themselves on participating to the full in native ways and who consider it their professional duty to do so. In 1798, a Pole, Joseph Kopeć, a literary figure of some standing, tried the mushrooms (but not the urine) and published his remarkable im-

1. Philip Johan von Strahlenberg, *Das Nord- und Östliche Theil von Europa und Asia, in so weit solcher das gantze Russische Reich mit Siberien und der grossen Tartary in sich begreiffet. . . .* Stockholm, 1730. *Vide* [3], p. 234.

pressions of the experience.[1] That he was ill at the time and running a fever detracts from the value of his testimony.

In the ṚgVeda Soma also has *Two Forms*, expressly so described in IX 66[2, 3, 5]:

IX 66[1-2]
Cleanse thyself, O [thou] to whom all peoples belong, for all wondrous deeds, the praiseworthy god, the friend for the friends.
With those two Forms [*dual, not plural*] which stand facing us, O Soma, thou reignest over all things, O Pávamāna!

pávasva viśvacarṣaṇe 'bhí víśvāni kā́vyā
sákhā sákhibhya ī́ḍyaḥ
tā́bhyāṃ víśvasya rājasi yé pavamāna dhā́manī
practīcī́ soma tasthátuḥ

IX 66[3]
The Forms [*plural, not dual*] that are thine, thou pervadest them, O Soma, through and through, O Pávamāna, at the appointed hours, O Wonder-worker!

pári dhā́māni yā́ni te tvám̐ somāsi viśvátaḥ
pávamāna ṛtúbhiḥ kave

IX 66[5]
Thy shining rays spread a filtre on the back of heaven, O Soma, with [thy] Forms [*plural, not dual*].[2]

táva śukrā́so arcáyo divás pṛṣṭhé ví tanvate
pavítraṃ soma dhā́mabhiḥ

In the Soma sacrifice the *First Form* is drunk by Indra and his charioteer, Vāyu, who are impersonated by high functionaries in the rite. The Vedic commentators, knowing nothing of the fly-agaric, have reached a consensus that the *First Form* is the simple juice of the Soma plant, and the *Second Form* is the juice after it has been mixed with water and with milk or curds. The commentators are agreed on this, the Vedic mythologies are so written, the matter is considered settled.[3]

1. *Vide* [9], pp. 243 ff.
2. Sanskrit and Vedic possess three numbers, singular, dual, and plural. In IX 66[2] the dual number is used speaking of the *Two Forms*. This is natural as the poet faces two vessels containing, one the juice of Soma presumably mixed with milk, *etc.*, the other Soma urine. In verses 3 and 5 he speaks of all Soma's forms, the celestial, the plant, the juice, the Soma urine, and therefore uses the plural.
3. *Vide, e.g.*, A. A. Macdonell, *The Vedic Mythology*, London, 1897, pp. 82, 106.

While in default of any other explanation it is easy to see how this was arrived at, it is unsatisfactory because it flies in the face of the ṚgVeda text. Indra and Vāyu are repeatedly drinking the juice of Soma mixed with milk or curds, and, saving error on my part, they are never in the ṚgVeda drinking juice expressly described as not mixed with anything.

I will cite four instances where Indra and Vāyu drink Soma mixed with milk or curds. The first instance, V $51^{4\text{-}7}$, is crucial because there is no mention of curds and the reader might think the poet was speaking of the unmixed juice, exactly as the commentators contend, until suddenly in verse 7, just before the end, it seems that all along the poet took for granted the curds! In the many instances where the poet does not mention the milk or curds, the omission seems accidental. The fly-agaric was often, perhaps usually, dried up when it was used in the sacrifice, and initially it had to be soaked in water, reinflated so to speak. Here are verses V $51^{4\text{-}7}$:

I

V $51^{4\text{-}7ab}$ . . . Here is the Soma [that] pressed in the vat is poured all around inside the cup, he dear to Indra, to Vāyu.

. . . O Vāyu, arrive hither for the invitation, accepting it, to share in the oblation! Drink [of the juice] of the pressed stalk, up to [thy full] satisfaction!

. . . Indra and thou, Vāyu, ye have a right to drink of these pressed [stalks]. Accept them, immaculate ones, for [your full] satisfaction!

. . . Pressed for Indra, for Vāyu, have been the Soma plants requiring a mixture of curds.

ayáṃ sómaś camū́ sutó 'matre pári ṣicyate
priyá índrāya vāyáve
vā́yav ā́ yāhi vītáye juṣāṇó havyádātaye
pibā sutásyā́ndhaso abhí práyaḥ
índraś ca vāyav eṣāṃ sutā́nāṃ pītím arhathaḥ
tā́ñ juṣethām arepásāv abhí práyaḥ
sutā́ índrāya vāyáve sómāso dádhyāśiraḥ

The following three quotations say expressly that Indra and Vāyu drink Soma mixed with milk or curds:

2

IX 11^{2} The Atharvans have mixed milk with thy sweetness, longing for the god, the god [Soma] for the god [Indra].

IX $11^{5\text{-}6}$ Cleanse the Soma, pressed out by the hand-worked stones; dilute the sweet one in the sweetness [milk or water].

Approach with reverence; mix him with curds, put the Soma juice into Indra.

abhí te mádhunā páyó 'tharvāṇo aśiśrayuḥ
deváṃ devā́ya devayú

hástacyutebhir ádribhiḥ sutáṃ sómaṃ punītana
mádhāv ā́ dhāvatā mádhu

námaséd úpa sīdata dadhnéd abhí śrīṇītana
índum índre dadhātana

3

IX $62^{5\text{-}6}$ The beautiful plant beloved of the gods, [the Soma] washed in the waters, pressed by the masters, the cows season [it] with milk.

Then like drivers [urging] on a horse, they have beautified [the Soma], the juice of liquor for drinking in common, for the Immortal One [Indra].

śubhrám ándho devávātam apsú dhūtó nṛ́bhiḥ sutáḥ
svádanti gā́vaḥ páyobhiḥ

ā́d īm áśvaṃ ná hétāró 'śūśubhann amṛ́tāya
mádhvo rásaṃ sadhamā́de

4

IX 109^{15} All the gods drink this Soma when it has been mixed with milk of cows and pressed by the Officiants.

IX $109^{17\text{-}18}$ The prize-winning Soma has flowed, in a thousand drops cleansed by the waters, mixed with the milk of cows.

O Soma, march ahead toward Indra's bellies, having been held in hand by the Officiants, pressed by the stones!

píbanty asya víśve devā́so góbhiḥ śrītásya
nṛ́bhiḥ sutásya
sá vā́jy àkṣāḥ sahásraretā adbhír mṛjānó
góbhiḥ śrīṇānáḥ
prá soma yāhī́ndrasya kukṣā́ nṛ́bhir yemānó
ádribhiḥ sutáḥ

If then the traditional view of the first and second 'Form' of Soma is to be replaced, what evidence do I adduce in favor of my interpretation?

Only the words of the ṚgVeda. In the hymns to Soma there comes a time when the religious emotion reaches a climax, an intensity of exaltation, that is overwhelming, and that after 3,000 years, in a world of utterly different orientation, even in translation, cannot but move any perceptive reader. These hymns are in Maṇḍala IX, from say 62 through perhaps 97, the mood then tending to ease off. The 74th hymn, in particular, consists of an enumeration of phrases that we have learned by now to recognize when they occur singly, clearly numinous phrases for the contemporary believers, tense phrases piled one on the other, Pelion on Ossa, in portentous sequence, until we suddenly read, at the end of verse 4, a phrase not met with before and not to be met again:

> IX 74[4]
> Soma, storm cloud imbued with life, is milked of ghee, milk. Navel of the Way, Immortal Principle, he sprang into life in the far distance. Acting in concert, those charged with the Office, richly gifted, do full honor to Soma. The swollen men piss the flowing [Soma].
>
> *ātmanván nábho duhyate ghṛtám páya ṛtásya nā́bhir amṛ́taṃ ví jāyate*
> *samīcīnā́ḥ sudā́navaḥ prīṇanti táṃ náro hitám áva mehanti péravaḥ*

If the final clause of this verse bears the meaning that I suggest for it, then it alone suffices to prove my case.

Renou renders the final phrase of this verse 4 as follows:

> Les [Maruts] seigneurs à la vessie pleine compissent [le Soma] mis-en-branle.

> The [Maruts] lords with full bladders piss [Soma] quick with movement.

Renou had lived with the Ṛg Veda text for a lifetime and knew everything that had ever been said by scholars about it. He discerned that the 'swollen' men had full bladders and that they were urinating Soma. But to give meaning to the sentence he introduced the gods of rain, the Maruts. Certainly there are precedents for the clouds' 'urinating' rain. But in this verse and at this point in the hymns the Maruts are out of place. From IX 68 to 109 there are 24 other citations of *nṛ́* in the plural (men) and in every instance they are the officiants at the sacrifice.[1] So are they in 74^4. It is noteworthy that Grassmann translates *nṛ́* in this verse by 'men serving . . .' *etc.*, in conformity with his third definition of *nṛ́*, 'men serving the gods, such as singers and sacrificers'. He does not translate it by his 6th definition, which would include gods. The priests appointed to impersonate Indra and Vāyu, having imbibed the Soma mixed with milk or curds, are now urinating Soma. They in their persons convert Soma into the *Second Form.* When Renou translated this verse, he had never heard of the Siberian use of the fly-agaric. Roger Heim and I apprised him of the facts when we dined with the Renous in the middle of April 1966.

Let us pause for a moment and dwell on a rather odd figure of speech. The blessings of the fertilizing rain are likened to a shower of urine. The storm-clouds fecundate the earth with their urine. Vedic scholars have lived so long with their recalcitrant text, and so close to it, that they remark no longer on an analogy that calls for explanation. Urine is normally something to cast away and turn from, second in this respect only to excrement. In the Vedic poets the values are reversed and urine is an ennobling metaphor to describe the rain. The values are reversed, I suggest, because the poets in Vedic India were thinking of urine as the carrier of the Divine Inebriant, the bearer of *amṛta.* This would explain the rôle that urine – human and bovine – has played through the centuries as the medico-religious disinfectant of the Indo-Iranian world, the Holy Water of the East.

1. *Vide* IX $68^{4,7}$; $72^{2,4,5}$; $75^{3,5}$; 78^{2}; $80^{4(2)}$; $86^{20,22,34}$; 87^{1}; 91^{2}; 95^{1}; 97^{5}; 99^{8}; 101^{3}; 107^{16}; 108^{15}; $109^{8,15,18}$.

The words of this hymn IX 74 are redolent with a most holy mystery, the handling of the miraculous Soma planted on the mountain tops by the gods. Those charged with the Office are the Guardians-of-the-Meaning, the Guardians-of-the-Melodies, the Guardians-of-the-Mystery. The Pressing Stones, the woolen Filtre, the mingling of the Soma juice first with water, then with milk or curds in the vessels – all this is set forth clearly. Then the details of the Mystery are hidden from us. This is not in my opinion deliberate. Every party to the proceedings knew every detail. But when an event takes place that stirs people to their depths, a hush naturally falls, a feeling of awe and terror and adoration mingle. They speak in a whisper, as the rubric directs the clergy to do at the climax of the Mass. The details of the Mystery are certainly not to be put into Hymns. Thus we do not know what the dose of the juice was, nor how much water was added, nor how much milk or curds, nor what the effect was. We do not know whether the effect of the *First Form* and the *Second Form* was identical. Chemists say it could well be different, the juice being one thing and the metabolite another. Or it might be the same, the juice developing its marvelous properties only after it has been converted into the metabolite. There is a further possibility. In modern experience the fly-agaric causes nausea. If the agent that provokes vomiting is not the same as the one that leads to ecstasy, the former might be eliminated in the digestive track and the urine be thus freed from this inconvenience. We do not know how the metabolite was taken, whether neat or with water or milk or curds or honey. In the ṚgVeda we are not told who shared in the divine beverage. Afterwards only the priests, or some of them, were privileged to imbibe, but must there not have been a primitive age when others who participated in the rites shared in the drink? We know that centuries later the Śūdras, and outcastes, were not permitted even to hear the words of the ṚgVeda hymns, so holy were these.

Bhawe calls attention to three passages in the ṚgVeda that seem to him to stress the skill needed in the mixing of the Soma juice.[1] This

1. S. S. Bhawe: *The Soma-Hymns of the ṚgVeda*, Part III, p. 176, comment on 70[8d]; also 47[1a] and 99[7a].

might refer only to complicated ritualistic gestures and postures that had to be executed with precision. But in his judgement it is more likely that the blending of the ingredients had to be just right. Perhaps there was a secret recipe, the fruit of esoteric experience passed on from one generation of priests to the next. This recipe might be able to reduce or eliminate the initial nausea provoked by the Soma juice.

Let us pause for a moment and consider the probabilities. Some 3,250 years ago the Indo-Aryans living in the Indus Valley were worshipping a plant whose juice, pressed out and drunk immediately, seems to have had astonishing psychic effects, effects comparable to those of our Mexican hallucinogenic mushrooms, comparable but far different. The identity of that plant is not known. Hundreds of poets living over centuries in different centers speak of it in hallowed syllables but without mentioning leaves, roots, blossoms, or seed. Its stalk ('stipe' in mycologists' language) was obviously fleshy. How well these fit a wild mushroom! Nowhere in the thousand hymns is a dimension given of the 'stalk' that is incompatible with the stipe of a mushroom. It grew only high in the mountains. How well this fits the fly-agaric in the latitude of the Indus Valley! The poet says that the priests who have drunk the juice of this mysterious plant urinate the divine drink. In the traditions of Eurasia there is only one plant that supplies a psychotropic metabolite – the fly-agaric. Could any key unlock this combination save the fly-agaric?

If mine is indeed the interpretation to place on the *Second Form* of Soma in the ṚgVeda text, this should not be the only reference to the potable metabolite in Indo-Iranian literature and Parsi traditional practice. In the Avesta there is a verse in a famous Yasna, 48: 10, supposed to preserve the very words of the Teacher himself, which has never been satisfactorily explained:

> When wilt thou do away with this urine of drunkenness with which the priests evilly delude [the people] as do the wicked rulers of the provinces in [full] consciousness [of what they do]?
>
> [*Translation by R. C. Zaehner.* – The text below corresponds to the underlined part of the translation only.]
>
> *Kadā ajə̄n mūθrəm ahyā madahyā . . .?*

The learned commentators, not knowing of the *Second Form* of the Soma of the Iranians (called Haoma), have arbitrarily changed 'urine' to 'excrement' and have puzzled over the meaning. Surely Zoroaster meant what he said: he was excoriating the consumption of urine in the Soma sacrifice. If my interpretation be accepted, there is opened up a promising line of inquiry in Zoroastrian scholarship.

In the vast reaches of the *Mahābhārata*, the classical Indian epic, there occurs one episode – an isolated episode of unknown lineage – that bears with startling clarity on our *Second Form*. It was introduced into the text perhaps a thousand years after the fly-agaric had ceased to be used in the Soma sacrifice, and perhaps the editor did not know its meaning, which only today we are recovering. Here it is, as translated by Wendy O'Flaherty.

Mahābhārata, Aśvamedha Parvan, 14.54.12-35

> *Kṛṣṇa* had offered *Uttañka* a boon, and *Uttañka* said, 'I wish to have water whenever I want it.' *Kṛṣṇa* said, 'When you want anything, think on me,' and he went away. Then one day *Uttañka* was thirsty, and he thought on *Kṛṣṇa*, and thereupon he saw a naked, filthy *mātañga* [= *caṇḍāla*, an outcaste hunter], surrounded by a pack of dogs, terrifying, bearing a bow and arrows. And *Uttañka* saw copious streams of water flowing from his lower parts. The *mātañga* smiled and said to *Uttañka*, 'Come, *Uttañka*, and accept this water from me. I feel great pity for you, seeing you so overcome by thirst.' The sage did not rejoice in that water, and he reviled *Kṛṣṇa* with harsh words. The *mātañga* kept repeating, 'Drink!', but the sage was angry and did not drink. Then the hunter vanished with his dogs, and *Uttañka's* mind was troubled; he considered that he had been deceived by *Kṛṣṇa*. Then *Kṛṣṇa* came, bearing his disc and conch, and *Uttañka* said to him, 'It was not proper for you to give me such a thing, water in the form of the stream from a *mātañga*.' Then *Kṛṣṇa* spoke to *Uttañka* with honeyed words, to console him, saying, 'I gave it to you in such form as was proper, but you did not recognize it. For your sake I said to Indra, "Give the *amṛta* to *Uttañka* in the form of water." Indra said to me, "A mortal should not become immortal; give some other boon to him." He kept repeating this, but I insisted, "Give the *amṛta*." Then he said to me, "If I must give it, I will become a *mātañga* and give the *amṛta* to the noble descendant

> of *Bhṛgu* [*i.e., Uttaṅka*]. If he accepts the *amṛta* thus, I will go and give it to him today." As he continued to say, "I will not give it [otherwise]," I agreed to this, and he approached you and offered the *amṛta*. But he took the form of a *caṇḍāla*. But your worth is great, and I will give you what you wished: on whatever days you have a desire for water, the clouds will be full of water then, and they will give water to you, and they will be called *Uttaṅka*-clouds.' Then the sage was pleased.

We found the first reference to Soma-urine in the ṚgVeda at a point in the liturgy where the proceedings and the religious emotion called for frankness in utterance. This reference did not stand alone in the literatures of Iran and India. There should be yet others in the ṚgVeda itself, perhaps more veiled as befits a holy mystery. I believe there are many such, but for the orderly progress of my argument I will defer their discussion until I reach the third of the three 'filtres'.

PLATE I · THE SOMA OF THE ṚGVEDA

PLATE II · THE IMMORTAL HÁRI

RV IX 69^5: With unfading vesture, brilliant, newly clothed, the immortal *hári* wraps himself all around.

VIII

EPITHETS, CONCEITS, AND TROPES FOR SOMA

Until this very day no poet in the English language has ever sung the supernal beauties of the fly-agaric; nor I believe has any novelist or essayist paid obeisance to this remarkable fruit. The same can be said probably for all the Germanic and Celtic languages. The original 'toadstool' was the fly-agaric, if we judge by the French and Basque languages. In the conservative provinces of France the dialectal name of the fly-agaric is *crapaudin*, 'the toad,' and in the Biscay and Guipuzcoa country the Basques call it by the precise Basque equivalent of the French provincial name, – *amoroto*. In *Mushrooms Russia and History* (Pantheon Books, New York, 1957) my wife V. P. Wasson and I explored the folkloric and linguistic background of the fly-agaric throughout Europe, and showed the deep hold that it exerted at one time on the imagination of the north European peoples. It seemed that a shadow hung over the fly-agaric, an ancestral curse; yes, a tabu. In many West European languages there are childhood ditties dealing with it, but beyond the nursery no one dwells on it except to repudiate it whenever occasion demands. At present I leave to the reader to find his own explanation for this tabu.

In the fall of the year, hard by a birch or pine, one is apt to find the fly-agaric. The season in the temperate zone lasts two or at most three weeks, with the climax coming in the middle week. The fly-agaric emerges as a little white ball, like cotton wool. It swells rapidly and bursts its white garment, the fragments of the envelope remaining as patches on the brilliant red skin underneath. At first the patches almost cover the skin, but as the cap expands they are reduced in relative size and finally are nothing more than islands on the surface. In fact, under certain conditions, especially as a result of rain, they are washed off altogether and the fly-agaric then shines without blemish as a resplendent scarlet mushroom. When the plant is gathered it soon loses its lustre and takes on a rather dull chestnut hue. Such is the dominant fly-agaric in Eurasia and in Washington, Oregon, and

British Columbia. There is another variety found commonly in the rest of the United States, a variety that is a brilliant yellow, sometimes with a reddish tinge in the center of the cap. These two varieties are not mutually exclusive: occasionally specimens of each are found in the other's territory, and attention is then drawn to the oddity in mycological journals, perhaps in the miscellaneous notes for amateurs. No one knows what causes these two varieties in different sectors of the temperate zone, nor does anyone know whether the difference extends beyond the coloration and certain morphological features to the chemistry of the plant.[1]

When the fly-agaric is crushed and the juice milked out, the liquor comes forth a tawny yellow. As we shall see, in the ṚgVeda it is sometimes hard to say whether the poet, when he is speaking of Soma, has in mind the plant or the juice. The glowing adjectives of enhancement that he employs could describe either.

Now let us see how the poets of the RgVeda describe Soma.

A. INTRODUCTORY.

There are no words in the ṚgVeda that describe Soma as black, or gray, or green, or dark, or blue. All the great Vedic scholars from Burnouf to Renou seem to be in agreement on this.[2]

B. 'HÁRI' AND RED.

Hári is the most common of the colour epithets for Soma in the ṚgVeda. Numerically it far exceeds all the other colour words put together and rivals the epithet 'bull' that the poets never tire of applying to Soma. The word *hári* is cognate with *híraṇya* (golden) in Sanskrit and with χόλος (gall) and χλωρός (yellow) in Greek, and ultimately with the English 'gall' and 'yellow'.

Hári is the precise adjective that one would wish to employ in Vedic to describe the fly-agaric. *Hári* is not only a colour word: the intensity

1. My division of *A. muscaria* into two varieties will not satisfy the mycologists. Those interested in pursuing this matter should consult Roger Heim: *Un problème à éclaircir: celui de la tue-mouche*, in the *Revue de Mycologie*, Vol. xxx, fasc. 4, 1965, pp. 296 ff.

2. Occasionally in later times *hári* came to include 'green' among its meanings, but this usage seems not to be ṚgVedic, except possibly in the late hymns that we exclude from consideration.

PLATE III · TAWNY YELLOW PÁVAMĀNA

of the colour is also expressed by it. It is dazzling, brilliant, lustrous, resplendent, flaming. In colour it seems to have run from red to light yellow. The mythological horses of the sun-god were *hári*: in this context the word is usually rendered by 'bay' or 'chestnut', but one doubts whether any mundane colour such as 'bay' would describe the steeds of the sun. They are flaming and full of brio. *Hári* is of course a term of enhancement, and by being linked together, Soma and the sun-god's horses are mutually enhanced. How well the breathtaking fly-agaric fits into this picture.

Some other colour adjectives are used from time to time for Soma. Thus on one of the many occasions when Soma is called a bull, the bull is 'red,' *vṛ́ṣā śóṇo* (IX 97[13a]). Others are:

> *aruṇá*: This according to Grassmann means reddish, bright brown, golden-yellow, red, the red of morning.
>
> *aruṣá*: Again Grassmann: red, fire-colour, applied especially to fire, the sun, lightning, dawn, Soma, *etc.*
>
> *babhrú*: reddish brown, brown. Monier Williams gives 'tawny'.

The juice of the fly-agaric is tawny yellow. As we have said already, often we do not know whether the plant or the juice is being described. The dried fly-agaric is dull by comparison with the fresh specimen, *babhrú* rather than *hári*.

The poets of the ṚgVeda not only use the same adjective for Soma and the sun-god's horses. They compare Soma directly with the sun. The sun is a shining disc and thus a compelling metaphor for the fly-agaric, as compelling as it is inappropriate for any chlorophyll-bearing plant. Here is a selection of such figures:

> I 46[10ab]
> Light has come to the plant, a sun equal to gold . . .
> *ábhūd u bhā́ u aṃśáve híraṇyaṃ práti sū́ryaḥ*
>
> I 135[3b]
> This [Soma is] thy precise share, accompanied by the rays that are his in common with the sun
> *távāyáṃ bhāgá ṛtvíyaḥ sáraśmiḥ sū́rye sácā*

IX 2^{6c}
He [Soma] shines together with the Sun . . .
sáṃ sū́ryeṇa rocate

IX 27^{5ab}
Here he is, racing with the sun, Pávamāna in the sky . . .
eṣá sū́ryeṇa hāsate pávamāno ádhi dhávi

IX 28^{5ab}
He [Soma] has made the sun to shine.
eṣá sū́ryam arocayat pávamāno vícarṣaṇiḥ

IX 37^{4bc}
He has made the sun to shine.
pávamāno arocayat jāmíbhiḥ sū́ryaṃ sahá

IX 61^{8c}
He [Soma] joins forces with the sun's rays.
sáṃ sū́ryasya raśmíbhiḥ

IX 63^{7ab}
Purify thyself with this stream by which thou [Soma] madest the sun to shine
ayā́ pavasva dhā́rayā yáyā sū́ryam árocayaḥ

IX 63^{8ab}
Pavamāna has hitched Etaśa [the sun's steed] to the Sun . . .
áyukta sū́ra étaśaṃ pávamāno manā́v ádhi

IX 64^{7}
[the Soma's flowing liquor] like the rays of the sun.
pávamānasya viśvavit prá te sárgā asṛkṣata sū́ryasyeva ná raśmáyaḥ

IX 64^{9c}
thou hast whinneyed like the sun-god.
ákrān devó ná sū́ryaḥ

IX 71^{9b}
he has clothed himself with the fire-bursts of the sun.
ádhi tvíṣīr adhita sū́ryasya

IX 76^{4c}
he who has been cleansed by the sun's ray.
yáḥ sū́ryasyā́sireṇa mṛjyáte

IX 86[22d]
thou hast made the sun to mount the sky.
nṛ́bhir yatáḥ sū́ryam ā́rohayo diví

IX 86[29d]]
Thine, O Pávamāna, are the lights, the sun.
táva jyótīṃṣi pavamāna sū́ryaḥ

IX 86[32a]
He [Soma] wraps himself all around with the rays of the sun.
sá sū́ryasya raśmíbhiḥ pári vyata

IX 97[31d]
[once] born, thou [Soma] didst fill the sun with rays.
jajñānáḥ sū́ryam apinvo arkaíḥ

IX 97[33d]
O Soma juice, . . . go bellowing to the sun's ray.
krándann ihi sū́ryasyópa raśmím

IX 97[41d]
the juice has engendered light for the sun.
'janayat sū́rye jyótir índuḥ

IX 111[3b]
[The Soma] races against the rays [of the Sun], vehicle beautiful to see, celestial vehicle beautiful to see.
sáṃ raśmíbhir yatate darśató rátho daívyo darśató ráthaḥ

For the past century students of the ṚgVeda have been aware of a link that ties Soma to Agni, the god of fire. This tie is intimate and all pervasive, to the point where Bergaigne even went so far as to advance the hypothesis that the two had been interchangeable. Hymns addressed to Soma sometimes call him 'Agni'. (IX 66[19-21]; 67[23-24]) Soma is the child of the thunder-storm. The plant shares its liquid nature with the rain, its brilliance with the lightning (IX 22[2]), and the fire that lightning causes. Its inebriating potency is thought to rival the subtlety of flames. 'Make me to burn like fire started by friction', says the poet, addressing Soma. (VIII 48[6]) Some years ago I gathered together evidence indicating that a peculiar relationship existed in primitive man's men-

tality between the mushroom world and thunder-storms,[1] extending far beyond the Indo-European tribes but also including them.
This relationship exists between Soma and the thunder-storm, and in
fact it reaches an intensity no where else found. The flame-like plant,
child of the thunder-bolt, possesses inebriating qualities that harmo-
nize with its celestial appearance.

These observations on the colour of Soma may have failed to convey the full impression of radiance that marks Soma throughout the
ṚgVeda, radiance without a specific colour linked to it. Take for
example this verse:

> IX 69[5]
> With unfading vesture, brilliant, newly clothed, the immortal *hári*
> wraps himself all around. By authority he has taken the back [*i.e.*,
> the vault] of heaven to clothe himself in, a spread-cloth like to a
> cloud ...
>
> *ámṛktena rúšatā vāsasā hárir ámartyo nirṇijānáḥ pári vyata
> divás pṛṣṭháṃ barháṇā nirṇíje kṛtopastáraṇaṃ camvòr nabhasmáyam*

In the following verse the poet telescopes the life history of the fly-
agaric, and how delicately he does it! For the first time in millennia
the verse takes on meaning:

> IX 71[2]
> Aggressive as a killer of peoples he advances, bellowing with power.
> He sloughs off the Asurian colour that is his. He abandons his enve-
> lope, goes to the rendez-vous with the Father. With what floats
> he makes continually his vesture-of-grand-occasion.
>
> *prá kṛṣṭihéva šūṣá eti róruvad asuryàṃ várṇam ní riṇīte asya tám
> jáhāti vavríṃ pitúr eti niṣkṛtám upaprútam kṛṇute nirṇijaṃ tánā*

In the first line the poet reminds us of the extraordinary strength
displayed by a simple mushroom in forcing its way to the surface
against obstacles. 'Asurian' is not a colour: it is the radiance associated
with Asuras, which at this period in Indo-Aryan history meant the
divinities. The fly-agaric sloughs off the radiant envelope that is his to
start with, the 'universal veil', and prepares to meet with the Sky

---

1. *Vide* R. Gordon Wasson, 'Lightning-bolt and Mushrooms: an essay in early cultural exploration.'
Festschrift *For Roman Jakobson*, Mouton, The Hague, 1956, pp. 605-612.

Plate iv · SŪRYA: Sun.
RV IX 37^{4bc}: He has made the sun to shine.

(= Father). He dons of course his gorgeous apparel, his *nirṇíj*, his 'vesture-of-grand-occasion', what Renou calls his *robe d'apparat*. Nine times[1] in Maṇḍala IX the poets speak of the *nirṇíj* of Soma: this is his dazzling vesture-of-grand-occasion. Often his costume is linked with the milk of cows: this is the fly-agaric while still studded with plaques of snow, with tufts of snowy wool. There is one occasion on which the poet stoops to a banal simile:

> IX 86^{44c}
> Like a serpent he creeps out of his old skin.
> *áhir ná jūrṇā́m áti sarpati tvácam*

At least some of the poets knew their fly-agaric *in situ*, high in the mountains: could the last phrase in this verse have been written by anyone who did not know it?

> IX 70^{7}
> He [Soma] bellows, terrifying bull, with might, sharpening his shining [*hári*] horns, gazing afar. The Soma rests in his well-appointed birth-place. The hide is of bull, the dress of sheep.
> *ruváti bhīmó vṛṣabhás taviṣyáyā śṛ́ṅge śíśāno háriṇī vicakṣaṇáḥ*
> *ā́ yóniṃ sómaḥ súkṛtaṃ ní ṣīdati gavyáyī tvág bhavati nirṇíg avyáyī*

'The hide is of bull, the dress of sheep.' The red bull of IX 97^{13} supplies him with his skin; his dress is of fluffy tufts of white sheep's wool.

Often have I penetrated into a forest in the fall of the year as night gathered and seen the whiteness of the white mushrooms, as they seemed to take to themselves the last rays of the setting sun, and hold them fast as all else faded into the darkness. When fragments of the white veil of the fly-agaric still cling to the cap, though night has taken over all else, from afar you may still see Soma, silver white, resting in his well-appointed birth-place close by some birch or pine tree. Here is how three thousand years ago a priest-poet of the Indo-Aryans gave voice to this impression:

> IX 97^{9d}
> By day he appears *hári* [colour of fire], by night, silvery white.
> *dívā hárir dádṛśe náktam ṛjráḥ*

1. IX 68^{1}, 71^{2}, 82^{2}, $86^{26,46}$, 95^{1}, 99^{1}, 107^{26}, and 108^{12}.

Soma's scarlet coat dominates by day; by night the redness sinks out of sight, and the white patches, silvery by moon and starlight, take over.

C. THE BULL AND SOMA

The bull was the mightiest beast familiar to the Indo-Aryans. It was the symbol of strength in the ṚgVeda and it was the commonest metaphor for Soma. It exchanged attributes with Soma: both were seated high in the mountains, both were gazing afar off, both bellowed, both sharpened shining horns. Sometimes the image was taken from Soma, sometimes from the bull, thus:

> IX 70[7ab]
> He [Soma] bellows, terrifying bull, with might, sharpening his shining horns, gazing afar . . .
> *ruváti bhīmó vṛṣabhás taviṣyáyā śṛ́ṅge śíśāno háriṇī vicakṣaṇáḥ*

The hallucinogenic mushrooms of Mexico act passively: the subject seems to an outsider to be withdrawn in meditation. Soma, in addition to translating one to Elysium, seems to have possessed a kinetic potency, filling one with the joy of extraordinary physical and vocal activity. This is reported alike in the ṚgVeda and in Siberia, where, as we shall see, phenomenal displays of physical prowess sometimes attend inebriation by the fly-agaric.

It is important that the modern reader fix his attention on the sense of power that Soma gave to the poets of the ṚgVeda. They ring all the changes on this metaphor of a bull. (Sometimes Soma is a buffalo.) For the poets the bull is a creature that constantly sharpens his horns: there are many such references. We have just seen Soma compared to a bull 'sharpening his shining horns'. Once the poet resorts to synecdoche:

> IX 97[9c]
> Soma with sharpened horns [*i.e.*, Soma the bull] attains his [full] reach.
> *parīṇasáṃ kṛṇute tigmáśṛṅgo*

Some have deduced from this verse that Soma must have been a plant with thorns! But of course the 'sharpened horns' are nothing more than the familiar cliché for a bull, and the bull is Soma.

PLATE V · AGNI: Fire.

D. THE UDDER AND SOMA

The swollen hemisphere of the fly-agaric's cap naturally suggests an udder to the poet:

III 48[3ab]
Approaching his mother, he [Indra] cries for food; he looks toward the sharp Soma as toward the udder.
upasthā́ya mātáram ánnam aiṭṭa tigmám apaśyad abhí sómam ū́dhaḥ

IV 23[1ab]
What priest's sacrifice has [Indra] enjoyed, [approaching] the Soma as it were an udder?
kathā́ mahā́m avṛdhat kásya hótur yajñáṃ juṣāṇó abhí sómam ū́dhaḥ

VII 101[1ab]
Raise the three voices that are preceded by light and that milk the udder, which is milked of sweetness . . . [. . Soma]
tisró vā́caḥ prá vada jyótiragrā yā́ etád duhré madhudoghám ū́dhaḥ

VIII 9[19a]
When the swollen stalks were milked like cows with [full] udders . . .
yád ā́pītāso aṃśávo gā́vo ná duhrá ū́dhabhiḥ

IX 68[1]
The sweet juices have hurried to the god like milch cows [to a calf]. Resting upon the barhis, noisy, with full udders, they have made the red ones their flowing garment. [This is before the Soma plants are pressed. They are resting in an open space, on the ground, waiting to be pressed.]
prá devám áchā mádhumanta índavó 'siṣyadanta gā́va ā́ ná dhenávaḥ
barhiṣádo vacanā́vanta ū́dhabhiḥ parisrútam usríyā nirṇíjaṃ dhire

IX 69[1ab]
The thought is placed like an arrow upon a bow; like a calf to the udder of his mother he hastens. [The figure may not refer to Soma but the context suggests that it does.]
íṣur ná dhánvan práti dhīyate matír vatsó ná mātúr úpa sarjy ū́dhani

IX 107[5ab]
Milking the dear sweetness from the divine udder, he has sat in his accustomed place.
duhāná ū́dhar divyáṃ mádhu priyáṃ pratnáṃ sadhástham ā́sadat

E. THE STALK AND SOMA

Not only is the Soma plant likened to an udder; the stalk or *aṃśú* (literally a 'shoot', a perfect word for the stipe of a mushroom) is likened to a teat:

I 137[3ab]

The priests milk this *aṃśú* for you both [Varuṇa and Mitra, two gods], like the auroral milch cow, with the aid of stones they milk the Soma, with the aid of stones.

tā́ṃ vāṃ dhenúṃ ná vāsarī́m aṃśúṃ duhanty ádribhiḥ sómaṃ duhanty ádribhiḥ

II 13[1cd]

The first milk of the *aṃśú* is the best.

tád āhanā́ abhavat pipyúṣī páyo 'ṃśóḥ pīyū́ṣaṃ prathamáṃ tád ukthyàm

III 36[6cd]

Indra is farther than this seat when the milked *aṃśú*, the Soma, fills him.

átaś cid índraḥ sádaso várīyān yád īṃ sómaḥ pṛṇáti dugdhó aṃśúḥ

IV 1[19cd]

He has tapped so to speak the pure udder of the cows, rendering the milk clear as is the juice yielded by the *aṃśú*.

śúcy ū́dho atṛṇan ná gávām ándho ná pūtáṃ páriṣiktam aṃśóḥ

V 43[4]

The ten fingers, the two arms, harness the pressing stone; they are the preparers of the Soma, with active hands. The one with good hands [the priest] has milked the mountain-grown sap of the sweet honey [Soma]; the *aṃśú* has yielded the dazzling [sap].

dáśa kṣípo yuñjate bāhū́ ádriṃ sómasya yā́ śamitā́rā suhástā mádhvo rásaṃ sugábhastir giriṣṭhā́ṃ cániścadad duduhe śukrám aṃśúḥ

VIII 9[19ab]

When the swollen *aṃśú* were milked like cows with [full] udders ...

yád ā́pītāso aṃśávo gā́vo ná duhrá ū́dhabhiḥ

IX 72[6a]

They milk the thundering[1] *aṃśú* ...

aṃśúṃ duhanti stanáyantam ákṣitaṃ kavíṃ kaváyo 'páso manīṣíṇaḥ

1. The word for 'thundering' here, *stanáyantam*, from the verb *stan*, 'to roar' or 'to thunder', is probably related to the Vedic and classical Sanskrit word for breast, *stana*, thought also to be derived from *stan*, perhaps *via* the image of the cloud.

Plate vi · RV IX 71^{2bc}: He sloughs off the Asurian colour that is his. He abandons his envelope.

IX 95[4ab]
They milk the *aṃśú*, this bull at home on the mountain.
táṃ marmṛjānáṃ mahiṣáṃ ná sā́nāv aṃśúṃ duhanty ukṣáṇaṃ giriṣṭhā́m

IX 107[12cd]
With the milk of thy *aṃśú* . . .
aṃśóḥ páyasā madiró ná jā́gṛvir ácchā kóśaṃ madhuścútam

In the light of my fly-agaric hypothesis, the milking imagery that pervades the Soma passages in the ṚgVeda acquires new meaning. A chlorophyll-bearing plant, whether leafy or leafless, does not suggest the udder and milking. The dominance of the word *aṃśú* – stalk, stem, stipe – calls for comment. Over much of Eurasia certain important species of wild mushrooms are dried and strung together on strings by the stipes, hanging caps down. (The caps shrink more than the stipes and are more friable.) This may have been the practice in Vedic times and would explain, if explanation be needed, the emphasis on the stipe in the vocabulary.[1] Bogoraz, writing about the turn of the century, says expressly that in the Chukotka the fly-agarics were usually *strung up* in three's, this being it seems the trading unit. (*Vide* [22] p. 273)

F. SOMA'S 'HEAD'

In English we speak of the 'cap' of the mushroom, but in many other languages including the Vedic 'head' is used instead:

IX 27[3]
This bull, heaven's head [*mūrdhán*], Soma, when pressed, is escorted by masterly men into the vessels, he the all-knowing.
eṣá nṛ́bhir ví nīyate divó mūrdhā́ vṛ́ṣā sutáḥ sómo vàneṣu viśvavít

IX 68[4cd]
While Soma enters into contact with the fingers of the officiants, he protects his head [*śíras*].
aṃśúr yávena pipiśe yató nṛ́bhiḥ sáṃ jāmíbhir násate rákṣate śíraḥ

1. When I was a boy in Chesterfield County, Virginia, the farmers would call the leaves of the tobacco plant 'stems'. When the time came for gathering the leaves, they would string them together by the principal rib or 'stem', perhaps a score of them together, and take them to the tobacco barn to be dried by smoking, hanging each batch over rungs that stretched from beam to beam in the barn. Fires were then built in troughs on the ground. This is another example where handling practices led to a curious use of the word 'stem'.

IX 69[8cd]

For you are, O Soma juices, . . . the heads [*mūrdhán*] of heaven, carried erect, creators of vital force.

yūyáṃ hí soma pitáro máma sthána divó mūrdhā́naḥ prásthitā vayaskṛ́taḥ

IX 71[4cd]

On Soma's head [*mūrdhán*] the cows with a full udder mix their best milk in streams. [*i.e.*, milk is mingled with Soma juice.]

ā́ yásmin gā́vaḥ suhutā́da ū́dhani mūrdhā́ñ chrīṇánty agriyáṃ várīmabhiḥ

IX 93[3abc]

The udder of the cow is swollen; the wise juice is imbued with its streams. In the vessels the cows mix with their milk the *mūrdhán*.

utá prá pipya ū́dhar ághnyāyā índur dhā́rābhiḥ sacate sumedhā́ḥ
mūrdhā́naṃ gā́vaḥ páyasā camū́ṣv abhí śrīṇanti vásubhir ná niktaíḥ

G. FOUR POETIC CONCEITS

In speaking of Soma the poets of the ṚgVeda have recourse repeatedly to four conceits. They have never been adequately explained. They are not descriptive in any immediate sense: they express what the poets considered transcendental truths. Let us see how the fly-agaric fares with them.

1. The Single Eye.[1]

I 87[5ab]

We speak because of our descent from the ancient father; the tongue moves with *the eye of Soma.*

pitúḥ pratnásya jánmanā vadāmasi sómasya jihvā́ prá jigāti cákṣasā

IX 9[4]

Quickened by the seven minds, he [Soma] has encouraged the rivers free of grief, which have strengthened *his single eye.*

sá saptá dhītíbhir hitó nadyò ajinvad adrúhaḥ yā́ ékam ákṣi vāvṛdhúḥ

IX 10[8ab]

I have drunk the navel [*i.e., Soma*] into the navel [*i.e., stomach*] for our sake. Indeed, *the eye* is altogether with the sun. [*Bhawe rendering*]

nā́bhā nā́bhiṃ na ā́ dade cákṣuś cit sū́rye sácā kavér ápatyam ā́ duhe

1. In the Atharvaveda, XIII 1[45], there is another allusion to the single eye that is relevant here. In Wm. D. Whitney's translation, p. 717.

IX 10[9]
The sun [*i.e.*, Soma] looks *with the eye* towards the dear places and the highest place of heaven, . . . [*Bhawe rendering*]
abhí priyā́ divás padám adhvaryúbhir gúhā hitám sū́raḥ paśyati cákṣasā

IX 97[46c]
[Soma] who has *for eye* the sun
svàrcakṣā rathiráḥ satyáśuṣmaḥ

The one element in these verses, some of them difficult, that concerns us here is 'the single eye'. Does not the photograph, reproduced in Plate x, explain the image that the poet had in mind? How perverse this metaphor is if we have to do with a creeper, *vallī*. How meaningless if we deal with rhubarb.

2. Mainstay of the Sky.

IX 2[5]
The ocean [of Soma] has been cleansed in the waters; *mainstay of the sky*, the Soma in the filtre, he who is favourable to us.
samudró apsú māmṛje viṣṭambhó dharúṇo diváḥ sómaḥ pavítre asmayúḥ

IX 72[7ab]
In the navel of the earth [is situated the Soma], which is also the *mainstay of the sky* . . .
nā́bhā pṛthivyā́ dharúṇo mahó divò 'pā́m ūrmaú síndhuṣv antár ukṣitáḥ

IX 74[2ab]
Mainstay of the sky, well laid, the full *aṃśú* runs throughout everything . . .
divó yáḥ skambhó dharúṇaḥ svàtata ā́pūrṇo aṃśúḥ paryéti viśvátaḥ

IX 86[35cd]
. . . thou sittest in the vessels, having been pressed for Indra, inebriating drink, which inebriates, supreme *mainstay of heaven*, [Soma] who gazes in the far distance.
índrāya mádvā mádyo mádaḥ sutó divó viṣṭambhá upamó vicakṣaṇáḥ

IX 86[46ab]
He has spilled forth, *mainstay of the sky*, the offered drink; he flows throughout the world . . .
ásarji skambhó divá údyato mádaḥ pári tridhā́tur bhúvanāny arṣati

IX 87^{2cd}
... father of the gods, progenitor of the moving force, *mainstay of the sky*, foundation of the earth.
pitā́ devā́nāṃ janitā́ sudákṣo viṣṭambhó divó dharúṇaḥ pṛthivyā́ḥ

IX 89^{6ab}
Mainstay of the sky, foundation of the earth, all establishments are in the hand of this [Soma] ...
viṣṭambhó divó dharúṇaḥ pṛthivyā́ víśvā utá kṣitáyo háste asya

IX 108^{16}
Enter into the heart of Indra, receptacle for Soma, like rivers into the ocean, thou [O Soma] who pleasest Mitra, Varuṇa, Vāyu, supreme *mainstay of heaven*!
índrasya hā́rdi somadhā́nam ā́ viśa samudrám iva síndhavaḥ júṣṭo mitrā́ya váruṇāya vāyáve divó viṣṭambhá uttamáḥ

IX 109^{6a}
Thou Soma art the *mainstay of the sky*, ...
divó dhartā́si śukráḥ pīyū́ṣaḥ satyé vídharman vājī́ pavasva

We have given only a selection of the passages where Soma is a mainstay of the sky. Others have translated this by 'pillar' and 'fulcrum' of the sky.

What poet could conceive of a creeper, a climber, any vine – some species of Sarcostemma or Ephedra – as 'mainstay of the sky', 'foundation of the earth'? But the sturdy stanchion with its resplendent capital that is the fly-agaric lends itself well to this poetic conceit.

3. The Navel.

'Navel', *nā́bhi*, is one of the most important words in the ṚgVeda. In its primary meaning as the umbilicus it occurs only once, in a late hymn, X 90^{14}. As the 'hub' of a wheel it recurs three times; this use need not detain us. What we call 'blood kin' for the Vedic poets was 'umbilical kin'; in this sense we find it nine times. By far the most interesting citations of the word are the ones where it is used transcendentally, to express a mythological idea, in a reverential and sacred context. Soma is the Navel of the Way (= *Ṛtá*), says the poet. By *Ṛtá* he means the divine order of things, a word that

PLATE VII · RV IX 71^{2d}: He makes of milk his vesture-of-grand-occasion.

PLATE X · RV IX 9^{4c}: The single eye.

seems to convey somewhat the same idea as the *Tao* of Lao Tze.

This figurative use of the navel for Soma in Vedic times arrested my attention. In the fungal vocabularies of various unrelated Eurasian peoples I had come across the navel, and here it was playing a rôle in the India of the Aryans! Was this accidental convergence, or was it because Soma was a mushroom – the fly-agaric?[1] True, all words relating to childbirth play a bigger rôle in the vocabularies of primitive peoples than among others. Thus in ancient Greece the ὄμφαλος, a stone carving fixed in Delphi and now in the Delphi museum, was famed for centuries as the center of the world.

But it is a fact that peoples who know their mushrooms and live with them in their daily lives are apt to see parallels between mushrooms and the umbilicus. The lexicographer Dal′ reports that in Russia *pup* means 'navel' and the derivative *pupyri* is applied in the familiar language to fungal growths. In contemporary Cambodian the word *psət* means both navel and mushroom. This word's primary meaning is navel and it is clearly borrowed from the Malayo-Polynesian family of languages, where the hypothetical proto-Malayo-Polynesian word was **pusəg*, 'navel', a word with far flung progeny in languages spoken off the east coast of Asia, including *heso* in Japanese, whose medieval form was *feso*, stemming back to **pëso* in proto-Japanese. In standard contemporary Korean the word recurs in the form *p°sət*. There it means only 'mushroom', but in two southwestern dialects spoken only on the island of Tšedju, in the province of Cholla Namdo, words that may stem back to the same root mean the navel. In these dialects the navel takes the forms *pötoŋ* and *pötoŋ-ro*.[2]

In April 1966 Georges Dumézil introduced to me a young Turkish national, Orhan Alparslan, a student of architecture at the University of Paris. He came from a village, Zennun, situated off the highway

1. One of our valuable sources about the fly-agaric in Siberia is G. H. von Langsdorf. He found it in use as an inebriant among the Kamchadal. In discussing its fungal identification he writes: '... the Kamchadal mushroom has a cap with a *navel-like* protuberance in the middle, ...'! *Vide* [10], p. 247. The word in the German text is *nabelförmig*.

2. I am indebted to Dr. Johannes Rahder of Yale University for this information about the Korean dialects. He relied on the two volume work on Korean dialects compiled by the late Professor Shinpei Ogura, *Chōsen-go Hōgen no Kenkyū* (Studies of Korean Dialects); publishers, Iwanami, Tokyo, 1944, pp. 97-98 of Vol. 1.

running from Ankara to Samsun, at about the half-way point. He was a Circassian and his village was largely Circassian. He had been a shepherd boy in his childhood, and he possessed a remarkably clear memory. He described for me the various kinds of mushrooms that he had learned to know and gave me the names for them. He drew the shaggy mane – *Coprinus comatus* –, explained that it quickly turned into black ink but the black was tinged with violet, and gave its local name as *göbek mantari*, Turkish words meaning 'navel mushroom'.

Even in the scientific vocabulary of mycologists the navel has crept in. Many species of mushrooms are either 'umbilicate' or 'umbonate', depending on whether the 'navel' is a pocket in the middle of the cap or a protrusion. There is a genus, the Omphalia, whose name comes from the Greek ὄμφαλος, navel. True, these are neo-classical words devised in the last two centuries by scientists, but scientific names, especially those devised in the youth of the science (such as is here the case), often reflect facts lying deep in the consciousness of the race. The *muscaria*, for example, in *Amanita muscaria*, expresses the folk traditions of the Germanic race.

It is not my contention that these far flung analogies in vocabulary have influenced each other; quite the contrary. I believe a mushroom is apt to suggest the navel to the primitive observer, wherever he be, and that the Vedic people, obsessed as they were with the fly-agaric, applied to it the navel analogy, and imbued it with a multitude of transcendental meanings. Already we have seen a number of these figures of speech. We learned in IX 74[4] that Soma was the Navel of the Way. Here are others:

IX 79[4ab]
Your highest navel is attached in heaven; your fingers grow on the back of the earth.
diví te nā́bhā paramó yá ādadé pṛthivyā́s te ruruhuḥ sā́navi kṣípaḥ

We recall how the single eye of Soma was tied to the eye of the sun in IX 10[8]; in that same verse the navel now figures intelligibly:

IX 10[8]
I have drunk the navel [*i.e.*, Soma] into the navel [*i.e.*, stomach] for

Plates VIIIa & VIIIb

RV IX 97^{9d}: By day he appears *hári*. By night, silvery white.

our sake. Indeed, the eye is altogether with the sun. I have milked the child of the wise. [*Bhawe rendering*]

nā́bhā nā́bhiṃ na ā́ dade cákṣuś cit sū́rye sácā kavér ápatyam ā́ duhe

Again two hymns later Soma is called the all-seeing navel of the wise:

IX 12[4]

The sharp seer, in heaven's navel, is magnified in the woolen filtre, Soma the wise, possessed of good intelligence.

divó nā́bhā vicakṣaṇó 'vyo vā́re mahīyate sómo yáḥ sukrátuḥ kavíḥ

In the following verse Soma's navel is associated with the 'head' (= cap) of the mushroom:

I 43[9abc]

Thy descendants, O Immortal One, according to the supreme institution of the Way, receive them on thy navel, O Soma, thou who art the head [of heaven];

yā́s te prajā́ amṛ́tasya párasmin dhā́mann ṛtásya mūrdhā́ nā́bhā soma vena

Soma is repeatedly said to be in the navel of the earth, in the navel of heaven. It was a practice of the ṚgVeda poets to use epithets proper to one god when speaking of another. Agni and Soma, fire and the 'fire-agaric' (as I am tempted to call the fly-agaric) thus exchange epithets, and 'navel' is often applied to Agni. But I think it will not be disputed that the navel figure belongs to Soma; the frequency of its use with Soma and the elaboration of the uses make it peculiarly Soma's. As we have seen, pp. 39 ff, 'Agni' for the poet is a way of saying 'Soma'.

4. The Filtres.

In the ṚgVeda filtres figure prominently. One of them, a filtre of lamb's wool, presents no problem. After the Soma had been pounded with stones and mixed with water, it was forced through a filtre or strainer, which caught the pulp and fibrous elements and allowed the tawny yellow liquor to pass through and run down to the vats.

But the ṚgVeda speaks of two other filtres that have always baffled the scholars. If Soma is the fly-agaric they present no problem. The woolen filtre is in fact the second or middle filtre.

The First Filtre

In the order of their function, the verses speak first of a celestial filtre, the filtre that I offer to the reader in Plate xii, where the sun's rays, escorting Soma down from the sky, are caught and held on the fiery back of heaven (= the pileus of the fly-agaric).

In the first verse that I shall quote the poet says expressly that King Soma has the filtre for his chariot, and immediately thereafter cites the thousand studs, *bhṛṣṭí,* that carry him to fame; *i.e.*, the white patches on his cap.

IX 86^{40cd}
King, having the filtre for chariot, he has attained the victory prize;
a thousand studs, he conquers puissant renown.
rā́jā pavítraratho vā́jam ā́ruhat sahásrabhṛṣṭir jayati śrávo bṛhát

IX 10^{5}
(This verse has puns difficult to translate. Its meaning is clear. The Soma plants are called 'suns', *sū́rā,* a natural metaphor in the light of our various plates. The heavenly Somas spread the strainer of their (= the sun's) rays for *themselves* to come down).
āpānā́so vivásvato jánanta uṣáso bhágam sū́rā áṇvaṃ ví tanvate

How clear this would have been for Geldner and Bhawe if they had possessed the key. Bhawe, commenting on this verse in his Part I, p. 53, refers to the 'mysterious sieve through which the sun's rays pass'. Geldner before him had sensed that the Somas and the Suns are the same, and in his commentary had divined that the filtre straining the sun's rays is referred to elsewhere, notably in IX 66^{5}, 76^{4}, and 86^{32}.[1]

IX 66^{5abc}
Thy clear rays spread over the back of heaven, the filtre, O Soma, . . .
táva śukrā́so arcáyo divás pṛṣṭhé ví tanvate pavítraṃ soma dhā́mabhiḥ

As a poetic figure for the fly-agaric, there seems nothing to explain here. In this same hymn, verses 19-21, Soma is addressed as Agni,

1. Harvard Oriental Series, 35, p. 17, ftn. to 5c.

PLATE IX · RV IX 70[7d]: The hide is of bull, the dress of sheep.

i.e., the fly-agaric as fire. (We have already called attention to this on p. 39.) The fly-agaric is both fire and sun, it catches the sun's rays on its back and holds them there, where they filtre the Soma juice into the plant. The same theme recurs in

IX 67[22-25]
This Soma, which today circulates in the distance, which is a cleanser, may it cleanse us in the filtre!
The filtre that has been spread in thy flame, O Agni, with it, cleanse our song.
Thy filtre, O Agni, equipped with flames, may it cleanse us, cleanse us with the fruits of sacred songs!
With these both, the filtre and the fruits [of song], O God Savitṛ, cleanse me through and through!

pávamānaḥ só adyá naḥ pavítreṇa vícarṣaṇiḥ yáḥ potā́ sá punātu naḥ
yát te pavítram arcíṣy ágne vítatam antár ā́ bráhma téna punīhi naḥ
yát te pavítram arcivád ágne téna punīhi naḥ brahmasavaíḥ punīhi naḥ
ubhā́bhyāṃ deva savitaḥ pavítreṇa savéna ca mā́ṃ punīhi viśvátaḥ

Here Soma is addressed under the name of Agni. Metaphorically the miraculous plant seems to share every attribute of Agni, – flame-coloured, subtle, it purifies with its filtre as fire does with its flames.

In the following verse the strainer is not mentioned by name. Soma, Lord of the Universe, cleanses himself in the Sun's rays, the celestial filtre:

IX 76[4]
Monarch of everything that sees the sun-light, Soma cleanses himself. Triumphing over the Prophets, he made the Word of the Way to resound, he who is cleansed by the Sun's ray, he the father of poems, Master-Poet never yet equalled!
víśvasya rā́jā pavate svardṛ́śa ṛtásya dhītím ṛṣiṣā́ḷ avīvaśat yáḥ
sū́ryasyásireṇa mṛjyáte pitā́ matīnā́m ásamaṣṭakāvyaḥ

IX 83[1ab]
Thy filtre has been spread, O Bráhmaṇaspáti [Soma] . . .
pavítraṃ te vítataṃ brahmaṇas pate prabhúr gā́trāṇi páry eṣi viśvátaḥ

IX 83^{2abd}

The filtre of the burning [Soma] has been spread in heaven's home. Its dazzling mesh was spread afar ... They climb the back of heaven in thought.

tápoṣ pavítram vítatam divás padé śócanto asya tántavo vy àsthiran
ávanty asya pavītā́ram āśávo divás pṛṣṭhám ádhi tiṣṭhanti cétasā

In this verse the 'filtre' that we see in Plate XII is translated to heaven in the flame of the sacrifice. How easy it is for the poet to move from the earthly to the transcendental plane. How compact is the cosmology. Soma, fire, sun, sun's rays, the navel of the earth, the single eye, mainstay of the sky, celestial strainer, thunder-storm, aurora – they are all interlocked, meeting in our resplendent fly-agaric.

Here are further verses in the same tenor:

IX 86^{30ab}

As for thee, O Soma-juice, thou art clarified in the filtre so as to establish thyself [in] space for the gods.

tváṃ pavítre rájaso vídharmaṇi devébhyaḥ soma pavamāna pūyase

IX 86^{32ab}

The Soma envelops himself all around with rays of the sun, ...

sá sū́ryasya raśmíbhiḥ pári vyata tántuṃ tanvānás trivṛ́taṃ yáthā vidé

IX 91^{3cd}

... By a thousand paths free of dust, Soma, armed with verses, knowing the Word, the Sun passes the filtre.

sahásram ṛ́kvā pathíbhir vacovíd adhvasmábhiḥ sū́ro áṇvaṃ ví yāti

It seems that in this verse the Sun is a metaphor, standing for Soma.

The Third Filtre

A third filtre is mentioned in two verses of the ṚgVeda, and this brings us back to the discussion of the *Two Forms* of Soma on pp. 25 ff, which we promised to resume at a later point in the argument.

IX 73^{8ab}

The Guardian of the *Ṛtá* [Soma] cannot be deceived, he of the good inspiring force; he carries three filtres inside his heart.

ṛtásya gopā́ ná dábhāya sukrátus trī́ ṣá pavítrā hṛdy àntár ā́ dadhe

IX 97[55]
Thou runnest through the three filtres stretched out; thou flowest the length, clarified. Thou art Fortune, thou art the Giver of the Gift, liberal for the liberal, O Soma-juice.

sáṃ trī́ pavítrā vítatāny eṣy ánv ékaṃ dhāvasi pūyámānaḥ ási bhágo ási dātrásya dātā́ 'si maghávā maghávadbhya indo

Let us assume the fly-agaric surmise is well founded. Then the third filtre becomes clear: the Soma juice that is drunk by 'Indra' and 'Vāyu' in the course of the liturgy is filtered in their organisms and issues forth as sparkling yellow urine, retaining its inebriating virtue but having been purged of its nauseating properties.

That the priceless ambrosia was filtered down from the celestial sphere on the sun's rays into the plant is clear. That the Soma juice was filtered through the lamb's wool into the vessels at the place of sacrifice is also clear. What happened next? 'Indra' and 'Vāyu' consumed the liquor mixed with milk or curds and it would appear that their condition was a matter of considerable anxiety. How else are we to explain the poets' preoccupation with the Soma as it passed through their organisms? The poets do not stress the inebriation of the priests. Instead they take us with Soma into Indra's heart, into his belly or bellies, into his entrails. If these verses do not mean that in the Vedic ritual the priests were impersonating the gods, what do they mean?

IX 70[10bcd]
Purify thyself in Indra's stomach, O juice! As a river with a vessel, enable us to pass to the other side, thou who knowest; thou who battlest as a hero, save us from disgrace!

[arṣ]éndrasyendo jaṭháram ā́ pavasva
nāvā́ ná síndhum áti parṣi vidvā́ñ chū́ro ná yúdhyann áva no nidáḥ spaḥ

Or the preceding verse:

IX 70[9]
Clarify thou thyself, O Soma, for the invitation to the gods. Thou who art a bull enter into the heart of Indra, receptacle for Soma! Enable us to traverse the evil passages saving us from oppression!

For he who knows the country gives the directions to him who informs himself.

pávasva soma devávītaye vŕ̥séndrasya hā́rdi somadhā́nam ā́ viśa
purā́ no bādhā́d duritā́ti pāraya kṣetravíd dhí díśa ā́hā vipr̥cchaté

Obviously there was doubt about the outcome of the perilous passage. 'Indra' and his colleagues had to know their business, had to be experienced pilots. It looks as though Bhawe was justified in saying that skill was needed in mixing the Soma juice with milk or curds in the right way.

Is not the following verse imbued with new meaning, in the light of my interpretation, – the *human waters* being put into movement?

IX 63[7]
Clarify thou thyself by that stream by which thou madest the sun to shine, putting into movement the human waters!

ayā́ pavasva dhā́rayā yáyā sū́ryam árocayaḥ
hinvānó mā́nuṣīr apáḥ

Here are other references to Soma in the belly of Indra:

IX 72[2ab]
. . . the [Officiants] . . . draw the Soma by milking into the belly of Indra.

sākám̐ vadanti bahávo manīṣíṇa índrasya sómam̐ jaṭháre yád āduhúḥ

IX 72[5ab]
Spurred on by the two arms of the Officiant, in jets, the pressed Soma is clarified according to its nature, suitable for thee, O Indra!

nŕ̥bāhúbhyām̐ coditó dhā́rayā sutò 'nuṣvadhám̐ pavate sóma indra te

IX 76[3a]
O Soma Pávamāna, . . . penetrate into the entrails of Indra!

índrasya soma pávamāna ūrmíṇā taviṣyámāṇo jaṭháreṣv ā́ viśa

IX 80[2d]
O Soma, thou clarifiest thyself for Indra; . . .

índrāya soma pavase vŕ̥ṣā mádaḥ

IX 80[3a]
In the belly of Indra the inebriating Soma clarifies itself.

éndrasya kukṣā́ pavate madíntama

PLATE XI · RV IX 109^{6a}: Thou art the mainstay of the sky.

Here is a clear statement. As in the other two filtres, Soma is clarifying itself in the belly of Indra, in preparation for a further step.

IX 81[1ab]
The waves of Soma Pávamāna advance into the belly of Indra.
prá sómasya pávamānasyormáya índrasya yanti jaṭháraṃ supéśasaḥ

IX 85[5cd]
Cleansed like a winning race horse, thou hast spilled thyself in the belly of Indra, O Soma!
marmṛjyámāno átyo ná sānasír índrasya soma jaṭháre sám akṣaraḥ

The 86th hymn is climactic for Indra as Soma's filtre:

IX 86[2]
Thy inebriating drinks, swift, are released ahead, like teams running in divers directions, like the milch cow with her milk towards her calf, so the Soma juices, waves rich in honey, go to Indra, thunder-bolt carrier.
prá te mádāso madirā́sa āśávó 'sṛkṣata ráthyāso yáthā pṛ́thak dhenúr
ná vatsáṃ páyasābhí vajríṇam índram índavo mádhumanta ūrmáyaḥ

IX 86[3]
Like a race horse launched in movement for the victory prize, flow, O Soma, thou who procurest the light-of-the-sun for heaven's vat, whose mother is the pressing stone; thou, Bull, seated in the filtre above the calf's wool, clarifying thyself, thou Soma, that Indra may have his pleasure!
átyo ná hiyānó abhí vā́jam arṣa svarvít kóśaṃ divó ádrimātaram
vṛ́ṣā pavítre ádhi sā́no avyáye sómaḥ punāná indriyā́ya dhā́yase

IX 86[16a]
He advances to the rendez-vous with Indra, the Soma juice . . .
pró ayāsīd índur índrasya niṣkṛtáṃ

IX 86[19cd]
By the action of the streams he has made the utensils resound while penetrating into the heart of Indra.
krāṇā́ síndhūnāṃ kaláśāñ avīvaśad índrasya hā́rdy āviśán manīṣíbhiḥ

IX 86[22]
Clarify thyself, O Soma, in the celestial structures of thine essence, thou who hast been released roaring into the vessel, in the filtre.

Lodged in the belly of Indra, roaring with vigour, held in hand by the Officiants, thou hast made the sun to mount the sky.

pávasva soma divyéṣu dhā́masu sṛjānā́ indo kaláśe pavítra ā́ sī́dann índrasya jaṭháre kánikradan nṛ́bhir yatáḥ sū́ryam ā́rohayo diví

IX 86[23ab]

Pressed by the pressing-stones, thou clarifiest thyself in the filtre, O Soma-juice, when penetrating into the entrails of Indra!

ádribhiḥ sutáḥ pavase pavítra ā́ṅ índav índrasya jaṭháreṣv āviśán

★

IX 108[15]

For Indra, that he may drink, clarify thou thyself, O Soma, held in hand by the Lords, well armed, inebriating ...

índrāya soma pā́tave nṛ́bhir yatáḥ svāyudhó madíntamaḥ pávasva mádhumattamaḥ

IX 108[16]

Enter into the heart of Indra, Soma's receptacle, like the rivers into the ocean, thou, [O Soma,] who pleasest Mitra, Varuṇa, Vāyu, O thou supreme Mainstay of the Sky!

índrasya hā́rdi somadhā́nam ā́ viśa samudrám iva síndhavaḥ júṣṭo mitrā́ya váruṇāya vāyáve divó viṣṭambhá uttamáḥ

IX 109[18]

O Soma, advance into the belly of Indra, having been held in hand by the Officiants, pressed by the stones!

prá soma yāhī́ndrasya kukṣā́ nṛ́bhir yemānó ádribhiḥ sutáḥ

H. 'Tongue of the Way'.

The second verse of IX 75 begins with 'Tongue of the Way' (*ṛtásya jihvā́*), and the poet continues to apostrophize Soma as the source of eloquence. 'The Way' is *Ṛtá*, the divine order of things. Abel Bergaigne[1] remarked that the expression, 'Tongue of the Way', was picturesque and said it meant prayer. Caland and Henry,[2] on the other hand, were baffled. 'Tongue of the Way' could not be translated in any other way, but how could Soma be a tongue? These scrupulous scholars did not visualize the fly-agaric: its cap, the full blown red tongue, held the clue to the little mystery.

1. Abel Bergaigne, *La Religion Védique*, Vol. III, p. 241.
2. W. Caland and V. Henry, *L'Agniṣṭoma*, Paris, 1906, Para. 221, p. 338, ftn. 7.

Plate xii · RV IX 86[40d]: With his thousand knobs he conquers mighty renown.

Plate xiii · RV IX 75[2a]: Tongue of the Way.

I. The 'Knots' or 'Knobs' or 'Studs' or 'Spikes' on the Fly-Agaric.

Already we have seen how well the spots on the fly-agaric serve as the first of the three 'filtres' for Soma. In a more mundane sense these spots are called in Vedic *bhṛṣṭí*, a word that is used for the knobs or studs on a cudgel, as on the cudgel of Indra. With his thousand knobs or studs Soma conquers potent fame: so say the hymns in two places, IX 83^{5d} and 86^{40d}.

There is yet another passage pertinent to this theme, and it illustrates well the kind of problem that working in a language as remote as Vedic leads one into. Early in my inquiries I came upon a verse, IX 15^{6}, that seemed to present an obstacle to my fly-agaric thesis. According to the poet of this hymn, when the Soma plant is pressed and then run through the woolen filtre, it leaves behind in the filtre its 'knots' or 'nodules'. (The Vedic word, in the plural, is *páruṣā*.) There has been some difficulty with the sense of the sentence, but agreement on this particular word. Now mushrooms have no knots or nodules, which are characteristic of shrubs and trees. Here was a hurdle to cross or I was in trouble.

The latest translation of this hymn is S. S. Bhawe's (1957), and I found that he had devoted two pages of concentrated commentary to all the words that attend these 'knots'.[1] Suddenly, without his knowing the full import of what he was saying, he cleared up my difficulties. The 'knots', it seemed, had been 'sticking to' (*pibdanā́*) Soma's body, and they were shining (*vásūni*) also! His verbal analysis is original and seemingly sound. Without knowing of our fly-agaric thesis, he comes out with a sense that fits the fly-agaric perfectly. The shining 'rays' (as the Vedic poets are always referring to them), the scales or white patches or knobs or warts sticking to the cap (as we say) of the fly-agaric, are left behind in the strainer. Whether the Vedic word *párus* covers a semantic area broad enough to embrace not only the knots in wood but also the 'knobs' on the fly-agaric's cap, or whether the poet was resorting to a metaphor as we do in English, I leave to Vedic scholars to determine. For the Indo-Aryans the white spots of the fly-agaric were shining (*vásūni*) knobs or studs (*páruṣā*)

1. Part 1, pp. 71-3; *vide* p. 6, ftn. 1.

sticking (*pibdanā́*) to the cap. This example illustrates the delicacy of the task of the translator, the rare prescience of gifted students who are oft-times groping in the dark, and finally the help it will be to the scholar when at last he knows the identity of Soma and can familiarize himself with all its characteristics and properties.

As an instance of this help I will cite the word *sahásrapājas*, an epithet for Soma that occurs twice in the ṚgVeda in almost identical passages. (ṚgVeda IX 13^3 and 42^3). The first part of this word offers no difficulty: *sahásra-* means 'thousand'. But what is the meaning of a thousand *-pājas*? On this there has been much comment but no agreement. Some have thought the word meant 'forms', others have suggested 'colours', and yet a third commentator sees in it 'rays'. But in the light of our Plate XII and our discussion of the First Filtre are we not simply viewing the thousand 'studs' from a different metaphoric angle? This is consonant with the imagery of the First Filtre, the rays escorting Soma down from their heavenly abode and then filtering the divine inebriant with midwifely solicitude into the plant. We do not yet know the precise meaning of *-pājas*, nor the anatomy of its poetic associations, but we do localize it and, for lack of a better, 'rays' will do. The Vedic poets see the white spots on the fly-agaric either as 'studs' or as sparkles of the divine light. (We in the English-speaking world see them as warts disfiguring a repulsive toadstool.) If I am right in identifying Soma with the fly-agaric, then in re-studying the whole of the ṚgVeda we must at all times be alive to the numinous glow of this awesome plant, a plant with miraculous inebriating virtue fully matched by its vesture-of-grand-occasion. When still in the dark as to its identity, great scholars like Renou have felt that Soma was the heart and soul of the ṚgVeda. If we know now what Soma was, like the holy *óṣadhi* itself the ṚgVeda with its thousand hymns – **sahásrarc* – is certainly destined to glow again with a rebirth of radiance from its thousand facets.

IX

SOMA AND THE FLY

Does the fly-agaric help us in understanding this verse, which has troubled the translators? It reads, in part, as follows:

> I 119[9ab]
> To you, O Aśvins, that fly betrayed the Soma.
> *utá syā́ vāṃ mádhuman mákṣikārapan máde sómasyauśijó huvanyati*

The word for 'fly', *mákṣika*, might mean bee. The word for Soma is *mádhu*, 'honey', a frequent metaphor for Soma. The verse might well be interpreted as meaning that the bees have betrayed their honey to the Aśvins. This is grammatically and semantically unexceptionable, but it is banal to the point of inanity. On the other hand, if a fly betrayed Soma to the Aśvins, we are plunging to the depths of Indo-European folklore. Throughout northern Europe, wherever the fly-agaric is well known, there is a folk belief that the fly-agaric is linked to flies. Here we find the statement that the fly betrayed Soma to the Aśvins. Did the fly lead the Aśvins to the fly-agaric?

I have conducted experiments with the fresh Eurasian fly-agaric, splitting the stipe (stem, *aṃśú*) lengthwise and letting it bleed. Where flies have access to such stipes, the flies are drawn to them, suck the juice, and collapse helpless into a stupor. They do not die from the juice. On the contrary they recover completely in a matter of hours or one or two days. While in the stupefied state they may of course be killed by their enemies or be blown away to their deaths in non-viable surroundings.[1]

1. *Vide* also *Nature*, Vol. 206, No. 4991, pp. 1359-1360, June 26, 1965: 'Constituents of *Amanita Muscaria*', K. Bowden, A. C. Drysdale, G. A. Mogey.

X

WORDS USED FOR SOMA IN THE ṚGVEDA

The name Soma is derived from the root *su*, meaning 'to press'. Soma is the pressed one. Another word for Soma is *ándhas*. Both Soma and *ándhas* are used in the ṚgVeda to designate the plant and its juice, *ándhas* being probably cognate with ἄνθος, the Greek word for 'flower'. (It is as though the Aryans called Soma *the* flower.) In the ṚgVeda 'Soma' and *ándhas* are used exclusively for the sacred plant and its juice, but the plant must have carried a name before it was elevated to its high station as a god and before it could have been called by a name derived from the liturgical act of pressing; this early name is lost. The Vedic lexicographer Grassmann translated *ándhas* by the word *Kraut*, 'herb', such as 'food for cattle'. In classical Sanskrit one of the common words for 'herb' is *tṛ́ṇa*, and, surprisingly, in the earliest Sanskrit dictionary, the *Amarakośa* (*ca.* A. D. 450), it is defined not only as 'herb' but as 'mushroom'. In the ṚgVeda *tṛ́ṇa* occurs from time to time but no translator had ever found a passage where he said it meant 'mushroom'. But there is a hymn directed to the Ṛbhus that says:

> I 161 ^{11ab}
> In the uplands you have created *tṛ́ṇaṃ* for this people, in the lowlands ingenious waters.
> *udvátsv asmā akṛṇotanā tṛ́ṇaṃ nivátsv apáḥ svapasyáyā naraḥ*

Because Soma is repeatedly associated in the hymns with the heights, if I am right that Soma is the fly-agaric, this sentence should be reviewed to consider whether *tṛ́ṇa* does not mean here the divine plant, the fly-agaric. Another word often used for Soma in the ṚgVeda is *óṣadhi*, and the lexicographers tell us that it also meant 'herb'. But in the Vedic mind the plant categories did not correspond precisely to our own, and the three terms for 'herb' seem to have embraced all small fleshy plants. Similarly, as we shall see, the Chinese term *chih* 芝 originally meant a fleshy plant, but as time passed its meaning came to be confined to a particular mushroom.

Many other names are applied to Soma in the RgVeda, all of them metaphors stressing one or another of its aspects. In passages where the drink receives the highest homage it is sometimes called *amṛ́ta*, cognate with 'ambrosia', the liquor of immortality. That which is pressed is *aṃśú*, 'stalk'. Sometimes Soma is called simply *vīrúdh*, 'the plant'. The juice is *pávamāna*, from the root *pu*, 'clear flowing', or *índu*, the 'bright drop', or sometimes *drapsá*, the 'drop', or *rása*, the 'fluid', or *pitú*, the 'beverage', or *máda*, 'inebriation', or *mádhu*, 'honey'.

The rich Soma vocabulary reflects the importance of the plant and its sacred rôle. The ordinary word for a common or garden mushroom occurs only once in the ṚgVeda, in a hymn addressed to Indra, where the poet says:

> I 84[8ab]
> When will he trample upon the godless mortal as upon a *kṣúmpa*?
> *kadā́ mártam arādhásaṃ padā́ kṣúmpam iva sphurat*

Yāska, the earliest of the commentators on the ṚgVeda, who lived not later than the 5th century B. C., said *ksúmpa* meant a mushroom, *ahichatrikā*, and Geldner accepts this identification. The word survives in contemporary Hindi as *khumbi* and in the market place of Old Delhi one buys under that name a wild mushroom belonging to the genus Phellorina, close to or identical with *Phellorina Delestrei* Dur.

It is not surprising, on the contrary it is to be expected, that the sacred mushroom should be in a category by itself, segregated from the rest of the mushroom world. This is what happens among the Mazatec Indians of Mexico, in whose language there are two words, one that embraces mushrooms belonging to sacred species, and the other that includes the rest of the fungi. Among the Mazatecs the two words neatly divide the fungal world between them.

'Mushroom' in classical Sanskrit is *chattra*. In the sense of 'mushroom' this word may not have existed in Vedic times; certainly the Aryans did not bring it down from the North. The word itself comes from the root *chad*, 'to cover', and its primary meaning is 'parasol'. For southern peoples the parasol, furnishing protection from the sun, is of importance, and from Cambodia to Ethiopia it is a symbol of authority, in India the mark of the Kṣatriya caste, the rulers, the rajputs. Until recently the northern peoples have not known the parasol or the umbrella. When the Aryans invaded Iran and India, they gave to this newly discovered utensil an Aryan name, *chattra*, and later extended the meaning of that name to embrace the fleshy capped fungi.

In India today mushrooms do not play an important role. Only in the Northwest, in Kashmir and the Punjab, are they much relished, as well as among Muslims and Sikhs in the Punjab and Delhi. *Agaricus hortensis* (our cultivated mushroom, the *champignon de Paris*) is almost unknown, as well of course as *Agaricus campestris*. The mushroom that enjoys some popularity for the table is *Pleurotus Eryngii* in various of its subspecies. In Delhi large quantities of *Pl. Eryngii* s. sp. *nebrodensis* are imported from Afghanistan for sale in the market, where they are called *dhingri*. In Kashmir there is a smaller variety, *Pl. Eryngii* Fr. ex D. C. forma *tesselatus*, that goes by the name of *hĕḍar*, *hĕnḍa*. The aristocrats of the mushroom world in Kashmir are the morels, called *kana-gᵘ̈ch*, the 'ear-morel'; recently they have become too expensive for the people of Kashmir to eat, and today they are gathered only for export to the Punjab, New Delhi, and Paris. But they still figure in the folklore of Kashmir and must be served at wedding feasts, if there are means to buy them. For mushrooms considered inedible, *kukur-mut(t)ā*, 'dog's urine', and endless variations of the same expression, are the dominant word throughout India. In the languages derived from Sanskrit of modern India I have found no trace of a sacred mushroom.

The Laws of Manu, Chap. V 5, place a tabu on mushrooms for the three upper castes of Hindus (= 'twice-born men'): 'Garlic, leeks and onions, mushrooms [*kavaka*] and all plants springing from impure substances are unfit to be eaten by twice-born men.' It is impossible to say whether this prohibition is related to a sacred use of the fly-agaric in Vedic times; probably not. The hermit or ascetic lies under a similar inhibition, Chap. VI 14: 'Let him avoid honey, flesh, and mushrooms growing on the ground [*bhūmi kavaka*], *bhūstṛṇa*, . . .' Elsewhere *bhūstṛṇa* is said to mean, among other plants, mushrooms that grow on the ground, the same as *bhūmi kavaka*. Clearly the Hindus are mycophobes. In the Punjab, Kashmir, and the Northwest, where the population has been Muslim and Sikh for centuries – in the Indus Valley, south of the Hindu Kush – there is a deep-seated mycophilia running counter to the general Hindu attitude. This is the area where the ṚgVeda was composed.

There is one episode conspicuous in the religious annals of India in which mushrooms played, or may have played, a decisive rôle: the death of Buddha at the age of 80 in the middle of the sixth century B. C. He was making his way with his followers, talking and preaching as he went, through the kingdom of Magadha, the present state of Bihar, and had paused for the night at a place called Pāvā. One Chunda, a metal worker, asked him for dinner and he accepted. Among other dishes Chunda served *sūkara-maddava*, called in one recension *sūkara-mansa*. (These words are from the Buddhist scriptures written in the Pali language, a tongue spoken a century or more later than the events we are describing and farther to the west. The Buddha himself must have spoken Magadhi Prakrit, also closely related to Sanskrit.) The two terms mean the same thing, but what is that meaning? The first word means 'swine', 'boar', cognate with the Latin *sus*, English 'swine'. The second element means soft flesh. 'Soft flesh of swine?' 'Swine's soft flesh?' *i.e.*, the soft flesh of which pigs are fond? A subjective genitive or an objective genitive? Buddhism has never made much of this, but the two great schools, Mahayana and Theravāda, have disagreed on it. The Mahayana school has held that the Buddha ate pork, it proved to be bad, and he died of the effects, after dysentery. The Theravāda school believes, on the other hand, that he died from mushrooms, a food of which pigs are fond. A Chinese translation of the *Dīgha Nikāya*, including the Book of the Great Decease, made in the beginning of the fifth century A. D., rendered *sūkara-maddava* by *mu erh*, 'tree mushroom'.[1] I record the extraordinary circumstances of Buddha's death because they bring in mushrooms, but I am not pre-

1. *Vide* (1) '*Sūkara-Maddava* and the Buddha's Death', by Fa Chow, Silver Jubilee Vol. of *Annals* of Bhandarkar Oriental Research Institute, Poona, 1942, pp. 127-133; and (2) 'Nourriture du Dernier Repas du Buddha', by André Bareau, *Mélanges d'Indianisme*, Editions E. de Boccard, Paris, 1968, pp. 61-71.

pared at this time to advance an explanation for the myth that would link it with Soma: the sources – Chinese, Sanskrit, and Pali – need to be re-examined with minute care. Nor, so far as I now know, are the ṚgVeda and Soma to be associated with the triple-tiered *chattra* ('parasol' and 'mushroom') that surmounts the great stupa at Sanchi, one of the earliest and most awe-inspiring Buddhist structures that survive; nor with the megalithic 'mushroom-stones' found in great numbers in Kerala, and less often in Nepal.

XI

MISCELLANEA

George Watt, the botanist who devoted his life to the study of the flora of India,[1] is said to have declared that 'no plant is known at present [1884] which would fulfil all the requirements [of Soma], and he lays particular stress on the fact that the vague and poetical descriptions given of the Soma make any scientific identification almost impossible.'[2] Watt did not read Sanskrit, much less Vedic. He did not allow for the deficiencies in the translations of the ṚgVeda on which he depended: he may have been unaware of them. The early renderings of the ṚgVeda made the ancient poets sound like upper class Europeans of the 19th century, frock-coated, with their inhibitions and pruderies, their etiolated religious outlook. Even today one hears it said that the ṚgVeda is vague and contradictory about Soma. This recalls the blind men who were defining the elephant by feeling each one a different part of the huge beast. Their reports were contradictory indeed. Who is to say that the ṚgVeda is contradictory without first knowing what Soma is?

The hymns of the ṚgVeda fit the fly-agaric like a glove. True, one must possess some awareness of the psychotropic plants of the world and their rôle in primitive religion. Given that familiarity, a reading of Geldner, Renou, and Bhawe leads straight to the fly-agaric.

Indians and Westerners who reverence the ṚgVeda for its religious feeling will perhaps be revolted by the dual forms of Soma and will even experience a visceral resistance to this solution of the enigma. Furthermore, a few Vedic scholars may know a momentary pang of regret. For almost a century and a half Soma has been the great unknown of Vedic studies, until this unknown had come to be considered a permanent built-in feature of this remote area of learning. Playing the Vedic game with aces wild has had its charms: it allowed individual

1. Watt's monumental work, *Dictionary of the Economic Products of India*, 1889-1896, edited and partly written by him, in many volumes, is a major legacy of the British rule in India, of lasting value.
2. Quoted from Max Müller, *Collected Works*, London, Vol. x, 1888, p. 223. The observations attributed by Müller to Watt do not appear in Watt's published statements.

leeway in reading difficult passages. It permitted even the great Bergaigne, in an unguarded moment, to suppose that Soma was merely some plant or other expressly selected by Aryan priests who went looking high in the mountains for this purpose (an 'Ecclesiastical Commission', we would call it), that the myth of Soma's mountain origin might be fulfilled;[1] as though the myth came first and the plant was an afterthought! This was possible only because Soma remained a blank. If Bergaigne was right, the poets of the ṚgVeda, and especially of Maṇḍala IX, must have been engaging in make-believe. His remark flouted all that the poets had to say in praise of the marvelous properties of the plant and the divine inebriation that resulted from drinking its juice. Their religion was founded on a hoax.

On the other hand the identification of Soma will give impetus to Vedic studies. The religion of the ṚgVeda now assumes body, fresh colour, a sharp bite. If I be right, the whole corpus of hymns, and the Avesta as well, must be re-read in the light of the discovery that a divine mushroom was at the center of these religions, was the focus of these poets. How astonishing that we can still draw parallels with the fly-agaric cult in Siberia, where as we shall see in *Part Three* it lingers on, in the last stages of degeneration among the peripheral tribes of the extreme north.

In Siberia the fly-agaric is utilized by the shamans. In the Indus Valley we associate it with an organized priesthood. This priesthood may have characterized Indo-European society in their homeland, but are we safe in assuming so? May not a shamanistic religion have acquired an hieratic structure under the pressures of a tough war of conquest lasting centuries? In a world of enemies the shamans may have found it in their own interest and the interest of the community to close their ranks, to band together and organize a tribal priesthood, as a weapon of political power. The assembling of the hymn book, about which Vedic scholars know something from internal evidence, may have kept step with the organizing of the priesthood.

1. Abel Bergaigne, *La Religion Védique*, Vol. 1, p. 183. The French text reads: 'Enfin le choix que les Aryas védiques faisaient d'une plante croissant sur les montagnes pour en tirer le breuvage du sacrifice, ne leur avait-il pas été suggéré par le mythe du Soma, venu de la montagne suprême, IX 87^8, c'est-à-dire du ciel et particulièrement des nuages du ciel, I 187^7?'

Why was Soma so soon abandoned in India, perhaps even before the forms were closed on the canon of the ṚgVeda? For one thing, questions of supply, which must always have been awkward, became impossible when the Indo-Aryans spread out over all of India. The mushroom crop in the Hindu Kush and the Himalayas was each year a fixed quantity. Of course for a time the priests could make do with insufficient fly-agarics (as they had had to do many times in seasons of short supply), stretching out the Holy Element by utilizing the *Second form*, and by mixing the precious fluid with ever more water and milk until only a symbol of the Real Presence remained. Possibly the *Second form* fell from favour and came under the condemnation of the priesthood, as it seems to have done in some quarters in Iran. The silence among later Indian writers about the true Soma would indicate that the decision to abandon it was deliberate and universal, and extended even to the discussion of it. But the memory of Soma must have stayed alive in the inner circles of Brahmans, perhaps for centuries. India is a land where the incredible sometimes comes true, and I should be delighted, but not altogether surprised, to discover that there are still circles privy to the knowledge of the true Soma.

As the substance of the Sacrifice became diluted and finally vanished, as the Divine Inebriant was reduced to a fading sacerdotal memory, inevitably more and more emphasis was placed by the priests on the efficacy of pure liturgy, and sacerdotalism proliferated to a point that the world has never seen equalled. The Brahmans of those days, profoundly moved by the legacy they had received from their ancestors and not yet possessed of an alphabet, set up, and copper-riveted in place, a method of preserving by sheer memory the words and melody of the original hymns that would withstand the vicissitudes of time. Would it be too much to say that the psychotropic mushrooms stirred them to make this supreme effort? With the passing of the generations Soma, the Divine Plant, no longer a part of Hindu experience, became sublimated in later Hindu mythology and took its place in the heavenly firmament as the Moon God. Dr. Stietencron has asked whether the Soma-Moon equivalence may not have been suggested by several of the attributes of the fly-agaric – its successive phases as it passes from the egg stage through to its ultimate end, all in a period comparable with the monthly cycles of the moon, though quicker; its pock-marked face; its white flesh; its white rotundity before it breaks through the white veil; the crescent that appears when the cap is bisected at a certain stage of its growth. A people richly endowed with poetic imagination would need only these features to see in the moon a transcendental fly-agaric.

The sublime adventure of religious contemplation, the mystical experience, which the priestly caste (and perhaps others) of the Indo-Aryans had known through the mediation of the fly-agaric, could now only be achieved through regulated austerity and mortification of the flesh, and the Hindus, who had known to the full the bliss that contemplation can give, made themselves the masters of these techniques: the price being counted as nothing compared with the prize.

XII

MANI, MUSHROOM, AND URINE

Possibly we can adduce evidence that will show vestiges of the cult of the sacred mushroom surviving in esoteric circles under Iranian influence down to the 11th century of the Christian era. The evidence is tenuous but in the light of my fly-agaric thesis it is tantalizingly suggestive. For the first time our trail leads us to China.

Of the sacred texts of Iran only a part survives, and much of what survives is corrupt and confusing. A succession of religions marked the history of ancient Iran – the primitive religion that Zoroaster undertook to reform, Zoroastrianism, Mithraism, Manichæism, some others, each of them preserving features of its predecessors and changing the emphasis, but all of them laying stress on the dualism of this world, the conflict between light and dark, between good and bad.

For some years before his conversion to the Christian faith in A.D. 386, St. Augustine was a follower of Mani, though he never visited Iran. Immediately thereafter he wrote his attack on the Manichæans, in which there is a passage, seldom noticed, condemning them for eating mushrooms, as well as other delicacies.[1] The Latin text reads as follows:

> Quid porro insanius dici aut cogitari potest, hominem boletos, orizam, tubera, placentas, caroenum, piper, laser, distento ventre cum gratulatione ructantem, et quotidie talia requirentem, non inveniri quemadmodum a tribus signaculis, id est a regula sanctitatis excidisse videatur.

Writing to confound the Manichæans, St. Augustine speaks of those who stuff themselves every day, to gratify their appetites, with *boletos* (a class of mushroom), rice (in Rome at that time an expensive

1. St. Augustine, *De Moribus Manichæorum* ('About the Ways of the Manichæans'), Chap. 13, Para. 30. In imperial Rome *boletus* was the name applied to what we call the genus Amanita, including both the edible and the toxic amanitas. *Vide*: *Mushrooms Russia and History*, Wasson & Wasson, New York, 1957, Chap. IV, Mushrooms for Murderers. So great was St. Augustine's influence on later churchmen that I think his mycophobic utterance may have had a part in shaping the virulent diatribes against mushroom-eaters of St. François de Sales and Jeremy Taylor. *Vide* our *MR&H*, pp. 21-2, 353.

luxury), truffles (or perhaps underground mushrooms of the genus known today as Terfezia, common in North Africa), flat cakes, sweet wine boiled until thick, pepper, and silphium (a spice, highly esteemed, from a plant now allegedly extinct), and he asks whether anyone can suppose that such people do not lose their standing in holiness.

Only an ethno-mycologist might remark, on coming across this sentence in the great Doctor of the Church, that of the seven cates bringing down his censure, two belong to the fungal tribe, that the first of the seven is the Amanitas to which the fly-agaric belongs, and that he is writing against the Manichæans, an Iranian religion founded by Mani about a century earlier. Even an ethno-mycologist could hardly do more than make a mental note of this; which note, however, would spring to life on reading about the Manichæans in China.

The Iranians introduced the religion of Mani into China in A. D. 694 and 719, and during the rest of the T'ang dynasty and down through the Sung this Iranian sect of Manichæans played something of a rôle in the religious life of the Chinese, especially in their impact on the Taoists.[1] They won a victory when in A. D. 763 they gained as a convert the Khan of the Uighurs, a powerful Mongolian people. Much later an unfriendly Chinese official, Lu Yu 陸游 (A. D. 1125-1209), wrote two reports on the activities and practices of the devotees of Manichæism in Fukien province. Among their evil ways he lists their practice of eating certain mushrooms:

> What they eat is always the *red mushrooms, hung hsün* 紅蕈

The two distinguished French sinologues who edited this text, Messrs. Chavannes and Pelliot, could not be expected to know that it is

1. The information about the activities of the disciples of Mani in China comes from *Un Traité manichéen Retrouvé en Chine, traduit et annoté par Ed. Chavannes et P. Pelliot*, Paris, 1912, pp. 292-340, especially 302-5 and 310-314, including the footnotes. This study was first published in the *Journal Asiatique*, November-December 1911. On p. 304 the French scholars, translating the passage from St. Augustine, render *boletos* by *cèpes*, thereby falling into error. In Antiquity, as we have said, the *boletus* was an amanita and the *cèpe* was the *suillus*, the Italian *porcino*. Linnæus created the confusion: when he was naming the genera and species he resorted to the ancient Greek and Latin words but paid not the slightest heed to their ancient meanings.

precisely in Fukien province that there grows an abundance of an edible red mushroom, which is gathered there and widely eaten not only in Fukien province but throughout China. Chinese mycologists call it *Russula rubra* (Kromb.) Bres., and anyone conversant with the province would assume that this is the mushroom of which Lu Yu speaks. How astonishing was this charge of mycophagia, which Lu Yu hurled against the little sect of Mani disciples. The Chinese know their mushrooms and consume wild species in quantities. It is not as though a mycophobic Englishman had leveled the heinous accusation against, say, a band of Gypsies. To be guilty of this offense the Manichæans must have been gluttons for mushrooms, and the local red mushrooms to boot, and clearly, since he was speaking about their religion, the practice was a part of their religion. Were not these mushrooms another substitute for Soma, a substitute more appropriate, in that it was *red mushrooms*, than any known to have been used in India?

The second report of Lu Yu, probably submitted in A.D. 1166, is just as interesting.

以溺為法水用以沐浴

> They [the Manichæans] consider urine as a ritual water and use it for their ablutions.

MM. Chavannes and Pelliot observe in a footnote that 'the use of cow's urine is known in Brahmanism and, on certain occasions, in Mazdaism [Zoroastrianism]; but the context here hardly permits one to think of anything except human urine, and analogies are more rare on this point'. Lu Yu goes on to complain that as the Manichæans eat so many Russula mushrooms, the price of mushrooms of the Russula kind goes up. For 'mushrooms', he here uses the two characters 蕈 *hsün*, 菌 *chün*. The first Chinese character is the same as he used before. The second one is the other standard Chinese character for mushroom. In the second half of the sentence he repeats the two characters but in reverse order:

食<u>蕈菌</u>則<u>菌蕈</u>為之貴

As they eat fungi mushrooms, therefore mushrooms fungi grow dear.

Mani was born in Iran perhaps a millennium or more after Zoroaster and at least 1,500 years after the period when the ṚgVeda was being composed in the Indus Valley. If anyone takes the attitude that, because of the lapse of time, the practices of the Manichæans are irrelevant to my Soma-Haoma-fly-agaric thesis, I will not argue with him. In a part of the world and in a period where religious practices and beliefs showed a marked underlying stability and persistence, it seems to me that the use in this Iranian cult of urine, apparently human urine, and of mushrooms, *red* mushrooms, is more than chance. In esoteric religious circles the ancient practices may have lived on with modifications for many centuries.

The religion of Zoroaster still survives in the community of Parsis, largely centered in Bombay. It is pertinent to my argument that they still drink urine in their religious rites, though only in token amounts and only the urine of a bull.[1] As I have already mentioned, throughout the area that stretches from Iran to India cow's urine is used as a disinfectant, a religious or ceremonial disinfectant, paralleling somewhat our historical use in the West of Holy Water.

So much emphasis is laid on cows in the ṚgVeda and on the urine of bulls in the religion of the Parsis that the question naturally presents itself whether cows consume the fly-agaric and whether they are affected by it, along with their urine and milk. I cannot answer this. In early October 1966 with three Japanese friends, all mycologists,[2] I visited Sugadaira, in Nagano prefecture, five hours west of Tokyo by fast train. There were many birch trees scattered over the mountains, and an abundance of fly-agarics growing at their feet. A herd of heifers was grazing in the lush highland pasture, and we took advantage of the opportunity to offer them some of the fly-agarics that we had in our baskets. Two of the animals ate them from our hands avidly; two or three turned away with indifference. We did not carry our

1. Jivanji Jamshedji Modi: *The Religious Ceremonies and Customs of the Parsees*, 1923, 2nd ed. 1937; Bombay. *Vide* p. 93 (in the 2nd edition) and other references indexed under 'Gaomez', 'Nirang', and 'Nirangdin'.

2. These friends were Yoshio Kobayashi, Rokuya Imazeki, and Masami Soneda, all well known mycologists. We were guests of Shigeyoshi Iwasa at the lodge of the Meguro High School, Tokyo; unfortunately Professor Iwasa could not join us. We were most hospitably cared for by the hosts at the lodge, Mr. and Mrs. Mikiyoshi Motai.

experiment further, as the animals were of fine Frisian breed and the owner was unaware of our scientific activities. We did not have an opportunity to catch the urine of the heifers that had eaten the fly-agarics; besides, they would have eaten many, many more, had we supplied them.

In 1965 and again in 1966 we tried out the fly-agarics repeatedly on ourselves. The results were disappointing. We ate them raw, on empty stomachs. We drank the juice, on empty stomachs. We mixed the juice with milk, and drank the mixture, always on empty stomachs. We felt nauseated and some of us threw up. We felt disposed to sleep, and fell into a deep slumber from which shouts could not rouse us, lying like logs, not snoring, dead to the outside world. When in this state I once had vivid dreams, but nothing like what happened when I took the Psilocybe mushrooms in Mexico, where I did not sleep at all. In our experiments at Sugadaira there was one occasion that differed from the others, one that could be called successful. Rokuya Imazeki took his mushrooms with *mizo shiru*, the delectable soup that the Japanese usually serve with breakfast, and he toasted his mushroom caps on a fork before an open fire. When he rose from the sleep that came from the mushrooms, he was in full elation. For three hours he could not help but speak; he was a compulsive speaker. The purport of his remarks was that this was nothing like the alcoholic state; it was infinitely better, beyond all comparison. We did not know at the time why, on this single occasion, our friend Imazeki was affected in this way.

In the cultural history of Eurasia has the reindeer played a part in the attitude toward mushrooms and also toward urine, owing to certain traits peculiar to the reindeer? The northern forest and tundra folk live in an intimacy with the reindeer that is hard for us to imagine, an intimacy that amounts almost to a symbiotic relationship. Reindeer manifest two addictions, two passions, one to urine especially human urine, and the other to mushrooms including the fly-agaric. When human urine or mushrooms are in the vicinity, the half-domesticated beasts become unmanageable. All reindeer folk know of these two addictions (how could they not know about them?), though not all know that the mushroom addiction embraces the fly-agaric since the fly-agaric does not always grow over the areas where reindeer abound. Reindeer, like men, suffer (or enjoy) profound mental disturbances after eating the fly-agaric. In *Part Three* and the Exhibits we shall give the testimony of travelers on these points. I now call attention, for the first time in this context, to the odd traits of the reindeer because of their possible bearing on the fly-agaric complex in the religious life of the Siberian peoples. In the give and take of the human species and the reindeer may not the human race have learned from the reindeer to esteem urine and the inebriating qualities of the fly-agaric also, and finally the combination of the two? It has been said that in Vedic times – *ca.* B. C. 1500-1000 – reindeer had not been domesticated. How conclusive is the evidence for this in northern Siberia? It is thought that the ancestors of the Aryans did not live in the reindeer latitudes. There is linguistic evidence that may show the Indo-Europeans discovered the virtues of the fly-agaric before the Siberian tribes, and we are beginning to perceive the extent of the trade relations that always existed across those vast land expanses of Siberia. In that heyday of the fly-agaric cult may the Indo-Europeans have mastered techniques that encouraged cows to eat the resplendent heavenly mushroom? All these questions are speculative and at present unanswerable.

XIII

THE MARVELOUS HERB

1

In the Shahnameh of Firdousi there is an episode that bears on our quest for the genuine Soma. The Shahnameh ('Book of Kings') is the great Persian epic of 60,000 rimed couplets, a repository of all Iranian history and all national legends known to the poet. Firdousi completed his prodigious work just after A. D. 1000, and the episode that interests us is attributed to the 6th century, at the court of the outstanding Sassanid king Khosru I, called Anushirvan ('the Blessed'), who reigned from A. D. 531 to 579. Whether the episode as recounted by Firdousi took place as told or whether he merely passes on an embroidered version of what had become a legend is immaterial: its terms are what interest us,[1] whether true or false. Here is my summary of the passage pertinent to our inquiry. It will be apparent later why I have put a sentence in italic.

> At the court of King Khosru there was an outstanding physician, one Bursōē, an elderly man who loved to talk and who was reputed to be versed in every branch of knowledge. One day he presented himself before the King at the hour of audience and said: 'O King, friend of learning, you who explore science and who keep it in memory, today I have perused in a serene spirit an Indian book. It is said therein that on a mountain in India there grows a plant brilliant as Byzantine satin. If a skillful man gathers it and mixes it cannily, and *if then he spreads it on a dead man, the dead man recovers the power of speech without fail and forthwith.* If the King permits, I am going to undertake this difficult quest. I shall make use of all that I know to guide me and I hope to accomplish this marvel. It would be only just that the dead return to life, since the world has for king the Blessed One.' Whereupon the King replied: 'It is not likely that this will be, but perhaps we must try.'
>
> The King added that as Bursōē would probably need a guide, he should go amply supplied with gifts for the Indian Rajah. [Later in the story the Rajah is identified as having his capital at Kanauj, the

1. Firdousi: *Shahnameh*, verses 3431-3568; I have relied on the French translation of Jules Mohl.

> meager ruins of which still exist in Uttar Pradesh, fifty miles from Kanpur, on the Ganges far south of the Himalayas.]
>
> Bursōē set out on his journey and arrived in the presence of the Rajah, who welcomed him and read the letter delivered to him. He assured Bursōē that all the Brahmans living in the mountains would help him. Bursōē went everywhere in the mountains on foot, and he chose herbs dry and fresh, faded and others in their full glory. After crushing them he spread them over corpses, 'but those herbs did not bring back a single one to life.'

The rest of the tale, which recounts how Bursōē extricated himself as well as he could from his embarrassing predicament, does not concern us.

It is noteworthy (1) that the book that Bursōē had read was Indian; (2) that according to the Indian book the herb shimmered like satin of Byzantium; (3) that in India Bursōē had recourse to the mountains; and (4) that he was promised Brahmans living in the mountains for his guides. The herb grew in the mountains of India – no other place is mentioned. Bursōē had been reading an Indian book – we are not told what one nor when it was written. In the episode do we not hear a clear though fading echo of the Soma of the ṚgVeda, of the marvelous herb that grew high in the mountains and for which the Brahmans alone held the key?

2

In Indian literature the Purāṇas contain legends and tales of the olden days, all of them religious and in verse. The different Purāṇas took their present shape at different times, running from perhaps the 4th to the 16th centuries.

A story is told in several of the Purāṇas that may bear on the Bursōē episode in the Shahnameh. The version that I shall give is from the *Padma Purāṇa*, Part 2, Book 6, Chapter 8, verses 40-63. Scholars tell us that it assumed its present form between A. D. 800 to 1000. A certain Mount Droṇa figures in it: '*dróṇa*' (from *dru*, Sanskrit, 'made of wood') is the word commonly used in the ṚgVeda for the wooden vessels that contain Soma. Here the word has been transferred to that other receptacle for Soma, the mountain out of which the marvelous herb grew.

> Jalandhara was the king of the demons and he was waging war against the gods. The spiritual preceptor of the demons, their guru, was named Śukra; he had a magic incantation that he had received from Śiva by means of which he revived all the demons as they fell in battle. Viṣṇu, who was leading the army of the gods, said to Bṛhaspati, who was the guru of the gods, 'Śukra has revived all the demons. Why do you not revive the gods?' Bṛhaspati said, 'I will revive the gods with herbs.' Then Bṛhaspati went to the great Droṇa Mountain beside the ocean of milk, and he took the herbs that grew there and by means of yoga he revived the gods. Seeing this, Jalandhara said to Śukra, 'How can the gods be revived without your magic?' Śukra said, '*There is a mountain named Droṇa beside the ocean of milk, and herbs grow there that revive the dead. The guru of the gods went there and took the herbs and revived the gods who had fallen in battle.*' Hearing this, Jalandhara went to the ocean of milk and he went to the mountain Droṇa and beat it with his fists until Droṇa said, 'I am your slave; protect me.' Jalandhara then commanded the mountain to go down to the nether world below the earth, and Droṇa went there; as he went, all the herbs cried out. Then Jalandhara returned to the battlefield and fought with Viṣṇu and conquered him. – *Translated by Wendy Doniger O'Flaherty.*

In the Sanskrit text the word for herb is *oṣadhi*, one of the terms often used for Soma in the ṚgVeda. In Sanskrit there was a poetic convention that precious herbs (*oṣadhi*) grew on the Himalayas. They were brilliant and shone like lamps at night. There are hundreds of such references, of which perhaps the most famous is Kālidāsa's *Kumārasambhava* I 10:

> Where the magic herbs (*oṣadhi*) cast their glow into the caves of mountaineers who lie there with their mistresses, lamps of the love-chamber that need no oil.[1]

1. I am indebted to Professor Daniel H. H. Ingalls for this reference. He also supplied me with a Sanskrit proverb, which may go back to a Prakrit original: 'On the snowy mountains grows the magic herb, but the snake is on your head.' (Snowy mountains = Himalayas) *Vide* translation by D. H. H. Ingalls: *An Anthology of Sanskrit Court Poetry*, Harvard Oriental Series 44, verse 791; also *idem* p. 522 for further references to the proverb.

3

All Chinese know the *ling chih* 靈芝, a conception that goes back in Chinese cultural history for thousands of years; just how far back is the question at issue here.

The *ling chih* is a symbol of happy augury, bespeaking good fortune, good health, longevity, even life with the immortals. Conventionally the phrase is translated by 'the divine fungus – or mushroom – of immortality'. The 'marvelous herb' would be a simpler rendering, or again the 'herb of spiritual potency'. It is one of that large family of Chinese expressions drawn from nature – animals, plants, mountains, clouds, *etc.* – which seem to have constituted in a sense the furniture of every Chinese mind. The Chinese would distill from the chosen objects a conventional meaning, and thereafter the objects were used in literature and art to convey this meaning, like pictographs. At an early date the Taoists captured the concept of a divine mushroom of immortality, and they exploited it fantastically in their writings in the first millennium of our era and even later, until the original idea was lost in a welter of imaginary divagations. Beginning with the Yüan Dynasty (A.D. 1280-1368) the *ling chih* has been endlessly represented in art – in paintings, carvings in jade and deer's antlers, furniture and carpet designs, balustrades, jewelry, lady's combs, perfume bottles, in short wherever the artistic urge found an outlet. It has become a cliché in newspapers, novels, conversation. For two thousand years the idea of the *ling chih* has passed through various phases in the cultural history of China, and I hope someday to publish my notes on these, looking to the time when enough will be known for someone to write the biography of the miraculous fungal idea.

Up to now students of Chinese culture have always regarded the *ling chih* as indigenous to China.[1] I shall suggest that it came from India

1. *Vide, e. g.*, Michael Sullivan: *The Birth of Landscape Painting in China*, University of California Press, San Francisco and Los Angeles, 1962, p. 52. Sullivan's discussion of the *ling chih* (*vide* references in his index) is the best that has appeared in recent times. Also, Marcel Granet: *La Pensée Chinoise*, Paris, 1934, 'L'art de la longue vie', pp. 507 ff.

and was a 'literary' reflection of Soma, the miraculous mushroom of the ṚgVeda. Certainly the fungal idea, once it reached China, found there fertile soil: an 'elixir of immortality' has bemused many peoples, such as the West Europeans in the 16th century, but the Chinese seem to have been singularly susceptible to this will-o'-the-wisp, always hoping to come upon a plant or mineral or juice that would prolong virility and longevity. The novelty of the *ling chih* is precisely that it was a mushroom, and this is India's contribution to our story.

Alfred Salmony in his *Antler and Tongue*[1] has called attention to the accumulating evidence of the influence of India on the China of the Late Eastern Chou, from the 7th to the 3rd centuries B.C. He has pointed out that the influence could be solely 'literary', *i.e.*, whereby the Chinese 'heard tell' of practices elsewhere and under the stimulus of such reports sought to duplicate them. This is what Joseph Needham calls 'stimulus diffusion',[2] what I would call 'idea diffusion'. Salmony cites a number of examples and supplies the bibliographical references. To this list I now propose to add the *ling chih*.

First a word as to the vocabulary for the divine, the miraculous, mushroom. This assumed such importance in Chinese culture that alternative expressions evolved. It is often called the 'miraculous *chih*', 神芝, or 'auspicious herb', 瑞草. With Chinese culture the idea spread to Korea and Japan, where however it has tended to be confined to the literate and intellectual classes. *Ling chih* in Japan, written with the same Chinese characters, becomes *reishi*, the *l* yielding to *r* in the customary way. But more widely used than *reishi* today is *mannentake* 萬年茸, 'ten-thousand-year mushroom'. Scholars say that in the archaic stage of Chinese culture the character for *chih* 芝 was written 㞢, a pictograph of 'herb', a small plant that was not woody. In Japanese this meaning is still current: where *chih* 芝 appears in contexts other than *ling chih*, it is pronounced *shiba* and means 'lawn'. Presumably it meant this in spoken Chinese in the early T'ang Dynasty (A.D. 618-906), when the Japanese probably borrowed the

1. Alfred Salmony: 'Antler and Tongue, an essay on ancient Chinese symbolism and its implications', *Artibus Asiæ*, Ascona, Switzerland, 1954, pp. 51-52.

2. Joseph Needham: *Science and Civilization in China*, Cambridge University Press, Vol. 1, 1961, pp. 244-248.

word. But in the specific sense of 'mushroom' it had already had a long history in China. In the *Li Chi*, one of the 'Thirteen Classics' of China, *chih* 芝 is listed as a ritual food of the emperor, thus figuring in the religious ceremonies that were an important part of his functions.[1] The early commentators gloss this as meaning an edible mushroom growing on trees, probably what is known in recent centuries as the tree-ear, *mu erh* 木耳, familiar throughout the temperate zone to those versed in mushroom lore as the famous *Auricularia Auricula-Judae*. The *Li Chi* comes down to us in a recension dating from the early Han (the two centuries before Christ), but the contents are supposed to have been assembled in the late Chou period, from the 7th to the 3rd centuries B. C.

Though *chih* in the sense of 'mushroom' seems already to have been familiar, the idea of *ling chih*, a supernatural mushroom with miraculous powers, appears first in the Ch'in Dynasty (B. C. 221-207) under the great Emperor Shih-huang, who assumed the title of the 'First Emperor', by which he has always been known, for it was he who unified all China for the first time and who built the Chinese Wall to keep the barbarians out. Until his reign we find not a single word about the marvelous plant: no mention in the Classics, in the inscriptions on the bronze vessels, on the oracle bones. There are no carvings in jade or deer's antlers, no representations in pottery or jewelry, on halberds or belt buckles. If the conception had existed, we must assume an implausible conspiracy of silence in this singularly articulate people concerning an idea that prompts thousands of tongues to wag.

Then, suddenly, there is a burst of talk about the wonder fungus, but only talk. People busy themselves looking for it, especially in the mountains, but no one finds it. The great historian Ssu-ma Ch'ien (B. C. 145-?87) is the source of most of our reliable information on this. There are many passages in his account of Shih-huang's reign telling of talks between the Emperor and his necromancers: a magical herb ('*chih*') existed, but where was it? Everyone was searching but it was nowhere to be found. Matters reached a point where the

1. *Li Chi*, Chap. 12. 禮記内則.

Plate xiv and Plate xv · Twin-leafed Ling Chih from above and from below.
(*Collection R. G. Wasson*)

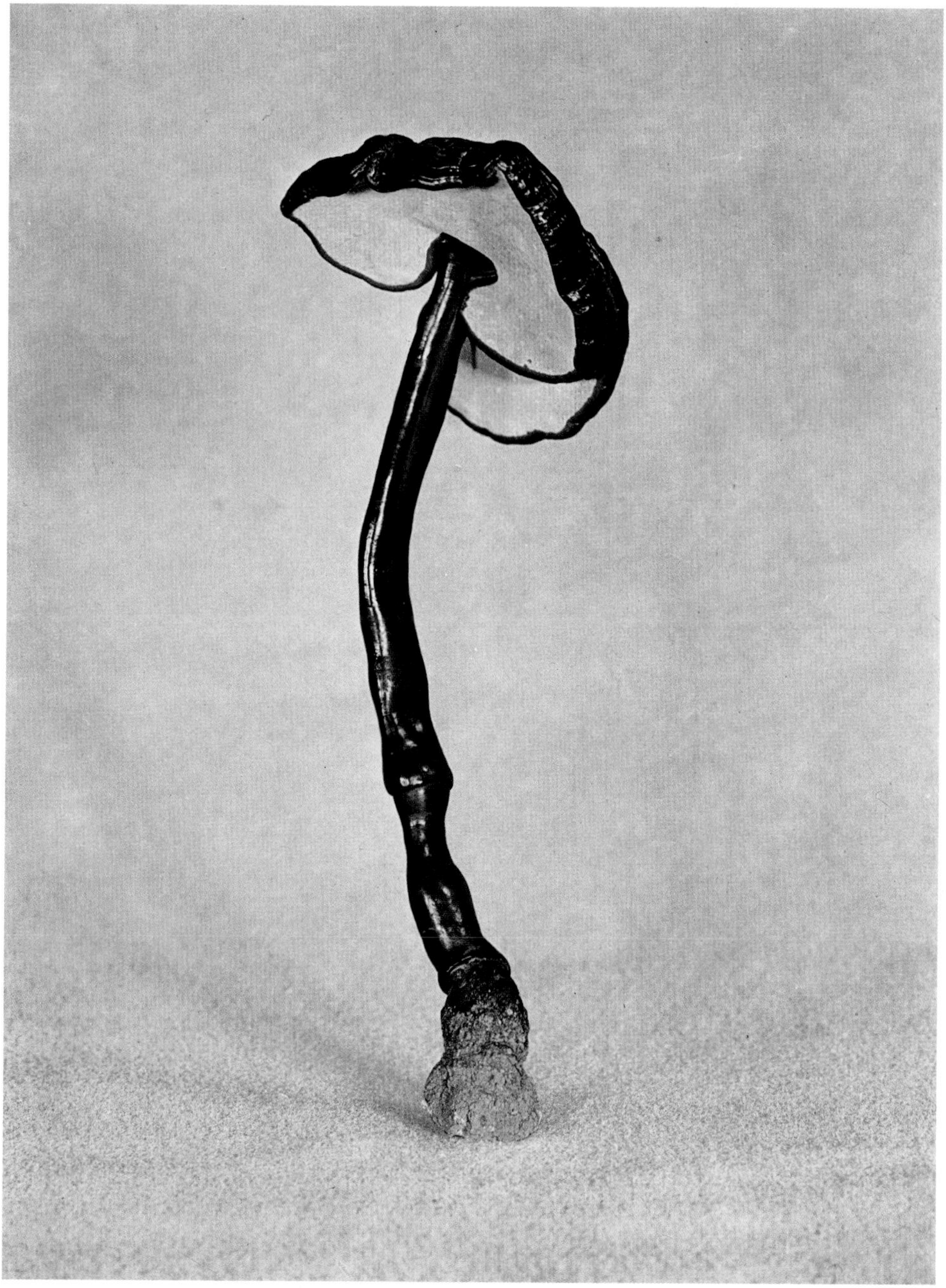

Emperor was advised to go into the mountains alone, and in disguise, obscurely, on tip-toe, and perhaps he might surprise it! 'We your humble servants are trying to find *chih*. It is one of the miraculous medicines that even the gods cannot come upon easily.' This is the manner in which the Emperor is addressed by Lu-shêng, one of the sages who was giving advice.[1] The Emperor has been told of a marvelous plant and, asking his necromancers to find it, gets only evasive answers. They believe (or pretend to believe) the reports of its properties, but have no idea what to look for. Addressing the 'master of humanity', who clearly believed the reports, they do not dare confess their ignorance by telling the truth.

Or again we read that Shih-huang sends vessels into the Eastern Sea to seek the mysterious fungus on islands off the coast and far away.[2] One sage (or necromancer, or shaman), Hsü Fu, seems to have rallied hundreds of youths of both sexes to embark with him and to have sailed away to the Southeast. The reports of this voyage are numerous. According to one account, they never returned, but Ssu-ma Ch'ien tells us that Hsü Fu came back and reported falsely about as follows: 'Your humble servants sailed east and southward and reached P'êng-lai [allegedly one of the three godly islands in the Seas]. There your humble servants saw palaces made of *chih*. There were servants, copper in colour and dragon in appearance. The palaces were so bright that they lit up the heavens.'[3]

I take it that the reports of Soma had reached the Emperor by the sea route: hence the voyages to establish contact with the source of information. I take it that the reports placed Soma high in the mountains: hence the futile excursions into the mountains to find the plant. The first reports about Soma probably reached China in the reign of

1. From Ssu-ma Ch'ien: *Shih chi* 史記, biography of Shih-huang.

2. The proponents of early deliberate trans-Pacific contacts with America must face these texts in considering the stage of development in navigation that the Orientals had reached as late as the Ch'in Dynasty. Contacts by sea between India and China involved only plane sailing, but the difficulties were staggering for the people of those times. The Japanese had hardly established contacts with China *via* Korea, much less the islands of the Pacific. As late as the middle of the first millennium A. D. voyages from Kyushu to Shantung were formidable adventures, exceedingly perilous.

3. *Op. cit.*, Vol. 118, 'The Biographies of Princes Huai-nan and Hêng-shan' 淮南衡山列傳; quotation from the biography of Prince Huai-nan, a grandson of the first Han emperor.

Shih-huang-ti, but possibly they came a few decades earlier, Shih Huang being the first to pursue the matter. In any case in China the conception of the miraculous *chih* probably does not antedate the 3rd century B.C.

Ssu-ma Ch'ien, the father of Chinese history, lived in the reign of the early Han Emperor Wu (B.C. 157-87), about a century after Shih-huang. Contemporary with Ssu-ma Ch'ien there was a famous gentleman-in-waiting at court by name Tung-fang Shuo 東方朔, a man renowned for his ready wit. In the Taoist canon there survives a work attributed to him, the *Shih chou chi* 十洲記, 'Notes on Ten Continents', that gives another account of the voyage of Hsü Fu. Scholars say that the attribution to Tung-fang Shuo is certainly spurious, and that the *Shih chou chi* could not have been written before the 4th or 5th centuries and probably several centuries later. In short it is a product of the second half of the first millennium, therefore contemporaneous *sensu lato* with Bursōē's journey and with the *Padma Pūraṇa* from which we have already quoted. In the following excerpt taken from the *Shih chou chi*[1] we have underlined phrases that will strike a familiar chord:

> The isle of Tsu is situated close by, in the Eastern Sea. There grows a never-dying plant, shaped like water-grass, with blades three to four *ch'ih* [feet] in length. *A man who has been dead three days revives immediately when this plant is laid on him.* When it is eaten it prolongs life. In the time of Shih-huang-ti of the Ch'in dynasty, when murdered people lay broadcast in the 'great preserve' and across the roads, *birds resembling crows or ravens appeared with this plant in their bills, and placed it on the faces of those corpses, with the effect that they sat up immediately, and revived.* The officers reported this to the Emperor. On this Shih-huang-ti sent out an envoy with a sample of this plant [草 *ts'ao*]; and he interrogated the doctor of the Spectre Valley, who lived near the north wall. 'This herb [*ts'ao*]', thus spoke the sage, 'is the herb of immortality of the Tsu island in the Eastern Ocean,

1. The Chinese text of this episode is given by J. J. M. de Groot: *The Religious System of China*, Brill, Leiden, Vol. IV, pp. 307-8. I have followed de Groot's translation except in a few particulars where on the advice of my Chinese friends I have emended it.

> where it grows in a red marble field. Some call it *chih* 芝 which feeds the *shên* 神 [spirit]. Its leaves grow luxuriantly, and one stalk suffices to give life to a man.' On these words Shih-huang-ti with enthusiasm spoke: 'Can it be fetched from there?' And he sent an envoy to the island, one Hsü Fu, with five hundred young people of both sexes, in command of a ship with decks. They put to sea to seek the island but they never came back.

Here then we have three tales that intermesh in an odd manner. The Iranian version conveys a clear echo of the Soma of the ṚgVeda. There has been added to the marvelous herb one fabulous property: it revives the dead. In the Indian tale an herb with the identical miraculous virtue grows on Mount Droṇa, the name of this mountain echoing the word in the ṚgVeda that designates the receptacle for the juice of Soma. In the late Chinese revision of the Hsü Fu tale the same fabulous virtue is attributed to an herb: it makes the dead to rise. This tale is a variant of one told by Ssu-ma Ch'ien perhaps six or seven hundred years earlier about the First Emperor, Shih-huang, and a *fungus or mushroom with miraculous properties.* Here is an example of idea diffusion, and the idea, I believe, had its source in the Soma of the ṚgVeda. The marvelous fungus of the Chinese, first appearing in the reign of Shih Huang-ti, must today be viewed in the light of Sino-Indian contacts and the evidence that Soma was a mushroom.

The next chapter in the history of the *ling chih* takes place in B. C. 109, a century after the time of the Emperor Shih-huang. The Han Dynasty (B. C. 206 - A. D. 220) now rules China and Wu-ti occupies the throne. The sources for our information about the episode that I am going to relate are the same Ssu-ma Ch'ien whom we have already quoted and who was contemporary with the later event, and Pan Ku (A. D. 32-92), the chronicler of the Han Dynasty and also a reliable historian. In the imperial Kan-ch'üan palace building operations had been under way in the first half of the year. Then, in late summer (probably August), in an inner pavilion of the palace, there appeared a fungal plant, a marvelous growth with nine paired 'leaves'. The Emperor Wu made the most of the event. He identified the plant with the famous *chih* that Shih-huang and his necromancers could not locate. He issued an edict and exclaimed that even the inner chambers of the palace were not discriminated against. He proclaimed an amnesty of prisoners, served beef and wine to a hundred families, and composed an ode for the occasion, the earliest poem about a mushroom (but far from the last and far from the best) that comes down to us in Chinese history. Pan Ku has preserved it for us:

齊房產草	My secluded dwelling produces an herb,
九莖連葉	Nine stems with twin leaves!
宮童効異	The palace pages busy themselves with this miracle;
披圖案諜	They lay out pictures and consult records!
玄氣之精	The essence of the Mysterious Breath,
回復此都	There it is, returned again to this residence,
蔓蔓日茂	Day after day this superb growth,
芝成靈華	This *chih*, which unfolds its beauties most marvelously!

Under the Emperor Wu taxes were levied harshly to meet his needs and grumbling was widespread. He was hardly popular and perhaps he seized on the unusual appearance of a fungal growth inside the palace chambers as a public relations ploy, bestowing on this plant the name *chih* to turn in his favor its good name and prestige. The episode is known to this day by everyone familiar with Chinese history, somewhat like the vase of Clovis in French history and King Alfred and the pancakes in English history. The ode is No. 13 in the nineteen ritual odes that were sung to music on the seasonal occasions of the Chinese year.

At last the unidentified *chih* of Shih-huang-ti had found an identity, an identity that it has retained to this day. Naturally this fungus had nothing to do with its Indian precursor, which Shih-huang had never succeeded in locating in the mountains of China. I think it is safe to assume that the plant found growing in the imperial palace in B.C. 109 was *Ganoderma lucidum* (Leyss.) Karst., the same species that has been represented in Oriental art – Chinese, Japanese, Korean – time

FIG. 1. *Ganoderma lucidum*. Normal and abnormal shapes.

without number in recent centuries. The definitive scientific identification of the fungus pictured in Chinese paintings since the Yüan Dynasty was made by two Japanese mycologists, by Iwao Hino in May 1937 and by Rokuya Imazeki in 1934 and 1939.[1]

1. Rokuya Imazeki: *Natural Science and Museums*, 1934, No. 61, pp. 11-15, 'Good Omen Plant, *Mannentake*'; also *Bulletin of the Tokyo Science Museum*, March 1939, No. 1, 'Studies on Ganoderma of Nippon'. Iwao Hino (Miyazaki): *Botany and Zoology*, May 1937, Vol. 5, No. 5, '*Reishi* and *Ganoderma lucidum* that grow in Europe and America: their Differences'. (All are in Japanese, with English summary.)

Ganoderma lucidum is a woody fungus, therefore inedible, that grows widely in the temperate zone. We see it in Europe and America, but we have never made anything of it. In appearance it can be of stunning beauty, with its rich lacquer-like finish, the concentric lines of its strange pileus, its odd stem. Those unversed in these matters often refuse to believe that the high polish of this mushroom has not been artificially applied. But its peculiar virtue is that it is protean. There is the normal form, and there is the endless variety of other shapes and consistencies that it assumes when it grows under abnormal conditions, as for example in the dark. The cap (or 'pileus') is always attached to the stem (or 'stipe') eccentrically, and it can give rise to the 'paired leaves' of Wu-ti, as we show in Plates XIV and XV. (*Vide supra*, poem by Wu-ti. In classic Chinese the pileus of a mushroom was a 'leaf' and in Japanese this use of 'leaf' has survived down to our own day. The Chinese character is 葉, in Chinese reading *yeh*, Japanese *ha*.) We know that there had been building operations going forward in the Imperial Palace earlier in the year. This is precisely the condition that might produce an abnormal *chih* inside the new structure: the wood was not seasoned and the growth made its appearance in the obscurity of the secluded chambers. The nine 'double-leaves' would be a rare curiosity, but it is certainly not at all unbelievable.

Pan Ku, the chronicler of the Han Emperors, composed a second ode to the *ling chih*, more than a century after the Emperor Wu's. The attentive reader will have noted that until now we have never quoted an ancient source for the name *ling chih*. Everyone has used either *chih* alone, or *ts'ao chih* 草芝, 'plant chih', or *chih ts'ao* 芝草, 'chih plant'. In the last line of the Emperor Wu's poem, the first character is *chih* 芝, the next to the last is *ling* 靈. *Ling* has a long and complicated history in Chinese. To start with, it is made up by the combination of three characters – 雨, 'rain'; 口口口 (three mouths) 'praying for'; and 巫, 'shaman'. Combined in a single character, they mean spiritual potency, a stirring of the soul. Pan Ku in his ode joined *ling* and *chih*, the two concepts that the Emperor Wu had associated together in

PLATE XVI · NOIN ULA TEXTILE (*here shown divided in two parts*). 1st century A. D. From Mongolian tomb. Contemporary connoisseurs of Chinese art suggest that the enigmatic plant toward which the birds are leaning is LING CHIH, the fabulous Fungus of Immortality. (*Courtesy of the Hermitage Museum, Leningrad*)

PLATE XVII · NOIN ULA TEXTILE (*detail*).

the same verse, thus minting the phrase that ever since has been in common currency. Here is Pan Ku's ode:

因露寢兮產靈芝	*Ling chih* grows with the settling dew,
象三德兮瑞應圖	The sign of the three virtues, happy omen's picture fulfilled.
延壽命兮光此都	It prolongs lives and glorifies the capital.
配上帝兮象太微	It accompanies the Emperor on high, Image of the Sky!
參日月兮揚光輝	Image of the sun and the moon, it throws out bursts of light!

The three Virtues of the Universe are the Heavens, the Earth, and Man. The pictures of happy omens were mythical documents from heaven depicting auspicious marvels such as the unicorn, the phoenix, the dragon, and the *ling chih*. Was not Pan Ku's ode intended to be used as an alternative to Wu-ti's in the liturgical year?

Lately the art historians have provided us with an unexpected development in the history of *ling chih*. A Russian archæological expedition in 1924-5 unearthed some remarkable artefacts in tombs at Noin Ula in what is now Mongolia. They date from the first century A.D., perhaps the early part of the century. Among the finds was a silk textile, reproduced in our Plates XVI and XVII.[1] A single motif is repeated several times. It consists of two rocky crags, a bird perched on each of the crags leaning down and outward, a graceful tree between the crags, and outside the crags, a plant with nine stalks terminating in what Michael Sullivan calls a 'poached-egg' design. The birds are so

1. The first and best description in English of the Noin Ula finds as a whole was that of W. Perceval Yetts: 'Discoveries of the Kozlov Expedition', *Burlington Magazine*, April 1926, pp. 168-176. The textile that interests us was discussed at some length by Michael Sullivan: *The Birth of Landscape Painting in China*, pp. 52-53 and Pl. 35; other references to *ling chih* in his index, with numerous reproductions of *bas reliefs* representing the fungus. William Willetts: *Chinese Art*, London and New York, 1958, also accepts the identification, pp. 290-292. The Noin Ula textile hangs in the Hermitage Museum, Leningrad, where the Museum authorities graciously photographed it for me.

placed that they are about to reach the 'poached eggs'. Perceval Yetts in his description of this textile refers to the plant as a 'clumsy fungoid form'. Sullivan has identified it with *ling chih.* Sullivan sees no parallel

FIG. 2. From *Sakikusa-kō* ('An Inquiry into the Happy Herb'), by Suigetsu Kan-ō. An essay published in 1850 in Japan. The inspiration for this representation of *Ganoderma lucidum* stems back almost 2,000 years to the specimen that the Emperor Wu of China discovered growing in the inner pavilion of his palace.

for the treatment of the rocky crags in either Han China or the Near East, and he suggests a derivation from India, the wall painting in Cave X at Ajanta dating from the first century B.C. showing a similar silhouette. Here is independent support for my thesis of an Indian origin for the *ling chih* notion.

PLATE XVIII · Rubbing of a Han Dynasty Stone Carving. Found in Szechwan Province. The two hags with pendulous breasts, having wings on their arms, are assumed to be Immortals. They are playing the board game, popular at that time, known as *liu-po*. Behind the figure on the right there is what Chinese art critics today interpret as a Han Dynasty conception of the LING CHIH, the fabled 'Marvelous Herb' or 'Divine Fungus of Immortality'.

(*Courtesy of Rolf Stein, Esq.*)

Plate xix · Maid of Honour 女侍, attending the Heavenly Emperor. In her left hand she carries a vase with Ling Chih in two of its shapes, the normal one and the one that suggests deer's antlers or coral branches. Painted c. A. D. 1300.

I suggest that the fungal form was the artist's effort to embody the Emperor Wu's *chih* with nine twin-leaves, his knowledge of it being only literary. (The textile, dating from the later Han, came more than a century after the event in Wu-ti's reign.) And if India inspired the crags, perhaps the birds reaching down to the fungoid forms re-enact the familiar Indian legend of the rape of Soma by a falcon from a celestial mountain top.[1] Since the Noin Ula finds art historians have assembled a number of *bas reliefs* representing the same conventionalized motif, the 'poached eggs', and the stalks or stems that support them. In Plate XVIII we show one of these, also having nine 'stems' or stipes.

If I am right, we see the genesis of the *ling chih* in the reign of Shih Huang-ti, when on the strength of verbal reports from India about a miraculous mushroom he sent his necromancers looking for it here there and everywhere in the mountains. We see Wu-ti fixing the identity of this elusive mushroom on *Ganoderma lucidum* thanks to its fortuitous appearance in an inner chamber of his palace – a changeling that he was successful in foisting on the Chinese people. We see artists in the later Han, obviously unfamiliar with the fungus itself, depicting from verbal reports a fungus every whit as fabulous as the phoenix, the dragon, and the unicorn.

If I failed to make good my case for the fly-agaric as Soma, the argument in this chapter has no base. If Soma was the fly-agaric, then I submit the probabilities favour Soma as the inspiration for the Chinese idea of a divine mushroom of immortality. That the idea lived on in Iran and India well into the first millennium of our era is certain: Firdousi, the *Mahābhārata*, the Purāṇas, Kālidāsa and the other poets – all testify to that. In these centuries the Chinese picked up the idea but only by hearsay. It will not have escaped the perceptive reader that when Wu-ti joined in a single verse the *ling* and the *chih*, and thus took a vital step in fathering on a certain fungus the vibrant

1. The birds in the textile are probably the demoiselle crane, whose distribution extends from China to Spain. This was the guess of Roger Tory Peterson when we visited him in his home in Old Lyme, Conn., at Christmastide 1966. The legs are those of a wader. The black wing feathers and crest complete the identification. But of course the artist may not have considered the species of bird important, so long as it was big and spectacular.

name of the Marvelous Herb, he chose one that, in addition to other remarkable properties, was deep *red* in colour and that *shimmered like a satin of Byzance*. The strange convolutions of its pileus suggested to the Chinese mind the cumulus where the Immortals dwell, and thus in the course of time there came about an artistic convention by which, in the Yüan Dynasty and later, the painters made the clouds of the Celestial plane and the Herb of Immortality resemble each other to a point where, in extreme cases, it is hard to tell them apart.

We have advanced one step toward a United Field throughout Eurasia for a religious origin underlying the supernatural folklore that centres on an hallucinogenic mushroom. In *Part Three* we will carry this argument into northern Eurasia. But meanwhile we invite our readers to divert their attention to Dr. O'Flaherty's account of the post-Vedic history of the Soma question. The lesson that one must draw from her narrative about the futility of much scholarship is humbling.

PLATE XX · Chinese Sage Contemplating LING CHIH.
Painted by Chên Hung-shou (1559-1652). (*Courtesy of Wango Wêng, Esq.*)

PART TWO

THE POST-VEDIC HISTORY OF THE SOMA PLANT

by Wendy Doniger O'Flaherty

THE history of the search for Soma is, properly, the history of Vedic studies in general, as the Soma sacrifice was the focal point of the Vedic religion. Indeed, if one accepts the point of view that the whole of Indian mystical practice from the *Upaniṣads* through the more mechanical methods of yoga is merely an attempt to recapture the vision granted by the Soma plant, then the nature of that vision – and of that plant – underlies the whole of Indian religion, and everything of a mystical nature within that religion is pertinent to the identity of the plant.

In place of such an all-inclusive study, the present essay attempts to summarize what has been written since Vedic times about the physical nature of the Soma plant and the substitutes for Soma. I have worked in the Bodleian Library and the Indian Institute at Oxford, the India Office in London, and the British Museum, and of necessity I have omitted the contributions of Indian scholars whose works are not available in those collections. Furthermore, I have dealt summarily with the Haoma of the Avesta, since the work done on the botanical identification of Haoma has been subsumed for the most part under the study of Soma.

I. THE *BRĀHMAṆAS* AND THE *ŚRAUTA-SŪTRAS*

After the era of the Vedas there came a period when the centre of intellectual activity moved from the valley of the Indus to the upper Ganges and the Yamunā. A spate of works, the direct outgrowth of the Vedas and preserving ancient traditions, many of them lengthy, arose as a kind of corpus of ritual textbooks. These are the *Brāhmaṇas*, prose works dating from about 800 B.C.

Súrā is generally believed to have been an alcoholic drink of some sort – wine or rice wine or fermented liquor or beer or even distilled spirits[1] – or else to refer to alcoholic drinks in general. The *Brāhmaṇas* say clearly that Soma was not *súrā*. The *Śatapatha Brāhmaṇa* declares: 'Soma is truth, prosperity, light; and *súrā* untruth, misery, darkness.'[2]

1. Monier-Williams' dictionary gives 'distilled liquor' as a primary meaning of *súrā*, but evidence is strong that the distilling process was not known in India until a much later era.
2. *Śatapatha Brāhmaṇa*, Chowkhamba edition, 5.1.2.10.

The *Taittirīya Brāhmaṇa* says, 'Soma is male and *súrā* is female; the two make a pair.'[1] The sharp distinction made by the text seems to rule out the possibility that Soma was simply another kind of alcoholic drink, and it would seem probable that *súrā* embraced all the fermented drinks that rated mention in the ṚgVeda.

The *Brāhmaṇas* are much preoccupied with the question of substitutes for Soma. They are books of ritual composed for the sacerdotal caste, and in places they seem to be deliberately obscure. If the priests knew what Soma was, they never stated it clearly, and their references to the Soma plant are ultimately of little help in establishing its botanical identity. It may be assumed that Soma was none of the plants expressly suggested as substitutes for it, though of course it may have resembled any of them in some particular and probably did so. But it is difficult to draw any sure conception from these negative hints, for the substitutes often bear little resemblance to each other, including as they do grasses, flowers, creepers, and even trees.

The *Śatapatha Brāhmaṇa* sets forth an order in which substitutes should be used. First comes the reddish-brown (*aruṇá*) *phālguna* plant, which may be used because it is similar to Soma (*sómasya nyañga*), but the bright red (*lóhita*) *phālguna* plant must not be used. If *phālguna* is unavailable, then the *śyenahṛta* plant may serve, for there is a tradition that Soma was once carried away by a falcon, and a stalk (*aṃśú*) fell from the sky and became the *śyenahṛta* plant. The third choice is the *ādāra* plant, which sprang from the liquor that flowed from the sacrificial animal when it was decapitated. Fourth comes the brown *dū́rvā* grass, which is similar to Soma, and last choice is yellow *kúśa* grass. This being least satisfactory, a cow must be given in atonement.[2]

The *Tāṇḍya Brāhmaṇa* says that the *putīka* is the plant which grew from a leaf (or feather, *parṇa*) that fell when Soma was carried through the air, and that it is therefore a suitable substitute.[3] In his commentary on this work, Sāyaṇa says, 'If they cannot obtain the Soma

1. *Taittirīya Brāhmaṇa*, Ānandāśrama edition, 1.3.3.2.
2. *Śatapatha Brāhmaṇa*, 4.5.10.2-6.
3. *Tāṇḍya Brāhmaṇa*, Bibliotheca Indica edition, 9.5.1.

whose characteristics are described in the sacred text, then they may use the species of creeper (*latā*) which is known as *putīka*; if they cannot find *putīka*, then they may use the dark grass (*śyāmalāni tṛṇāṇi*) known as *arjunāni*.'[1] Yet another ætiological myth is used to explain the substitution of the fruit of the *nyagrodha* (sacred fig or banyan tree): the gods once tilted over their Soma cups, and the *nyagrodha* tree grew from the spilt drops.[2] Elsewhere it is said that even when Soma is available, one should use the juice of the *nyagrodha* fruit for non-Brahmans to drink.[3] It is probable that this fruit, like the *dūrvā* and *kúśa* grasses, was accepted as a substitute for Soma more by virtue of its own sacred nature than for any resemblance to the Soma plant.

Various other substitutes for Soma appear in the *Brāhmaṇas*: *syāmaka* (cultivated millet, said by the *Śatapatha Brāhmaṇa* to be most like Soma of all the plants),[4] *muñja* grass (sacred in itself), *kattṛṇa* (a fragrant grass),[5] and *parṇa* (a sacred tree).[6] The European lexicographers – Wilson, Roth, Monier-Williams – struggle to identify all these plants in modern botanical terms but often arrive at conflicting conclusions, as do the *Brāhmaṇas* themselves. Certain pertinent facts emerge, however, from the Brāhmaṇic literature:

1. The colour red is consistently associated with the Soma substitute. Red is the colour of the *nyagrodha* flower; the colour of the *phālguna* plant;[7] the colour of the acceptable *dūrvā* grass;[8] and even the colour of the cow used in the purchase of Soma.[9]

2. There is a clear distinction between the identity of Soma and the identity of the substitutes. For Soma one must look to the ṚgVeda;

1. *Tāṇḍya Brāhmaṇa*, 9.5.3.
2. *Aitareya Brāhmaṇa*, Haug edition, 7.5.30.
3. *Kātyāyana Śrautasūtra*, Albrecht Weber, ed., Berlin, 1859, verse 7.8.13.
4. *Śatapatha Brāhmaṇa*, 5.3.3.4.
5. *Cf.* Hemacandra, *Abhidhānacintāmaṇi*, ed. Böhtlingk and Roth, St. Petersburg, 1849, 1191-1192.
6. *Cf. Kauṣītaki Brāhmaṇa*, Ānandāśrama edition, 2.2. and *Śatapatha Brāhmaṇa*, 6.6.3.7.
7. H. H. Wilson: *Sanskrit Dictionary*, 1832, defines *phālguna* as 'a red plant, *Arjunana pentaptera*.'
8. Rudolph von Roth, in his 1881 article in the *Zeitschrift der Deutschen Morgenländischen Gesellschaft*, 'Über den Soma', identified the red *dūrvā* grass with the *Cynodon dactylon* which the Indians used, according to Roxburgh, to make a drink, 'a very cheap kind of Soma' (*eine sehr billige Soma*), as Roth remarked.
9. *Taittirīya Saṃhitā*, Keith edition, 6.1.6; *Śatapatha Brāhmaṇa*, 3.3.1.15.

for the substitutes, the *Brāhmaṇas* are the earliest sources of importance, but they contain no passages about the authentic Soma of sure evidential value. They are concerned with the ritual and symbolic nature of the Soma plant, not with its botanical identity.

II. LATER SANSKRIT WORKS

The writers of the post-Brāhmaṇic period, Sanskrit lexicographers and Vedic commentators, continued to dwell upon a multiplicity of plants that served as Soma, but most of them agreed upon only one thing: that Soma was a creeper (*vallī* or *latā*). Yet nowhere in the ṚgVeda do these terms appear, Soma being there considered an herb (*óṣadhi*) or plant (*vīrúdh*). Amara Siṃha, the earliest of the Indian lexicographers (*ca.* A.D. 450) gives many synonyms for what he calls *somavallī*, all of which Monier-Williams has the courage to define as *Cocculus cordifolius*. Amara also describes a plant that he calls *somarājī*, which Monier-Williams says is *Vernonia anthelmintica*.[1] These are distinguished from *súrā*.[2] The later lexicographers generally imitate Amara in their discussions of Soma: Medinī refers to an herb, the Soma creeper,[3] which Yāska had mentioned as an herb that caused exhilaration when pressed and mixed with water,[4] while Sāyaṇa, the most famous of the Vedic commentators, refers to it as the Soma creeper.[5] Śabarasvāmi, another great commentator, also refers to Soma as a creeper, but one that yields milky juice;[6] this was to be retained as an acceptable attribute of Soma from then on through the European discussions, and the belief in the milky sap appears in the Hindu medical works as well.[7]

1. *Amarakośa*, Kielhorn edition, Bombay, 2.4.82-3 and 2.4.95.
2. *Ibid.*, 2.10.39. Amara gives as synonyms for *súrā* the terms *varuṇātmayā*, *halipriyā*, and *pariśrut*; the latter may refer to an intoxicating liquor made from herbs or to the Soma of the Vedas (ṚgVeda IX 1⁶, and *Śatapatha Brāhmaṇa* 12.7.1.7).
3. *Medinīkośa*, Nathalal Laxmichand edition, 36.
4. Yāska's *Nighaṇṭu and Nirukta*, Laksman Sarup, ed., University of the Punjab, 1927; 11.2.2. and 5.17. *prīṇāti nacamanena* is given as the etymology for *nicumpana*.
5. Commentary on ṚgVeda III 48², Max Müller edition.
6. Śabarasvāmi, commentary on the *Pūrva Mimāṃsa Sūtra*, Gaekwad edition, 2.2.17. Soma is called *latā kṣīriṇī*.
7. This ascription to Soma of a 'milky' quality was probably based upon the Vedic statements that Soma was mixed with milk, or that Soma itself became white when mixed with milk, or that the

One might expect the early medical treatises to be less fanciful than the *Brāhmaṇas*, but this is not the case. The *Dhanvantarīyanighaṇṭu*, a medical work of *ca.* A. D. 1400, says that the Soma creeper yields the Soma milk and is dear to Brahmans.[1] The *Rājanighaṇṭu* describes the properties of Soma: the *somavallī* has great clusters and is a bow-like creeper, yielding the Soma milk; the *somavallī* is acrid, pungent, cool, black, sweet, and it serves to dispel biliousness, to quench thirst, to cause wounds to dry up, and to purify.[2] A more detailed, but hardly more scientific, description appears in the *Suśrutasaṃhitā*, a medical text in verse, dating from perhaps the fourth century A. D., which reduces to its ultimate absurdity the passion for symmetry and classification that permeates these writings. It tells us that although there was originally created one kind of *somavallī*, it was then divided into twenty-four varieties: *aṃśuman* Soma smells like ghee; *muñjara* Soma has leaves like those of garlic; *garuḍahṛta* (= *śyenahṛta*) and *śvetākṣa* ('white-eyes') are pale, look like cast-off snake-skins, and are found pendant from the boughs of trees; *etc.*[3] Unfortunately, the enumeration of these varieties proves of limited value for botanical identification: one reads that all of them have the same qualities, all are creepers with milky juice, all are used in the same way, and all have fifteen black leaves, which appear one per day during the waxing moon and drop off one per day during the waning moon.[4]

juice of Soma was 'milk' in the metaphorical sense of the supreme liquid, or the liquid pressed out of a swollen container; for it must be noted that nowhere in the Vedas is the raw Soma juice itself described as white or milky, but always as yellow or brown or red or golden.

1. *Dhanvantarīyanighaṇṭu*, Ānandāśrama Series No. 33, Poona, 1896; verse 4.4.

2. *Rājanighaṇṭu*, Ānandāśrama Series No. 33; verses 3.29-30. The pertinent terms are: *mahāgulmā dhanurvallī*; *katuś śitā madhurā*; *pittadāhakṛt* (or alternate reading: *pittadāhanut*); *tṛsnāviśoṣaśamanī* (alternate reading: *kṛṣṇā viśoṣaśamanī*); and *pāvanī* (or *pācanī*).

3. *Suśrutasaṃhitā*, Education Press, Calcutta, 1834, chapter 29. I have been unable to find the quotation in other editions of this work, but it is quoted in full in the *Śabdakalpadruma* (Vol. v, p. 417), Calcutta, 1814, which attributes it to the *Suśruta*. It will be noted that several of the *Suśruta's* 'varieties' of Soma refer to substitutes mentioned in the *Brāhmaṇas*.

4. This lunar connection is an extrapolation of the mythological association of Soma with the moon in the Vedas, and it accounts for the name 'moon-plant' given to Soma by the Bengalis of the last century. Yet Edward Balfour, in his *Cyclopædia of India*, published in London in 1885, maintained that the Soma plant itself derived its name from the Sanskrit word for moon, *soma*, because 'it was gathered by moonlight.'

Folk medicine and medical science in India are known collectively as *Āyurveda* ('The Sacred Knowledge of Long Life'). There is a widely quoted Āyurvedic verse, found both in Dhūrtasvāmi's commentary on Āpastamba's *Śrautasūtra* of the black Yajur Veda and in the *Bhāvaprakāśa*, that Max Müller cited in 1855 as the earliest 'scientific' description of Soma that he knew. It describes the plant as a black creeper, sour, leafless, yielding milk, having fleshy skin, causing or preventing phlegm, causing vomiting, and eaten by goats.[1] In spite of the admittedly late origin of this description, and in spite of the many equally authoritative descriptions in earlier Sanskrit works which contradict it, it served European scholars as a peg on which to hang their favourite theories. Its sharp detail and 'scientific' tone contrasted favourably with such descriptions as that of the *Suśruta*, and it seemed to agree with the descriptions of Soma given to Westerners by Indians of that day.

III. EARLY EUROPEAN REFERENCES

The earliest non-Indian notices of Soma are in the Avesta, where the plant appears as Haoma, but these references are more obscure than those of the ṚgVeda; the question of the 'authenticity' of the Soma cult of the Avesta will be discussed below as it arises in the course of European discussions. The earlier parts of the Avesta – the *Yašts* – refer to Haoma as being strained for the sacrifice,[2] as the only drink which is attended with piety rather than with anger,[3] as a tall, golden plant[4] with golden eyes.[5] The *Yasna* devotes three full hymns to Haoma (9-11), which it describes as growing in the mountains,[6] pressed

1. Cited by F. Max Müller in 'Die Todtenbestattung bei den Brahmanen,' *Zeitschrift der Deutschen Morgenländischen Gesellschaft*, No. 39, 1855, pp. XLII ff. The text reads:

śyāmalā 'mlā ca niṣpatrā kṣīriṇī tvaci māṃsalā
śleṣmalā vamanī vallī somākhyā chāgabhojanam

2. *Yašt* fragment 21.9. This and the following are from the edition of James Darmesteter and L. H. Mills, *Sacred Books of the East*, Volumes 4, 23, and 31.
3. *Ashi Yašt*, 2.5, and *Yasna*, 10.8.
4. *Sirozah*, 2.30.
5. *Ashi Yašt*, 6.37, *Gos Yašt*, 4.17, and *Mihir Yašt*, 23.88.
6. *Yasna*, 10.3.

twice a day,[1] odoriferous,[2] possessing many trunks, stems, and branches,[3] and yielding a yellow juice which is to be mixed with milk;[4] Haoma is golden and has a flexible stem.[5] The *Yasna* seems to distinguish between Haoma and other intoxicating beverages, of which it disapproves, but it has been suggested that Zoroaster hated the Haoma drinkers as well as drinkers in general.[6] Beyond this general description, and the evidence that the Soma cult was at least Indo-Iranian rather than simply Indian in its origin, the Avesta sheds little light on the Soma problem.

Megasthenes said of the Indians, 'They never drink wine except at sacrifices,' distinguishing this sacrificial wine from their ordinary liquor – probably *súrā* – which he describes as 'composed of rice instead of barley.'[7] McCrindle suggests that this 'wine was probably Soma juice,' and as the passage is certainly pre-Tantric it must in fact refer to Soma. Thus Megasthenes – misled no doubt by the similar rites among his own people – must be held responsible for the origination of a misconception that continues to plague Vedic studies to the present day. Plutarch speaks of a plant that the Iranians dedicated to a religious use:

> For pounding in a mortar an herb called ὄμωμι they invoke Hades and darkness; then having mingled it with the blood of a slaughtered wolf, they bear it forth into a sunless place and cast it away.[8]

Bernadakis conjectured that this ὄμωμι was the same as the μῶλυ of the *Odyssey* x 305, a fabulous herb probably cognate with the Sanskrit *mūla* (root); Orientalists of note, including Paul Anton de Lagarde,[9] have suggested that ὄμωμι was none other than the Haoma of the Iranians, the Soma of the Indians. In 1929 Emile Benveniste dismissed this notion:

1. *Yasna*, 10.2.
2. *Yasna*, 10.4.
3. *Yasna*, 10.5; *vide supra*, pp. 19-21.
4. *Yasna*, 10.13.
5. *Yasna*, 9.16.
6. Henrik Samnel Nyberg: *Die Religionen des Alten Irans*, Leipzig, 1938, pp. 188, 244, and 288.
7. Megasthenes, *Indika*, translated by J. W. McCrindle, Calcutta, 1887 (Schwanbeck edition, Bonn, 1846), fragment 27, page 69. Quoted by Strabo, 15.1.53-56.
8. Plutarch: *De Iside et Osire*, Squire edition, 1744, p. 117; μζ.
9. *Vide infra*, p. 108.

> We must beware of correcting the text on this point, as the majority of editors have done following P. de Lagarde. The substitution of the plant μῶλυ for the enigmatic ὄμωμι is the device of harassed interpreters and is no better than the explanation of ὄμωμι by *hauma*; the first is arbitrary, and the second absurd . . . ὄμωμι is another name of *amomum* which is used in the cult of Ahriman as *hauma* is sacred to the cult of Ohrmazd.[1]

In 1771 A. H. Anquetil-Duperron brought out the first translation of the Avesta, after having spent some years in India in association with the Parsis. For the Haoma plant, he observed, the Parsis used a tree (*arbre*) which, they said, grew in Persia but not in India and resembled a vine but never bore any fruit. Anquetil-Duperron thought the plant resembled a kind of heather (*bruyère*), with knots very close together and leaves like those of jasmine. 'All these details lead me to believe that the *Hom* is the ἄμωμος of the Greeks and the *amomum* of the Romans.'[2] As the Indo-Iranian connection was not yet established, nor the Soma cult itself discovered, this suggestion was not pursued.

Then in 1784 the first translation of a Sanskrit work into English appeared, Charles Wilkins's rendering of the *Bhagavad Gītā*. In it he included a note: '*Sōm* is the name of a creeper, the juice of which is commanded to be drunk at the conclusion of a sacrifice.'[3] This is the earliest published citation of Soma in a European language that I have found. In his 1794 translation of the laws of Manu, Sir William Jones describes Soma as 'the moon plant (a species of mountain-rue).'[4] H. T. Colebrooke then published his translation of the *Amarakośa*, saying that the *somrāj* was *Vernonia anthelmintica* (= *Conyza anthelmintica*),[5] but cautioning his readers that commentators seldom describe

1. Emile Benveniste: *The Persian Religion: According to the Chief Greek Texts*, 1929, p. 74. The *amomum* is thought to be the Indian spice plant, *Nepaul cardamom*. It is mentioned in Aristophanes' *The Frogs*, 110, and in Theophrastes' *Historia Plantarum*, 9.7.2, where it is said to come from India as cardamom comes from Persia.
2. A. H. Anquetil-Duperron: *Zend-Avesta, traduit par Anquetil du Perron*, Paris, 1771, Vol. II, p. 535.
3. Charles Wilkins: *The Bhāgvāt Geeta*, Serampore, 1784, p. 80, note 42.
4. Sir William Jones: *Institutes of Hindu Law, or, the Ordinances of Manu*, Calcutta, 1794, verse 3.158.
5. H. T. Colebrooke: *Cosha or Dictionary of the Sanscrit Language, by Amara Sinha, with an English Interpretation and Annotations by H. T. Colebrooke, Esq.*, Serampore, 1808, p. 102, note on verse 4.3.14. It should be added that Amara distinguished this plant from the *somavallī* for which Colebrooke gives no botanical name.

the plants they mention and that 'a source of error remains in the inaccuracy of the Commentators themselves . . . the correspondence of Sanscrit names with the generick and specifick names in Natural History is in many instances doubtful.'[1]

In 1814, William Carey published his *Hortus Bengalensis*, which was a summary of the manuscript that William Roxburgh was to publish in 1832. Carey identified *Sarcostemma brevistigma* (= *Asclepias acida, Sarcostemma acidum, Sarcostemma viminale, Cynanchium viminale*) as the plant known in Bengali and Sanskrit as *somalutā*, and also remarked that *Ruta graveolens*, a rue, bore the same Sanskrit name.[2] He did not link them with the Vedas, but he did take occasion to observe that Himalayan plants do not grow in Bengal,[3] an observation that was ignored by the Vedic scholars who later used Roxburgh's work to identify Soma.

In his 1819 *Sanskrit Dictionary*, Horace Hayman Wilson identified *soma* as 'the moon plant (*Asclepias acidum*) [= *Sarcostemma brevistigma*],' *somavallī* as 'a twining plant (*Menispermum glabrum*) [= *Tinospora cordifolia*],' or 'a medicinal plant (*Serratula anthelmintica*) [= *Vernonia anthelmintica*]'; in the 1832 edition of the dictionary, he added *somarājin*, which he defined as *Serratula anthelmintica*, though he still gave this as an alternative for *somavallī* as well.[4]

Sir Graves Chamney Haughton, in his 1825 edition of Jones's *Ordinances of Manu*, had corrected Jones's 'mountain rue' to 'swallow-wort (*Asclepias acidum*) [= *Sarcostemma brevistigma*],' probably following Carey. Finally, in 1832 William Roxburgh published his *Flora Indica*; he identified *Sarcostemma brevistigma* with *somalutā* in Sanskrit and Bengali, a plant (as he said) of much milky juice of a mild nature and acid taste. He added that 'native travellers often suck the tender

1. Colebrooke, *op. cit.*, p. 10.
2. William Carey: *Hortus Bengalensis, A Catalogue of the Plants Described by Dr. Roxburgh in his Manuscript Flora Indica*, Calcutta, 1814, pp. 20 and 32.
3. *Ibid.*, p. xi.
4. Horace Hayman Wilson: *Sanskrit Dictionary*, Calcutta, 1819 and 1832. In identifying *somavallī* with *Menispermum glabrum*, Wilson was misled by Colebrooke, whose Amara Siṃha he cites for this reference; Colebrooke, though giving no definition of *somavallī*, had given *Menispermum glabrum* (= *Tinospora cordifolia*) for the plant *guḍucī* (*guricha* in Colebrooke's note) which appears in juxtaposition with *somavallī* in the *Amarakośa*.

shoots to allay their thirst.'[1] This is hardly the description of a plant that induces religious ecstasy, but Roxburgh identifies it with *sōm* mentioned by Wilkins as the sacrificial plant. He distinguishes it from *somrāj* (as Colebrooke had done in 1808), which is *Vernonia anthelmintica*, a plant with an acid taste,[2] and he further distinguishes *Calotropis gigantea* (= *Asclepias gigantea*) as the plant used by the natives for medical purposes.

From these beginnings down to our own time Soma has been identified with various species of Sarcostemma (= Asclepiads, related to the American milkweeds), of Ephedra, of Periploca, all of them leafless climbers superficially resembling each other, yet belonging to genera botanically far apart. Botanists in India would gather specimens, identify them with scientific names, and add the vernacular names that local helpers would give them, such as *somalutā*. Linguists and serious travellers would occasionally bring back plant names picked up in the various languages spoken from India to Iran that seemed to stem back to Soma or Haoma – *e.g.*, *hum, huma, yehma*; *um, uma*; *um, umbur*[3] – and, linking them to the plant to which they belonged, present to the world another candidate for Soma. Most of these plants were or had been at some time used as substitutes for Soma or Haoma: to be eligible, plants had to meet certain requirements, which may have changed from area to area and from century to century. R. G. Wasson's Brahman informants said to him that the substitute 'Somas' had to be small, leafless, and with fleshy stems, attributes that are common in varying degrees to the three genera listed above and to the traditional descriptions in the *Brāhmaṇas* and medical texts.

IV. MID-NINETEENTH CENTURY

For the next fifty years, Sanskritists and botanists alike merely elaborated upon Roxburgh's identification. Henry Piddington gives '*Sarcostemma viminale*' for the Bengali *soom* and '*Asclepias acidum*' for

1. William Roxburgh: *Flora Indica*, Serampore, 1832, Vol. II, p. 33.
2. *Ibid.*, Vol. III, p. 406.
3. George Watt: *Dictionary of the Economic Products of India*, Calcutta, 1890, cites these examples in his article on 'Ephedra.' *Vide infra*, p. 121.

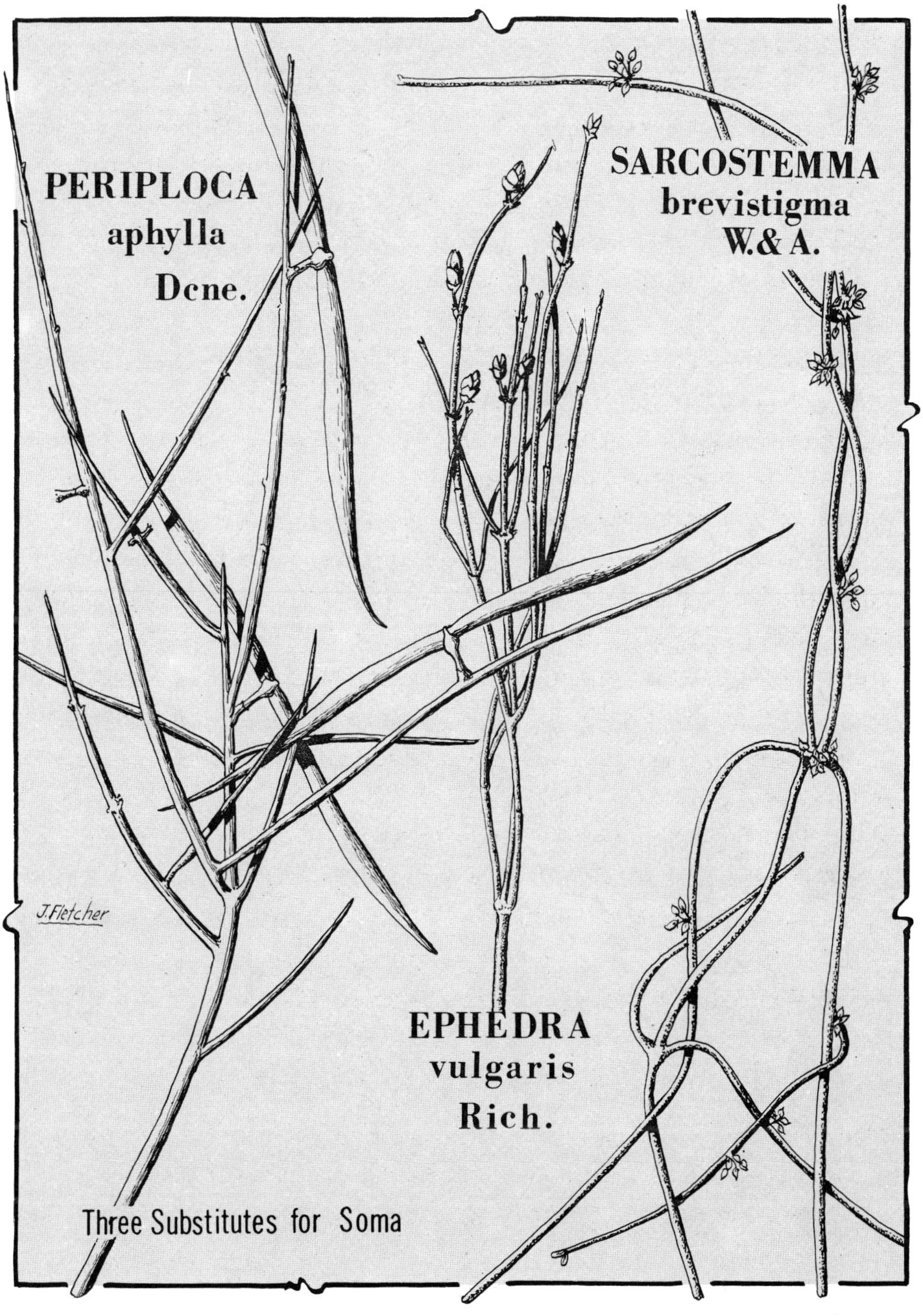

Three Substitutes for Soma

somalutā,[1] but these two names are now considered to represent the same species: *Sarcostemma brevistigma*. John Stevenson, in his translation of the *Sāma Veda*, says that Soma is '*Sarcostemma viminale*,' 'the moon-plant' and describes in some detail the method of its use, observing that according to the commentator it is pressed and mixed with barley and allowed to stand for nine days, 'but how many days precede [the ceremony] and how many follow, I do not know . . . the Soma, when properly prepared, is a powerful spirit . . .'[2]

Vedic studies had now begun to assume considerable importance in Europe. In 1844, Eugène Burnouf published the first of a series of articles on Haoma, wherein he said that the Haoma juice, obtained by trituration, was the same as the Soma of the Vedic sacrifice, but he did not venture a botanical identification.[3] In 1845, J. O. Voigt published his *Hortus Suburbanus Calcuttensis*, a catalogue of the plants in the H. E. I. C. Botanical garden. Once again *soma-rāj* is *Vernonia anthelmintica* and *shom-luta* is *Sarcostemma brevistigma*; Voigt also mentions that farmers use the *Sarcostemma brevistigma* to rid their fields of white ants.[4] He draws attention to the use of the Asclepiads in general as emetics and to their acrid and bitter milk, and points out that in the West Indies they are a popular remedy for worms in children, given in doses of a teaspoon to a tablespoon.[5]

Friedrich Windischmann thought that the Soma plant might be *Sarcostemma brevistigma*, but he doubted that the Indian Soma was the same as the Haoma of the Persians, as the plant might have changed with the change in location. Yet he considered that the Avesta preserved the tradition of Soma and the sacrifice better than the ṚgVeda did, and he called attention to the Persian belief (remarked long ago by Anquetil-Duperron) that Soma did not grow in India.[6] Christian Lassen

1. Henry Piddington: *An English Index to the Plants of India*, Calcutta, 1832, p. 79 and p. 9.
2. John Stevenson: *Sanhitā of the Sāma Veda*, London, 1842, p. iv.
3. Eugène Burnouf: 'Etudes sur la langue et sur les textes zends. IV: Le Dieu Homa,' in *Journal Asiatique*, 8th Series, No. 4, December, 1844, p. 468.
4. J. O. Voigt: *Hortus Suburbanus Calcuttensis, A Catalogue of the Plants in the Honourable East India Company Botanical Garden*, Calcutta, 1845, pp. 405 and 542.
5. *Ibid.*, p. 539.
6. Friedrich Windischmann: 'Über den Somakultus der Arier,' *Abhandlungen der Königlichen Baierischen Akadamie der Wissenschaften*, Munich, 1846, pp. 127-142.

added his assent that Soma was *Sarcostemma brevistigma*,[1] and William Dwight Whitney wrote that Soma was

> a certain herb, the *Asclepias acida* [= *Sarcostemma brevistigma*], which grows abundantly upon the mountains of India and Persia . . . which, when fermented, possesses intoxicating qualities. In this circumstance, it is believed, lies the explanation of the whole matter. The simple-minded Arian [*sic*] people . . . had no sooner perceived that this liquid had power to . . . produce a temporary frenzy . . . than they found in it something divine . . .[2]

Otto Böhtlingk and Rudolph von Roth accepted only tentatively *Sarcostemma brevistigma* as Soma, and perspicaciously added that it seemed to grow farther south than the Vedas indicate.[3]

Friedrich Max Müller in 1855 published an important article in which he quoted Dhūrtasvāmi's description of Soma[4] and said that this seemed to agree most strikingly with the accepted botanical descriptions of various Sarcostemmas. But he doubted whether the Vedic Soma would be found growing in Bombay (where the Sarcostemmas are found) rather than in the mountains of the North. Moreover, he asked, why would the Indians of the era of the *Brāhmaṇas* use *putīka* as a substitute for Soma if they could find 'Soma' itself – *i.e.*, Sarcostemma spp. – right in Bombay?[5]

For thirty years no notice was taken of this Āyurvedic source, nor of Max Müller's doubts. Major Heber Drury accepted *Sarcostemma brevistigma* as *shom-luta*,[6] and Walter Elliot referred *somalatā* to the same species.[7] Martin Haug believed that the Persians had probably replaced the original Haoma with something else, retaining the name,

1. Christian Lassen: *Indische Altertumskunde*, Bonn, 1847, Volume 1, p. 281.
2. William Dwight Whitney: 'On the main results of Vedic researches in Germany,' *Journal of the American Oriental Society*, No. 3, 1853, p. 299.
3. Otto Böhtlingk and Rudolph von Roth: *Sanskritwörterbuch nebst allen Nachträgen*, St. Petersburg, 1855-75.
4. *Vide supra*, p. 100.
5. F. Max Müller: 'Die Todtenbestattung bei den Brahmanen,' *Zeitschrift der Deutschen Morgenländischen Gesellschaft*, No. 39, 1855, p. xlii ff.
6. Major Heber Drury: *Useful Plants of India*, Madras, 1858, p. 385.
7. Walter Elliot: *Flora Andhrica, A Vernacular and Botanical List of Plants Commonly Met with in the Telegu District*, Madras, 1859, p. 169.

and he thought that the present-day Soma of the Indians was in fact a substitute, but that the substitute retained in a measure the qualities of the true Vedic plant, with no leaves and a bitter white sap: 'It is a very nasty drink, but has some intoxicating effect,' he wrote, having tasted it several times but being unable to take more than a few teaspoonfuls.[1]

Sir George Christopher Birdwood, in his *Catalogue of the Vegetable Productions of the Presidency of Bombay, including a list of the drugs sold in the bazaars of Western India*, included *Sarcostemma brevistigma* under 'drugs', identifying it as the *somalutā* of the Vedas, a 'fermented liquor . . . mixed with barley and ghee . . . This wine was drunk at all their religious ceremonies and was used as an intoxicant by the rishis . . . Water passed through a bundle of *somalutā* and a bag of salt will extirpate white ants from a field watered with it.'[2] Birdwood identified *Sarcostemma brevistigma* with *S. acida* and expressed the view that 'the Som of the Vedas and the Hom of the Zend Avesta' were perhaps 'the real plant . . . present to the mind of the writer . . . of the first chapters of Genesis,'[3] – an original contribution to the debate, and one of the few that is not echoed by anyone. In this year, Eugène Burnouf published his Sanskrit dictionary, wherein he described Soma as the juice of *Sarcostemma brevistigma*; the following year, J. Forbes Watson described the Soma plant as *Sarcostemma brevistigma*; the *somalatā* (Telegu) as the same species; and the *somalutā* (Sanskrit) as *Ruta graveolens*.[4]

Paul Anton de Lagarde (Paul Anton Boetticher) maintained that ὄμωμι was another word for μῶλυ or πήγανον, the mountain rue, and that it was a substitute used by the Greeks when they no longer had the *hōm* itself, the original sacred plant.[5] Drawing attention to the *Odyssey* verse describing the μῶλυ as a plant with a black root and a

1. Martin Haug: *Essays on the Religion of the Parsees*, 1861, pp. 219-222, and *Aitareya Brāhmaṇa*, Bombay, 1863, Vol. II, p. 489.
2. Sir George Christopher Birdwood: *Catalogue of the Vegetable Productions of the Presidency of Bombay*, Bombay, 1865, p. 53.
3. *Ibid.*, p. 209.
4. J. Forbes Watson: *Index to Native and Scientific Names of Indian and Other Economic Plants and Products*, London, 1866, p. 530.
5. Paul Anton de Lagarde: *Gesammelte Abhandlungen*, Leipzig, 1866, p. 174.

flower white as milk,[1] and to the mythological significance of rue,[2] Lagarde maintained that the description of the *harmal* (= *haoma* = ὄμωμι = μῶλυ) given by Arabian botanists described the plant precisely: a shrub with leaves like those of a willow and flowers like those of jasmine, with an intoxicating and soporific effect.[3] He added that Soma and Haoma were not the same plant, though they had the same name and use, and he linked the Soma of the ṚgVeda to rue by means of the Vedic epithet *sahásrapājas*, 'possessing a thousand *pājas*,' which he related to the Greek term for rue, πήγανον.[4]

But no matter how many botanists and scholars accepted the Sarcostemma thesis, there were always reservations held in some quarters. In 1871 John Garrett declared flatly that 'the Soma of the Vedas is no longer known in India'.[5] Nevertheless J. D. Hooker elaborated upon his earlier identification of Soma with *Sarcostemma brunonianum* by observing that this plant abounded in an acid milky juice and was 'hence eaten by the natives as salad, and sucked by travellers to allay thirst, thus forming a remarkable exception to the usually poisonous nature of the Asclepiadeous juices.'[6] Still there was room for new theories: Drury suggested that the 'moon creeper' might be *Calonyction muricatum* (*Ipomœa muricata*), a plant whose swollen pedicels are cooked as vegetables and whose seeds are used as purgatives.[7] Nevertheless, Hermann Grassmann, whose Vedic dictionary is the standard work to this day, accepted the general view that Soma was a Sarcostemma.[8]

In 1873 Rajendra Lala Mitra revived the case for Soma as an alcoholic beverage. The original Indo-Aryans drank 'soma-beer and strong spirits,' which, when they moved to the hot climate of India, tended to

1. *Odyssey*, X, 304-5.

2. Jacob Ludwig Grimm: *Deutsche Mythologie* (Göttingen, 1835), p. 962, had mentioned that rue was used in sacrifices to the devil.

3. Lagarde, *op. cit.*, p. 175.

4. The term *sahásrapājas*, occurring only twice in the ṚgVeda (IX 13³ and IX 42³), is generally translated as 'possessing a thousand forms,' or 'colours', or 'rays'.

5. John Garrett: *Classical Dictionary of India*, Madras, 1871, p. 594.

6. Joseph Dalton Hooker: *Curtis's Botanical Magazine*, London, 1872, Tab. 6002.

7. Major Heber Drury: *Useful Plants of India* (2nd edition), London, 1873.

8. Hermann Grassmann: *Wörterbuch zum Rig Veda*, Leipzig, 1873.

make them ill. 'The later Vedas, accordingly, proposed a compromise, and leaving the rites intact, prohibited the use of spirits for the gratification of the senses.'[1] Soma was 'made with the expressed juice of a creeper (*Asclepias acida* or *Sarcostemma viminale*) [both = *Sarcostemma brevistigma*] diluted with water, mixed with barley meal, clarified butter [ghee], and the meal of wild paddy [*nīvāra*], and fermented in a jar for nine days. . . . The juice of the creeper is said to be of an acid taste, but I have not heard that it has any narcotic property.' Mitra was of the opinion that the starch of the two meals – barley and wild paddy – produced 'vinous fermentation' and that the Soma juice promoted fermentation and flavoured the brew while checking the acetous decomposition, in the manner of hops.[2] In this way Mitra seemed to cover every likelihood – except for the possibility that Soma was not alcoholic at all, but a plant containing a psychotropic drug. It should also be pointed out that Mitra speaks of 'soma-beer and *strong spirits*.' Presumably he used this latter term in its customary sense, meaning distilled alcohol. This is an anachronism.

In 1874 Arthur Coke Burnell called attention to the fact that there were different 'Somas' in different parts of India, the Hindus of the Coromandel coast using *Sarcostemma brevistigma* in their rites, and those of the Malabar coast using *Ceropegia decaisneana* or *C. elegans*.[3] In the following year, Martin Haug elaborated on his previous descriptions by describing the Soma plant as a small twining creeper with a row of leafless shoots containing sour, milky sap; he identified the plant as *Sarcostemma intermedium*, mentioning *S. brevistigma* and *S. brunonianum* as related varieties, and he expressed his opinion that the plants denoted at the present time by the terms *somalatā* and *somavallī* were later substitutes.[4]

In 1878 Friedrich Spiegel reported that the Indian Parsis sent their priests to Kerman to obtain Haoma.[5] This was the year that saw the

1. Rajendra Lala Mitra: 'Spirituous Drinks in Ancient India,' *Journal of the Royal Asiatic Society of Bengal*, 1873, p. 2.
2. *Ibid.*, p. 21.
3. Arthur Coke Burnell: *Elements of South Indian Palæography*, Mangalore, 1874, p. 1.
4. Martin Haug: in a review of Grassmann's *Wörterbuch; Göttingische Gelehrte Anzeigen*, 1875, pp. 584-595.
5. Friedrich Spiegel: *Eranische Alterthumskunde*, Leipzig, 1878, Vol. III, p. 572.

publication of Abel Bergaigne's *La Religion Védique*; Bergaigne thought that Soma was a fermented drink and that milk was added to it to help the fermentation process. He calls it a spirituous liquor, the word spirituous (*une liqueur spiritueuse*) being applicable in a broad sense to any volatile inebriating drink.[1]

In 1879, Heinrich Zimmer acknowledged that present-day Hindus used a kind of Sarcostemma, citing Haug and adding that the nausea Haug had experienced was consistent with the *Brāhmaṇa* descriptions of Soma; but he expressed doubts whether this plant, found in Bombay and all over India, could have grown in the high site of Vedic civilization. He quoted various *Brāhmaṇa* references to substitutes to support his thesis that Sarcostemma was in fact just such a substitute, which had appeared in the course of time.[2]

In 1881 Mitra reiterated his beer theory,[3] and Kenneth Somerled Macdonald wrote, 'Soma is now admitted, we believe, by the best Sanskrit scholars to have been intoxicating. The numerous references in the ṚgVeda are consistent only with such an interpretation.'[4] He proceeded to repeat Mitra's theory in great detail, though he failed to give Mitra credit for it.

The early 1880's saw the most enthusiastic and intense period of the Soma debate. Scholarly tempers flared, new and important names began to appear, and ingenious theses were advanced. Rudolph von Roth began it with an article in which he reviewed all the recent theories about Soma as well as some of the *Brāhmaṇa* information; he concluded that although *Sarcostemma brevistigma* seemed to be well established as the present-day Soma plant, there was no assurance that this plant – which grew in the plains – was the Vedic plant, which must still grow in the mountains.[5] Though he believed that the Soma of the *Brāhmaṇas* was a substitute made necessary when the Vedic people moved away from their original home, he felt that the true

1. Abel Bergaigne: *La Religion Védique*, Paris, 1878, Vol. 1, pp. ix and 148.
2. Heinrich Zimmer: *Altindisches Leben*, Berlin, 1879, p. 275.
3. Rajendra Lala Mitra: *Indo-Aryans*, London, 1881, pp. 390-419.
4. Kenneth Somerled Macdonald: *The Vedic Religion*, London, 1881, p. 66.
5. Rudolph von Roth: 'Über den Soma,' *Zeitschrift der Deutschen Morgenländischen Gesellschaft*, No. 35, 1881, pp. 680-692.

Soma was probably some species of Sarcostemma or at least an Asclepiad, and he felt that any information about the present-day Soma might shed light on the Vedic plant.

In the following year, Edward William West expressed the opinion that the plants used for Soma in India and Persia at the present time were substitutes for the original Haoma-Soma plant, and he observed that the most common substitute in South and West India was the *Sarcostemma brevistigma*, 'a leafless bush of green succulent branches, growing upwards, with flowers like those of an onion,' and resembling *Euphorbia tirucalli* or thornless milk-bush when not in flower.[1] In that same year Angelo de Gubernatis, quoting Roth, expressed doubt whether the Soma plant could be identified at that time; he suggested that perhaps the Soma cult had shifted to wine in Persia, Asia Minor, and Greece, and that in India in later times its place was taken by a beverage offered to the gods and deliberately made unpalatable so that no mortal would be tempted to drink it.[2]

In 1883 Monier-Williams brought out his *Religious Thought and Life in India*. He defined Soma as *Sarcostemma brevistigma*, 'a kind of creeper with succulent, leafless stem,' but he added, 'And yet it is remarkable that this sacred plant [the original, true Soma] has fallen into complete neglect in modern times. When I asked the Brahmans of North India to procure specimens of the true Soma for me, I was told that, in consequence of the present sinful condition of the world, the holy plant had ceased to grow on terrestrial soil, and was only to be found in heaven.' Nevertheless a creeper 'said to be the true Soma' had been pointed out to him by Burnell in South India, where it was being used by orthodox Brahmans.[3]

In 1884 D. N. Ovsianiko-Kulikovskij published a book of some 240 pages devoted entirely to Soma and containing a great deal of learned and imaginative material; unfortunately, since it was published in Russian and never translated, it was not noted by subsequent scholars writing in other European languages, though Ovsianiko-Kulikovskij

1. Edward William West: *Pahlavi Texts, Sacred Books of the East*, Volume 18, 1882, p. 164.
2. Angelo de Gubernatis: *La Mythologie des Plantes*, Paris, 1882, Vol. II, pp. 350-369.
3. Sir Monier Monier-Williams: *Religious Thought and Life in India*, London, 1883, pp. 12-13.

was familiar with all of the significant literature in German, French, and English.[1] He maintained, on grounds of sound linguistic reasoning, that the word 'soma' applied first to the plant and later to the juice[2] and went on to discuss this plant:

> At the same time, the terms Soma and Haoma did not cease to signify also the divinity, whose gift or attribute the intoxicating beverage was considered to be. As concerns this last, it is presently obtained in India from a plant of the family *Asclepias acida*. Whether it was obtained from the same plant in antiquity, or from another, or if from another, from which particular one – all of these are questions that of necessity must remain for the time being without an answer. The description of the flower and the liquid of Soma in the ṚgVeda is not applicable to *Asclepias acida* and the beverage obtained from it.[3]

Ovsianiko-Kulikovskij described the sacrament of collecting Soma as taking place at night, 'in the light of the moon,' observing: 'Perhaps this feature has some link with the later identification of Soma with the moon.'[4] He continued:

> It was mixed with water, sour milk, and barley corn. Then it underwent fermentation, after which a strong, intoxicating beverage was obtained. The description of this procedure is met in a number of places in the ninth Maṇḍala of the ṚgVeda, but always in a fragmentary and often obscure form.[5]

This was the more or less conventional view of Soma, but then Ovsianiko-Kulikovskij went on to express his doubts that Soma had been a product of fermentation at all, and to suggest – in cautious but clear terms – that Soma might have been some sort of narcotic.

1. D. N. Ovsianiko-Kulikovskij: *Zapiski Imperatorskogo Novorosiiskogo Universiteta*, Volume 39, Odessa, 1884. Chast' 1: *Kul't bozhestva Soma v Drevnei Indii v epokhu Ved*. [Part 1: The Cult of the Deity 'Soma' in Ancient India in the Vedic Epoch] I am indebted to D. M. O'Flaherty for the translation.
2. *Ibid.*, p. 7.
3. *Ibid.*, p. 8.
4. *Idem.* This peculiar notion appears only one year later in the *Cyclopædia* of Edward Balfour (*vide supra*, p. 99), but it is most unlikely that he knew Ovsianiko-Kulikovskii's work.
5. *Ibid.*, p. 9.

This is by far the earliest reference to such a possibility, and it is significant that it appears in the writings of a Russian, who may have known of the Siberian cults:

> It is possible that the narcotic power of Soma was greater than the similar power of other drinks used in the time of the Veda; possibly it was of an essentially different nature. The action of Soma is always depicted in the most many-hued colours, as something fascinating, elevating, illuminating; on the other hand the action of common drink (*súrā*) is painted in far from such attractive colours . . . It is also quite possible that the superiority of Soma was entirely imaginary and rested solely on that religious sanction which fell to the portion of Soma (that is, not *Asclepias acida* or any other sort of Indian plant, but – Soma, as a religious and cultural psychological conception) as early as the most remote Indo-Iranian antiquity.[1]

Roth then brought out another paper, in which he emphasized his feeling that Soma must still exist.[2] But he acknowledged that Albert Regel, the botanist employed by the Russian government, who had searched for Soma at Roth's request in the Syr Darya and Amu Darya watersheds, had reported finding no trace of a plant meeting Soma's description. Regel had expressed his opinion that the closest thing to Soma that he had found had been rhubarb, though he admitted that rhubarb was not used by the natives to make any intoxicating drink. Roth had supplied him with Vedic descriptions of Soma, and from them it was clear that none of the Asclepiads, Euphorbias, Ferulas, yellow Compositæ, or *Cannabis sativa* (hemp) conformed to the formula. Wilkens, a zoologist specializing in South Turkestan, had reported to Roth that in his opinion Soma might be *Peganum harmala*,[3] belonging to the Rutaceæ, though it lacked the sweetness and juiciness that the true Soma was thought to have had. Nothing daunted, Roth ended up with a flourish: 'Usbekistani today may drink their kumiss in cups in which Soma once gave them cheer. . . . To

1. Ovsianiko-Kulikovskij, *op. cit.*, p. 12.
2. Rudolph von Roth: 'Wo wächst der Soma?', *Zeitschrift der Deutschen Morgenländischen Gesellschaft*, No. 38, 1884, pp. 134-139.
3. The πήγανον and *harmal* to which Lagarde had referred.

find the Soma one need not be a botanist; the plant will have to be recognized in all its juiciness [*Saftfülle*] by every eye.'

V. FILE NUMBER 118

Charles James Lyall, Secretary to the Chief Commissioner of Assam, translated Roth's papers and forwarded them, with his own remarks, to the Afghan Frontier Delimitation Commission. They in turn handed them over to the botanist George Watt, who published them with *his* added remarks, under the heading, 'A note upon Dr. Roth's suggestion regarding the Soma Plant', in a document issued as File No. 118 of the Government of India, Revenue and Agricultural Department, Simla, August 20, 1884.

Watt felt that Roth had produced no evidence that Soma was a species of Sarcostemma or any other Asclepiad; he considered it 'a great pity that Dr. Roth, instead of propounding his own theory at such length and in attempting to confute arguments against it, did not rather publish briefly the leading passages from Sanskrit literature descriptive of the plant . . . placing in the hands of the naturalist to the Commission a brief abstract from the Sanskrit authors, and thus leave his mind unbiased by any theories.' Watt was of the opinion that Soma was not necessarily a succulent, juicy plant, but that it might be rather a dry branch used in a decoction, 'either by simple maceration or boiling.' He went on: 'Can any one who has examined the bitter milky sap of the Ascledpiadæ . . . suppose that such a liquid could ever be used for more than a medicinal purpose, and still less become the Soma of the Vedas? It is much more likely that the oblong fruits of the Afghan grape . . ., imported into the plains, as they are at the present day, afforded the sweet and refreshing cup of which our Aryan ancestors became drunk while wrapt in the oblivion of religious enthusiasm.'[1] He did not pursue this suggestion, however, and considered the Compositæ or Umbelliferæ more likely than the Asclepiads as candidates for Soma.

1. That the Aryans knew of such a fruit and of an alcoholic drink made from it is established by Pāṇini's reference to the sweet grape juice of Kāpiśī, north of Kabul (Pāṇini 4.2.99).

In reply to this, Max Müller wrote to the *Academy* journal on October 20, 1884, misquoting Haug ('it was extremely nasty and not at all exhilarating', where Haug had described the Soma he tasted as 'a very nasty drink, but [it] has some intoxicating effect'), referring to the *Brāhmaṇa* substitutes and the tradition that Soma 'was brought by barbarians from the North,' and finally getting to the point – that he, Friedrich Max Müller, had published thirty years earlier the 'oldest scientific description of the Soma plant' that he knew of or had hope of finding, the Dhūrtasvāmi description of the dark creeper eaten by goats.[1]

On November 9, Roth replied in the same journal to 'the learned scholar': 'I did not, indeed, remember the passage referred to; but if it had been in my mind I should scarcely have mentioned it . . .' He said it was impossible to date the '*Āyurveda*', that Max Müller's passage sounded like descriptions of 'the later, even the latest date, especially in the so-called *Nighaṇṭus*,' taking exception to the adjective *vamanī* by insisting that 'it is not to be supposed that the Soma, or its principal substitute in later times, should have caused vomiting,' and, as a parting blow, correcting Max Müller's translation of *śleṣmalā* from 'destroying phlegm' to 'producing phlegm.' He concluded that the plant described was merely the 'Soma of later times which we know (that is, the *Sarcostemma acida* [*brevistigma*]), correctly described as bearing no leaves.' He still believed that the 'genuine original Soma' would bear great resemblance to its later substitute, and answered Watt's 'decoction' theory thus: 'I am sorry not to be able to conform my views to those of the distinguished botanist. The Aryans no more drank a decoction of the Soma plant than they drank tea or coffee. It would be, indeed, a disgrace to the interpreters of the Veda and Avesta if Dr. Watt were right.' He ended with the arch hope that the botanists of the Commission 'will not bring us home, as Soma, the Asafœtida, which there obtrudes itself upon one's notice, or any other Ferula.'

Max Müller replied on November 17 in another letter to the *Academy*, defending his Āyurvedic passage as the oldest *scientific* de-

1. *Vide supra*, pp. 100 and 107.

scription of the Soma, standing up for his translation of *śleṣmalā* as dissolving rather than producing phlegm, and keeping his original reading for *vamanī* rather than the *pāvanī* (purifying) suggested by Roth. He admitted that 'this oldest scientific description of the Soma plant' might refer to a later substitute, and concluded: 'As to the Soma which the Brahmans knew (ṚgVeda x 85³), I shall welcome it whenever it is discovered, whether in the valley of the Oxus or in that of the Neckar.'

At this point the botanists entered the forum. J. G. Baker of the Kew Herbarium wrote to the *Academy* on November 15, noting that Dr. Aitchison had been selected as the botanist to attend the Afghan Delimitation Commission and supporting Max Müller's faith in the Sarcostemma, which, he pointed out, is eaten by men and animals throughout Sindh, Arabia, and Persia. He said that the flowers of the *Periploca aphylla* are eaten by the natives of Baluchistan 'and taste like raisins.' W. T. Thiselton-Dyer of the Royal Kew Gardens then supported Watt's suggestion of the Afghan grapes: 'That the primitive Soma was something not less detestable than anything that could be extracted from a Sarcostemma I find it hard to believe. When, however, the original Soma was unprocurable, and the use became purely ceremonial, the unpalatableness of the Soma substitute was immaterial.'[1] The Sarcostemma, according to Thiselton-Dyer, was chosen when the Soma plant was forgotten (or unavailable in the hot plains) because 'there is a faint resemblance in texture and appearance, though not in form, between the joint of a Sarcostemma and an unripe green grape,' the Vedic Indians having had no word to distinguish a fruit from the stem of a plant, Thiselton-Dyer maintained.[2]

Max Müller answered this, on December 8, with a reference to the Vedic tradition that Soma juice was mixed with barley milk, a process that, he suggested, was incompatible with the grape hypothesis but not with a kind of hops; he added that 'a venturesome etymologist might not shrink even from maintaining that hops and Soma are the

1. Here he refers to de Gubernatis II, 352; *vide supra*, p. 112.

2. This is not true. *Phala*, 'fruit,' recurs in the ṚgVeda, referring to the fruit of a plant (not Soma) rather than to the stem. Cf. ṚgVeda III 45⁴, IV 57⁶, X 71⁵, X 97¹⁵ and X 146⁵.

same word,' deriving *hops* from Hungarian *komló*, mediæval Latin *humolo*, mediæval Greek χουμέλη, and ultimately Sanskrit *soma*, 'which, for a foreign word, brought from Persia into Europe, is tolerably near . . . Now hops mixed with barley would give some kind of beer. Whether milk would improve the mixture I am not brewer enough to know.'

Charles G. Leland then wrote to the *Academy* to support the hops hypothesis with information about the *soma* or *sumer* ('the pronunciation is not fixed') of the Romany tongue, apologizing that 'any confirmation of this, drawn from such a very disreputable source as gypsy, is, indeed, not worth much,' but pointing out that there was much Sanskrit in gypsy words and that *soma* in Romany meant 'scent or flavour . . . thus the hop gives the *suma* or *soma* to the beer, as the lemon to punch.' He added that the fact that hops do not grow in the present dwelling place of the Hindus confirms rather than disproves Max Müller's theory, since if it were still available there the Indians would be using it instead of the present substitute. At this point (December 20), A. Houtum-Schindler wrote from Teheran with a tale of a Sarcostemma that he had been shown by the Parsis in Kerman, a plant with a greenish white juice and a sweetish taste that caused vomiting when taken in amounts of more than a dozen drops, and that corresponded closely with Max Müller's Āyurvedic description – *if* it was viewed only 'several days after it had been collected,' by which time the stalk would have turned dark, the juice turned sour, and the leaves fallen off. To this ingenious postulate he added several descriptions of the *Hum* plant from various Persian dictionaries, including one of a deadly poison the fruit of which is much liked by partridges and which resembles a tamarisk tree – 'the latter qualities evidently refer to another plant.' W. T. Thiselton-Dyer then identified still another plant used by the Parsis in their rites – the *Ephedra vulgaris* which, he said, bore sweet red berries and somewhat resembled *Tamarix articulata*, like the plant mentioned by Houtum-Schindler: 'But he [Houtum-Schindler] also says it is "a deadly poison" (though apparently not to partridges). This does not agree with Ephedra, which is browsed by goats.'

On this note the *Academy* correspondence ended, but the argument continued. On January 25, 1885, Watt published his 'Second Note on the Soma Plant,' in which he said that a Dr. Dymock of Bombay had sent him a Haoma plant which was *Periploca aphylla* and had told him he thought that the plant was not used to obtain liquor but that a small portion of it was added to a liquor obtained from grain; he added that, according to the Parsis, the Haoma never decayed. This strengthened Watt's opinion that Soma was not a succulent plant, certainly not a Sarcostemma, an opinion which was further encouraged by a letter he had received from Rajendra Lala Mitra, who had suggested that Soma might have been used like hops as an ingredient in the preparation of a kind of beer, and that the Vedic phrase 'Soma juice' was merely a figure of speech. 'The word "sweet,"' Mitra wrote, 'which has so much puzzled the learned Professor von Roth, may be safely, nay appropriately, used in a poem in praise of bitter beer.' Watt was therefore convinced that the Haoma plant was the Soma after all, that 'the dry and bitter twigs' had been used to flavour some other beverages, 'much in the same way as Acacia bark is used throughout India.' In passing, Watt rejected a suggestion he had received from Benjamin Lewis Rice that Soma might have been 'sugarcane or some species of sorghum'; he concluded that *Periploca aphylla* – the Haoma of Dr. Dymock – was after all the most likely Soma plant, and he quoted Baker's description of *Periploca aphylla* as having 'flowers fragrant, eaten by natives, taste like raisins.'

Surgeon General Edward Balfour, in the third edition of his *Cyclopædia of India*, maintained that the Soma of Vedic times was a 'distilled alcoholic fluid' made from the *Sarcostemma brevistigma* which flowers during the rains in the Deccan. In the same article he referred to the Soma juice as 'a fermented liquor' and 'this beer or wine'; he believed that this liquid was used at all religious festivals and by the rishis at their meals; and he attributed to Windischmann the suggestion that the Soma plant may have been the gogard tree.[1]

1. Surgeon General Edward Balfour: *The Cyclopædia of India and of Eastern and Southern Asia, Commercial, Industrial, and Scientific; Products of the Mineral, Vegetable, and Animal Kingdoms, Useful Arts and Manufactures* (3rd edition), London, 1885, Vol. III, p. 703.

In 1885 Julius Eggeling published the second volume of his translation of the *Śatapatha Brāhmaṇa*, wherein he observed that the exact identity of the Soma plant was 'still somewhat doubtful,' but that every probability seemed to favour the *Sarcostemma brevistigma* or some other plant of the same genus. In answer to Watt's suggestion that the opinions of all the 'Sanskrit authors' be assembled for the botanists to use, Eggeling wrote: 'One might as well ask a Hebrew scholar to give accurate descriptions of the "lily of the valley" to enable the botanist to identify and classify the lovely flower which delighted the heart of King Solomon. It is exactly the want of an accurate knowledge of the nature of the Soma plant which prevents Vedic scholars from being able to understand some of the few material allusions to it.'[1] Undaunted, Aitchison wrote a letter to the Daily News (13 March 1885) expressing his opinion that Soma was wine after all. He modified this, however, when he returned from Afghanistan and published *The Botany of the Afghan Delimitation Commission*, wherein he informed his readers that the natives of North Baluchistan call *Periploca aphylla*, *Ephedra pachyclada*, and another *Ephedra* '*Hum*, *Huma*, and *Yehma*.' He said that the natives eat the small red fruit of these plants, but he hesitated to identify any of them with the original Soma.[2] Finally, in 1888, Max Müller republished the *Academy* correspondence in his *Biographies of Words and the Home of the Aryans*[3] with additional notes by Thiselton-Dyer, who supported Max Müller by saying that Soma 'was certainly in later times a fermented drink made from grain, to which the Soma plant itself was only added as an ingredient.' Observing that, according to Roxburgh, the Sarcostemma was 'not necessarily nauseous,' he nevertheless rejected it as a possibility for the Primitive Soma, rejecting as well Houtum-Schindler's *Hum* (which he identified as *Periploca aphylla*) and his tamarisk-like plant (which he identified as *Periploca hydaspidis*, indistinguishable, he said, from *Ephedra foliata* except when in flower). He concluded with a

1. Julius Eggeling: *Śatapatha Brāhmaṇa*, *Sacred Books of the East*, Volume II (1885), pp. xxiv ff.
2. I. E. Aitchison: *The Botany of the Afghan Delimitation Commission*, Transactions of the Linnæan Society, London, 1887, p. 112.
3. Max Müller: *Collected Works*, London, 1888, Vol. x, pp. 222-242.

belief that the Sarcostemma was used, like hops, to flavour the 'more effective ingredients' of fermented grain, and that it was used *not* 'as a ceremonial reminiscence of the grape' but in the absence of the original Soma plant – the hop. George Watt concluded this episode with his article on Ephedra in the *Dictionary of the Economic Products of India*, where he rejected, among other hypotheses, his own former suggestion of the Afghan grape, though he could not resist noting that wild grapes are called '*Um, Umbur*' in Kashmir.[1] He mentioned that *Ephedra vulgaris* grew in the Himalayas, that *Ephedra pachyclada* and *E. foliata* were found in Garhal and Afghanistan, that *Periploca aphylla* was used sometimes in Bombay, but that the Parsis' *huma* was usually an Ephedra. He said that *putīkas* (Basella spp.), when stripped of their leaves, would resemble Sarcostemmas; that *Vernonia anthelmintica* and *Pæderia fœtida* are known as *somaraj* in Hindustan; that Asclepiads are emetics and are eaten by goats; and that Sarcostemma is rare but Periploca plentiful in Central Asia. He rejected the Periploca, however, on the grounds that the Aryans would have recognized it in India and not have used the Sarcostemma in its place. And from all this he concluded that Soma refers to 'an early discovery of the art of fermentation' rather than to any plant in particular. All the sound and fury had proven nothing, after all.

VI. THE TURN OF THE CENTURY

Still the battle raged, if more quietly, over the same ground. Adalbert Kuhn said that Haoma and Soma were separate plants resembling each other in name and external appearance, the Soma of present-day India being *Sarcostemma brevistigma* (an identification that he chose to attribute to Roth), which was not however the original or at least the only plant from which Soma was taken.[2] John Firminger Duthie then added an odd piece of information that went unnoticed: The Marwara people call the spiked grass known as *Setaria glauca* Soma.[3] Darmesteter in his translation of the *Avesta* referred to Haoma as a yellow plant

1. George Watt: *Dictionary of the Economic Products of India*, Calcutta, 1890, 'Ephedra'.
2. Adalbert Kuhn: *Mythologische Studien*, Gütersloh, 1886, Volume 1, p. 106.
3. John Firminger Duthie: *The Fodder Grasses of North India*, Roorkee, 1888, p. 14.

with very close-set knots[1] 'like the Indian Soma'.[2] Monier-Williams' Dictionary (1891) simply gave *Sarcostemma brevistigma* for Soma, and J. Börnmuller mentioned that he had met with a Parsi priest in Yezd carrying *huma*.[3] At once he recognized it as *Ephedra distachya*, and thus he had solved at last the problem mentioned to him first by Max

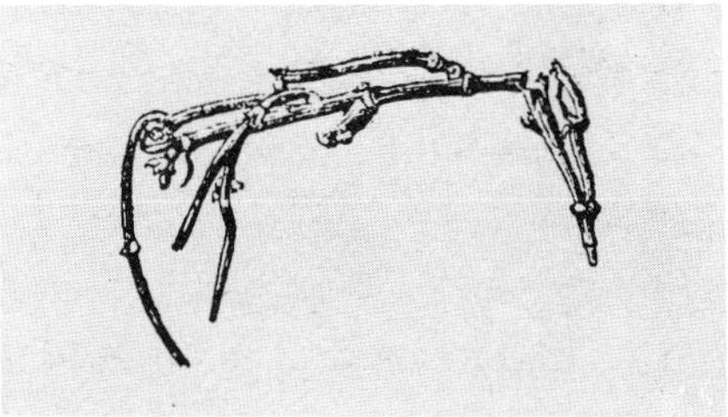

FIG. 4. 'Haoma' of the Persians. As pictured in James Darmesteter's translation of the Avesta, 1890-1892. Said to be life-size.

Müller in Oxford. He added that the plant grew all over Central Asia and that large quantities of it were dried and sent from Persia to Bombay every year.

Alfred Hillebrandt then produced his extensive summary of the recent Soma theories and introduced his own famous theory that Soma was the moon throughout the ṚgVeda. He explained that contemporary Soma plants were probably not the same as the original, that substitutes had been used as soon as the Aryans left their homeland in the Sindh, and that the Vedas themselves were self-contradictory, since they had been compiled in various times and places.[4] He pointed out that Soma was not a blossoming plant and therefore could not be hops as Max Müller had suggested; that Soma had a red stalk and reddish brown sap; and that the epithet 'with hanging branches' (*naicāśākhá*)[5] probably referred to the *nyagrodha* (*Ficus religiosa*) which was an important substitute. In fact, he concluded, even

1. *Vide supra*, pp. 59-60.
2. James Darmesteter: *Avesta*, translation into French, Paris, 1890-1892; introduction, p. lxv. He gives a picture of the plant as well (*vide* fig. 4).
3. J. Börnmuller: 'Reisebriefe aus Persien,' in *Mitteilungen des Thüringischen Botanischen Verlags*, 1893, p. 42.
4. Alfred Hillebrandt: *Vedische Mythologie*, Breslau, 1891, Vol. I, pp. 1-18.
5. This word, however, occurs only once in the ṚgVeda (III 53.14) and its meaning is uncertain; Grassmann thinks that it may refer to the name of a sage.

at the time of the ṚgVeda itself, various plants were already in use, and he cited Burnell's evidence that different Somas were used simultaneously in different parts of India.

Hermann Oldenberg devised a new theory, suggesting that the original Soma was itself a substitute, not for wine but for the Indo-European honey drink, mead or hydromel.[1] Of course it had long been recognized that the Sanskrit *mádhu*, the Greek μεθυ, and English *mead* were cognate, and that *mádhu* was applied to Soma in the ṚgVeda (whence all the trouble over the bitterness of Sarcostemma), but *mádhu* was a general term applied to milk and rain as well as to Soma, and the simple identification of Soma with *mádhu* had still left open the botanical identification of Soma as a plant whose sap was known as honey. Oldenberg, however, avoided this difficulty by postulating mead as a forerunner and Soma as a later substitute: this distracted attention from the Soma plant itself (since it was no longer to be regarded as the *original*, the *Ur*-plant, that everyone sought) but it did not, of course, add anything at all to what was known about the Indian Soma.

This theory was to become well known and widely accepted, but it had little immediate impact. Edmund Hardy maintained that Soma was neither any form of honey nor *súrā* but most probably *Sarcostemma brevistigma* after all.[2] Vedic scholars generally clung to the Asclepiad hypothesis[3] or they hazarded even less: P. Regnaud called Soma 'une liqueur enivrante,' and cautioned against taking the Vedic texts literally when they spoke metaphorically.[4] Rustomjee Naserwanyil Khory covered some *very* old ground by considering a *Brāhmaṇa* substitute to be the original Soma;[5] he identified Soma (*somavallī* in Sanskrit, *amṛtavallī* in Bengali) as a climbing shrub called *Tinospora cordifolia*, the extract of which (called *gurjo, gilo,* etc.) is used as an aphrodisiac, a cure for gonorrhea, and a treatment for urinary diseases.

1. Hermann Oldenberg: *Die Religion des Veda*, Berlin, 1894, pp. 366 ff.
2. Edmund Hardy: *Die Vedische-Brahmanische Periode*, Münster, 1893, pp. 152-153.
3. Zenaïde A. Ragozin: *Vedic India*, London, 1895, p. 171; W. Caland: *Altindische Zauberritual*, Amsterdam, 1900, p. 188.
4. P. Regnaud: 'Remarques sur le IXème Maṇḍala du Ṛg Veda,' *Revue de l'Histoire Religieuse*, XLIII, 1902, pp. 308-313.
5. Rustomjee Naserwanyil Khory: *Materia Medica of India and their Therapeutics*, Bombay, 1903.

Christian Bartholomæ in his famous *Altiranisches Wörterbuch* (1904) described Haoma as a plant used in medicine, for magic, and as an alcoholic drink, but he refrained from identifying the plant. W.W. Wilson thought Soma was the σίλφιον (Latin *laserpitium*) mentioned

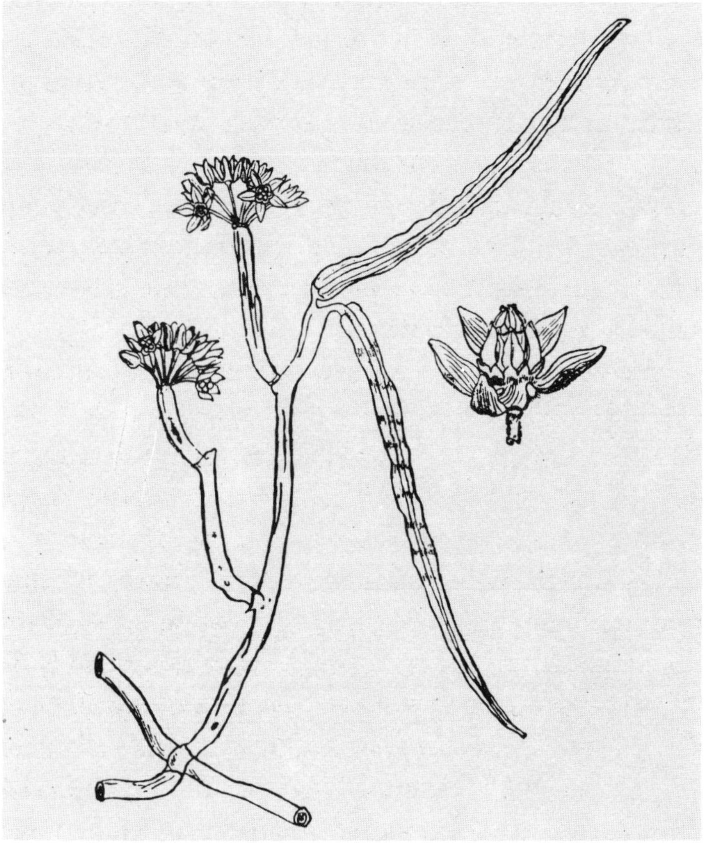

FIG. 5. 'Soma'. As pictured in Zenaïde A. Ragozin: *Vedic India*. New York, 1895.

in a fragment of Alcman as a plant that has wonderful properties, grows on mountains, is golden, and is plucked by birds.[1] He supported this argument with complex Indo-European linguistics but was cautious enough to note: 'It is not improbable that even at the time of the Vedas, use was made of more than one kind of plant.' W. Caland and

1. W. W. Wilson: 'The Soma offering in a fragment of Alkman,' *American Journal of Philology*, No. 30, 1906, pp. 188-195. Roth in his 1884 essay had suggested that Soma in India had occupied a sacred position similar to that held by the now extinct σίλφιον plant of Ancient Greece, but he had not considered the two plants identical.

Victor Henry considered the question of the identification of Soma insoluble.[1] Henry maintained that the Indo-Aryans had imported Soma ever since they entered the area of Northwest India, and that the present-day Soma of India did not correspond to the Vedic descriptions of its taste or characteristics.[2] Still the old definitions persisted: K. L. Bhishagratna in his 1907 edition of the *Suśruta* defined *somalatā* as *Sarcostemma brevistigma* and *somarājī* as *Vernonia anthelmintica*. Maurice Bloomfield referred to Soma merely as 'an intoxicating drink . . . regarded as the tipple of the gods,'[3] and A. A. Macdonell and A. B. Keith concluded that the ṚgVeda descriptions were 'inadequate to identify the plant . . . It is very probable that the plant cannot now be identified.'[4] Then Hermann Brunnhofer came forward to defend Lagarde's 'mountain rue' (originally Sir William Jones's theory, though Lagarde did not say so and probably did not know of Jones's opinion). Brunnhofer maintained that Lagarde had already solved the Soma question forty years previously, 'though the Vedists still ignore him.'[5] He elaborated upon Lagarde's theory by mentioning Pliny's description of a plant known as ἀμβροσία possessing characteristics that suggested a kind of rue to Brunnhofer.[6] He then called attention to the ṚgVedic verse (IV 3^9c^), perhaps an oblique reference to Soma, that speaks of a black cow giving white milk; Brunnhofer considered this a perfect description of the μῶλυ, a dark plant with white milk.

In 1911 Carl Hartwich published *Die menschlichen Genussmittel*, in which he discussed various forms of tobacco, alcohol, and drugs.[7] He included 'Soma – Haoma' in his study, but was uneasy about its status as a drug or stimulant ('*Genussmittel*'); having described various ac-

1. W. Caland and Victor Henry: *L'Agniṣṭoma*, Paris, 1906-7, p. 471.
2. Victor Henry: *Soma et Haoma*, Paris, 1907, p. 52.
3. Maurice Bloomfield: *The Religion of the Veda*, New York, 1908, p. 145.
4. Arthur Anthony Macdonell and Arthur Berriedale Keith: *Vedic Index*, London, 1912, Vol. II, p. 475.
5. Hermann Brunnhofer: *Arische Urzeit*, Bern, 1910, p. 297.
6. *Ibid.*, p. 300. The quotation from Pliny (*Historia Naturalis* 27.28.11) merely says that of several plants bearing the name of αμβροσία, one has leaves around the bottom of the stem resembling those of rue (*foliis rutae circa imum caulem*). Elsewhere (*Historia Naturalis* 14.40), Pliny speaks of αμβροσία as a kind of grape.
7. C. Hartwich: *Die menschlichen Genussmittel*, Leipzig, 1911, p. 806 ff.

cepted botanical 'Somas' (*Sarcostemma brevistigma, Periploca aphylla, Basella cordifolia, etc.*), he went on to say: 'Although all of these have been tested – though not altogether thoroughly – none of them is known to have stimulating or sedative or other characteristics which could mark it as a drug or stimulant.' He therefore considered it possible that they might have been used as supplements, perhaps as spices, to the other known ingredients of the Soma drink (meal, milk, and whey) in which alcohol might have been produced.[1] The possibility that Soma might have been some other plant which *did* have 'stimulating or sedative characteristics' does not seem to have occurred to him, though he granted that the Persians used for their Haoma a different plant (*Ephedra vulgaris*) containing a drug (ephedrine) which caused dilation of the pupil.[2] One of the plants that Hartwich gave as an alleged Soma was *Vitex negundo*, whose bitter leaves and root were used against fever. Since the plant was originally native to tropical America, Hartwich dismissed it as a possible Soma, but observed in conclusion that in India it was known as *Indrahasta* and *Indrasurā* ('the hand of Indra' and 'the wine of Indra').[3] It may be that this association with Indra – together with Indra's fondness for Soma (to which Hartwich refers) – recommended the *Vitex negundo* to Hartwich, for he gives no Sanskrit names for any of the other plants he mentions. Yet in this context it is surprising that he did not mention *Indrāśana* ('the food of Indra')[4] which was a common name for hemp, *Cannabis sativa*, the leaves of which were dried and chewed, supplying just the *Genussmittel* that Hartwich seemed to seek.

Keith reassembled the old evidence but still considered the Soma problem insoluble,[5] and L. H. Mills produced an odd throwback in which he insisted that Haoma and Soma grew independently from the

1. Hartwich, *op. cit.*, p. 809
2. *Ibid.*, p. 808.
3. *Indrasurā* (also known as *Indrasurasa* and *Indrasurisa*) is mentioned in the *Amarakośa*, the *Suśruta Saṃhitā*, and the *Śabdakalpadruma*, but is nowhere associated with Soma.
4. *Indrāśana* as a name for hemp (perhaps by confusion with *Indra-śaṇa*, which would mean the hemp of Indra) is mentioned in the *Śabdamālā* (quoted in the *Śabdakalpadruma*) and appears in its prākrit form — *Indrāsana* — in *Dhūrtasamāgama* 90.8.
5. A. B. Keith: The *Taittirīya Saṃhitā*, Harvard Oriental Series, Volumes 18 and 19, 1914, introduction and p. 119.

same original, and that 'nothing humorous, let us remember, attached to the idea of [alcoholic] stimulus at first in those early days'.[1] The *Oxford History of India*[2] merely observed that the Parsis say that Soma is Asclepias, and E. W. Hopkins concurred but questioned whether this was the plant referred to in the Vedas, the Avesta, and Plutarch, suggesting that the names might have been retained when substitutes for the plant were used.[3] Chapman Cohen stated without reference or explanation his belief that the Soma drink 'is prepared from the flower of the lotus,' an idea that was bound to occur to someone sooner or later, in view of the *Odyssey* tradition of the lotus-eaters and the sanctity of the lotus in India, but which seems to have attracted no supporters after this.[4] E. B. Havell then suggested that Soma was *Eleusine coracana* or *rāgi*, the common millet, an idea that he supported on the basis of Vedic characteristics (shaped like udders, tawny, growing in the mountains) and the *Śatapatha Brāhmaṇa* reference to *dūrvā* and *kuśa* as substitutes, both of which resemble *rāgi*, 'the common millet still used in the Eastern Himalayas for making the intoxicating drink called *Marua*.'[5] He mentioned that the Brahmans, while preparing Soma, 'sang a song which reminds one of a good old Aryan sailors' chantey, with a refrain, "Flow, Indu, flow for Indra".'[6] Havell admitted that, 'whether fermentation took place before or after it was so used is a point which is not very clear,' but he was clear on its history: *rāgi* had been the Aryans' principal food and drink until they moved to the Ganges Valley and switched to rice, 'and, perhaps under the influence of Buddhism, gradually gave up intoxicating liquors, or "went dry".' Then they used substitutes for the *rāgi-Soma*, forgot its name and retained contact with it only as the food of the lowest caste, the Śūdras. Havell said that *Marua* when kept too long is nauseating and evil-smelling, in keeping with certain descriptions of

1. L. H. Mills: 'The Avestic H(a)oma and the Vedic Soma,' *Asiatic Review*, No. 8, 1916, p. 315.
2. Vincent Smith: *Oxford History of India*, 1919, p. 23.
3. E. W. Hopkins: 'Soma,' in Hastings' *Encyclopaedia of Religion and Ethics*, Vol. 11, Edinburgh, 1920, p. 685.
4. Chapman Cohen: *Religion and Sex*, Edinburgh, 1919, p. 57.
5. E. B. Havell: 'What is Soma?', in the *Journal of the Royal Asiatic Society*, 1920, pp. 349 ff.
6. Ṛg Veda IX 113.

Soma, but when fresh 'it is an exhilarating drink that easily intoxicates the uninitiated.' He then administered the coup-de-grâce: the Indo-European myth of the bringing of Soma from heaven by an eagle is explained by the simple fact that birds lined their nests with *rāgi*. One would hardly think that this argument merited rebuttal, but it did in fact stimulate an answer that was to become the last significant theory of Soma: Mukherjee's theory that Soma was hemp, *Cannabis sativa* (= *Cannabis indica*), known to the initiate as *bhang* (the Hindi term derived from Sanskrit *bhañgá*, m., or *bhañgā*, f.), ganja, hashish, marijuana, or pot.

VII. MUKHERJEE AND THE *BHANG* THEORY

Braja Lal Mukherjee[1] picked unlikely grounds on which to challenge Havell's flimsy theory, which he said was based upon considerations 'which may be supplemented by others of a more important character,' *e.g.*, that there was no reference to cows' udders in the Ṛg Veda.[2] He then supplied an elaborate 25-point argument to show that Soma was in fact not *Eleusine coracana* but *bhang*: *Śatapatha Brāhmaṇa* 5.1.1.12 says that Soma is *uśānā*;[3] Soma was 'originally amongst the Kirāṭas,' a mountain tribe; amongst the Kirāṭas, *u* and *a* were articular prefixes; therefore *uśānā* = *śaṇa*; *śaṇa* is hemp; the Tanguts call hemp by the name *dschoma*; hemp = Greek *kanna* [sic] = Sanskrit *śaṇa*; the Tibetan word for hemp is *somaratsa*; the preparation of Soma is similar to that of *bhang*; the deity Mahādeva (Śiva) is a lover of *bhang*; *bhang* is used by the modern representatives of the Vedic people in the celebration of worship of the goddess Durgā, which is a Soma sacrifice.[4] The final link in the argument is this: '*Bhang* is sacred to Hindus by tradition.' In sum, 'May we not conclude that the weight of evidence is in favour of the identification of Soma with *Cannabis* (*bhang*)?' This strange argument, combining linguistic reasoning with the purest

1. Braja Lal Mukherjee: 'The Soma Plant,' *Journal of the Royal Asiatic Society*, 1921, p. 241.
2. This is untrue. *Vide supra*, p. 43.
3. This verse does not refer to Soma at all, but verses 3.4.3.13 and 4.2.5.15 of the Mādhyandina recension of the *Śatapatha Brāhmaṇa* speak of a plant called *uśānā*, from which Soma is made.
4. The Durgā celebration, the use of *bhang*, and the tradition that Śiva uses *bhang*, are all late characteristics of Bengal Śaivism.

twaddle, was further developed by Mukherjee in his book, *The Soma Plant*,[1] which was reviewed in a brief paragraph by L. D. Barnett, who said blandly that Mukherjee made out a good but not always convincing case for hemp as Soma.[2]

Sir Charles Eliot expressed 'considerable doubt' that Soma could be identified. He said it was a plant with 'yellow juice of a strong smell, fiery taste, and intoxicating properties,' and that the Parsis of Yezd and Kerman used Asclepiads.[3] N. B. Pavgee rebutted Havell without proposing any alternative, using the argument merely as a foil for his own hypothesis – that Soma was indigenous to the Saptasindhu region (*i.e.*, India proper) and was not 'brought in' by 'Aryans.' The Indo-Aryans were autochthonous in India, he wrote, and 'had not immigrated'; the Iranian Soma is, of course, spurious, but kept the old name. Soma could not be any kind of liquor, for liquor is an evil, while Soma – the true Soma – was 'exhilarating yet slightly intoxicating' and 'gave moral elevation'. Pavgee does not explain how this was done, nor does he identify the Soma plant.[4]

In 1922 Jakob Wilhelm Hauer published a work that lent a kind of peripheral support to the *bhang* theory, for he referred to the Soma cult as the most highly developed form of the use of narcotics to induce ecstasy, calling particular attention to the late Vedic hymn (x 136) that describes a long-haired sage who drinks poison with Rudra.[5] Hauer believed that Soma was the most important toxic means of inducing ecstasy, but not the only such means, and he suggested that the Vedic references might be traces of a primitive ecstatic practice of hallucination caused by certain plants.[6] Whether these 'plants' were Sarcostemmas, Afghan grapes, or hemp, Hauer neglects to say, but his final remark is more suggestive of a hallucinogenic plant than of anything else: opium is no longer used in a religious context, but every time we light up a good cigar we experience a faint reflection

1. Braja Lal Mukherjee: *The Soma Plant*, Calcutta, 1922.
2. L. D. Barnett: Review in *Journal of the Royal Asiatic Society*, 1923, p. 437.
3. Sir Charles Eliot: *Hinduism and Buddhism*, London, 1921, Vol. 1, p. 69.
4. N. B. Pavgee: *The indigenous far-famed Soma and the Aryan Authochthones in India*, Poona, 1921, and 'Soma juice is not liquor,' Third Oriental Conference, 1924, pp. 70-79.
5. Jakob Wilhelm Hauer: *Die Anfänge der Yogapraxis im Alten Indien*, Berlin, 1922, p. 137.
6. *Ibid.*, pp. 57, 59, 61.

of the splendour of the rapture of the primeval ecstatic.[1] Though this ancient Yogic cult is clearly non-Vedic, presumably pre-Vedic, Hauer does not say whether he believes the Soma plant to have been the same as the drug of the 'primeval ecstatic', or perhaps a substitute for it. Yet his remarks are provocative in the context of the search for a hallucinogenic Soma plant.

VIII. LATER RESEARCHES IN THE TWENTIETH CENTURY

In 1924, Gilbert Slater advanced a novel hypothesis: that *amṛta* (Soma) was Egyptian beer, a fermentation of date juice or palmyra palm or coconut palm, brought to India from Mesopotamia.[2] This theory was noted but not accepted. Georges Dumézil maintained that Soma was native to India, and that it was the Indian substitute for the Indo-European 'sacred barley beer' rather than for the Indo-European mead that Oldenberg had postulated.[3] The beer was replaced by wine in Greece and by Soma in India, and the one word for this beer, or for a beverage with a cereal base which must have preceded it, was the sacred ritual name: ambrosia, the Greek ἀμβροσία and the Sanskrit *amṛta*, the elixir of immortality. Barley itself must have been Indo-European, as the linguistic evidence indicates: Greek χριθή, Latin *hordeum*, Armenian *gari*, and various Celtic words for beer, *e.g.*, *cervesia* (Gallic) and *cuirm* (Irish). As for the sacred position of barley in India, Dumézil referred to the tradition that barley (*yava*) had stayed with the gods when all the other plants had left them, thus enabling the gods to conquer their enemies.[4] The *Mahābhārata* relates that the gods once churned the ocean in order to obtain the *amṛta*, but a terrible poison emerged and would have destroyed them all had Śiva not swallowed it and saved them; in the light of Dumézil's theory that the *amṛta* is barley beer, this myth is an expression of excessive fermentation that must be arrested.[5]

1. Hauer, *op. cit.*, p. 62.
2. Gilbert Slater: *The Dravidian Element in Indian Culture*, London, 1924, p. 78.
3. Georges Dumézil: 'Le Festin d'Immortalité', *Annales du Musée Guimet*, No. 34, 1924, p. 279.
4. *Śatapatha Brāhmaṇa*, 3.6.1.8-9.
5. Dumézil, *op. cit.*, p. 285.

Louis Lewin's *Phantastica: Narcotic and Stimulating Drugs* included Soma in the chapter on alcohol rather than narcotics, mentioning *Periploca aphylla, Sarcostemma brevistigma, Setaria glauca,* and *Ephedra vulgaris* as plants that had been identified with Soma, but he added: 'None of these plants is able to give rise to such effects as have been attributed to Soma. . . . I regard Soma as a very strong alcoholic beverage obtained by fermentation of a plant.'[1] Elsewhere he suggested that the yogis might have used some sort of narcotic such as Indian hemp or scopolamine, but he did not identify this practice with the cult of Soma.[2]

The Sarcostemma theory returned yet again, this time in a paper by L. L. Uhl, who maintained that Sarcostemma, and not Ephedra, was the original Soma, saying that he had found *Sarcostemma brevistigma* frequently at latitude 15° in South India, where it is called Soma and is used in sacrifices.[3] Arthur Berriedale Keith expressed the view that the Soma problem, though insoluble, had led to 'interesting investigations, but to no sure result, and the only thing certain is that the plant, which has been used in modern India as the Soma plant, is one which would not be considered by modern tastes as at all pleasant in the form of pressed juice mixed with water.'[4] Then, somewhat echoing de Gubernatis' reasoning, Keith went on to say that although one can't be sure what was pleasant to a Vedic Indian, nevertheless it is likely that 'the drink was originally a pleasant one; in the course of time the long distance from which the shoots had to be brought may easily have made it less attractive, as it certainly encouraged the use of various substitutes described in the ritual text books.' Keith agreed with Hauer's suggestion that ṚgVeda x 136 might be a reference to the use 'of some poison to produce exhilaration or hypnosis,'[5] relating

1. Louis Lewin: *Phantastica: die betäubenden und erregenden Genussmittel*, Leipzig, 1924; published in London, 1931, as *Phantastica: Narcotic and Stimulating Drugs*, from which I cite the above information, p. 161, note.
2. *Ibid.*, p. 117.
3. L. L. Uhl: 'A contribution towards the identification of the Soma plant of Vedic times,' in *Journal of the American Oriental Society*, No. 45, 1925, p. 351.
4. A. B. Keith: *Religion and Philosophy of the Vedas and Upanishads*, Harvard Oriental Series, Volumes 31-2, 1925, pp. 172, 283-4, 482.
5. *Ibid.*, p. 402.

this to a verse about hemp (*śaṇa*) in the *Atharvaveda*,[1] but not to Soma. Elsewhere he observed that the Avesta does not call Haoma mead, and suggested that mead was the Indo-European drink and that Soma was identified with it when Soma was discovered, in India, and found to produce 'a juice pleasant to drink or at least intoxicating.'[2]

In 1926 G. Jouveau-Dubreuil wrote an article suggesting that Soma was none other than a species of Asclepias after all.[3] He had discovered that the Nambudri Brahmans (on the Malabar coast, in Kerala) of Taliparamba, who practiced 'pure Vedic ceremonies,' sent to a Raja in Kollangod at the foot of the high mountains to obtain their Soma plant, a leafless, milky, green creeper that was an Asclepiad. *Plus ça change* ... Yet neither the Sarcostemma theory nor the theory that Sarcostemma was merely a substitute was incompatible with the beer theory of Max Müller and Mitra (for the beer could be made with the Sarcostemma or the plant for which it was a substitute), nor were these or the wine theory incompatible with Oldenberg's mead theory (for the mead could have been replaced by any of the above in India), and so the theories continued to live side by side. Otto Schrader and A. Nehring maintained that the Indo-European honey was replaced by the Soma plant – still called *mádhu* – in Aryan times, while wine and beer were later substituted for it throughout Europe.[4] P. V. Kane was unconvinced by any of the theories, but mentioned that in the Deccan a plant called *rānsera* was used as a substitute for Soma.[5]

In 1931 Sir Aurel Stein published a paper entitled, 'On the Ephedra, the Hum Plant, and the Soma,'[6] describing a cemetery in Central Asia filled with packets of Ephedra twigs. Recalling the evidence for Ephedra as Haoma and Soma and quoting Wellcome's *Excerpta Thera-*

1. *Atharvaveda*, ed. Roth and Whitney, Berlin, 1855, IV.5.
2. Keith, *op. cit.*, p. 172.
3. G. Jouveau-Dubreuil: 'Soma', translated by Sir R. Temple in the *Indian Antiquary*, No. 55, 1926, p. 176.
4. Otto Schrader and A. Nehring: *Reallexikon der Indogermanischen Altertumskunde*, Berlin, 1929, Vol. II, p. 139.
5. P. V. Kane: *History of Dharmaśāstra*, 1930-62, Vol. II, part 2, chap. 33, pp. 1202-3.
6. Sir Aurel Stein: 'On the Ephedra, the Hum Plant, and the Soma,' *Bulletin of the London School of Oriental Studies*, 6/2, 1931, pp. 501 ff.

peutica for evidence that the Chinese use an Ephedra called *Ma-huang* to get an alkaloid drug (ephedrine), he then followed this with the assertion that Ephedra could *not* have been the original Soma, for it was bitter and Soma sweet, and Ephedra was not mountainous. He then concluded: 'The only result of these inquiries has been to direct my attention more closely to . . . the wild rhubarb,' which grows in the mountains, has a fleshy stalk, and can be made into rhubarb wine, though Stein admits that the Indians do not do so.[1] Granting that the Vedic descriptions of the Soma plant could apply to Asclepias or rhubarb – or to anything else, for that matter – Stein nevertheless maintained that the descriptions of the Soma *juice* best applied to rhubarb. He added that the juice might be mixed with milk to facilitate fermentation, 'which alone could endow a juice like that obtained from the rhubarb with the exhilarating and exciting effect so clearly indicated in the Vedic hymns.'[2]

Perhaps the strangest episode in the history of Soma research came in 1933, with a truly Twentieth Century theory of Soma. Dr. Paul Lindner of Berlin published an article entitled, 'The Secret of Soma'.[3] He referred to a statement by E. Hubers that Soma was merely a decoction of barley or millet, into which the juice of the Soma plant had been added as a catalyst for fermentation (*Gärungserreger*), though it was unclear what kind of fermentation took place.[4] Lindner's own studies of the micro-organisms of the Agave in Mexico, particularly of *Thermobacterium mobile*, had brought him to the conclusion that yeast played only a secondary role in the fermentation, after the fermentation-bacteria had prepared the field. He went on to say that *Thermobacterium mobile* can produce alcohol in grapes and in cane sugar, and concluded that since the 'purity of fermentation' took precedence over the material fermented, even *Sarcostemma brevistigma, Calotropis gigantea,* or *Ephedra distachya,* which had been considered as 'the holy

1. Sir Aurel noted that Albert Regel had suggested rhubarb in 1884. *Vide supra*, p. 114.
2. Stein, *op. cit.*, p. 513.
3. Paul Lindner: 'Das Geheimnis um Soma,' *Forschungen und Fortschritte*, No. 5, 1933, p. 65.
4. E. Hubers: 'Schilderungen der Bierbereitung im fernen Osten,' in *Mitteilungen der Gesellschaft für Bibliographie und Geschichte des Brauwesens*. Lindner also cited J. Arnolds' *Origin and History of Beer and Brewing*.

Soma plant,' might have yielded the juice 'which was enjoyed by young and old.' Lindner was convinced that *Thermobacterium mobile* was the most important fermenting agent of the tropics, and that the songs of praise to Soma really must have been 'dedicated to this *Bacterium.*' He suggested that the tropical explorers try '*TM*' in the juice of the Indian butter tree, *Bassia latifolia,* or the Ceylonese cowtree, *Gymnema lactiferum,* which have milky saps, the former sweet, the latter bitter. When Lindner himself had tried a spoonful of *TM* it had resulted in 'undiminished feeling of well-being and almost odourless excrement.'[1] Dr. Leo Kaps had treated patients in the Wilheminenspital in Vienna with *TM* beer and obtained extraordinary results, and a Swedish firm and a Viennese brewery were about to produce a *TM* beer with small alcoholic content. I have found no evidence that Soma-TM-beer was in fact manufactured nor any further reference to the Lindner theory after this initial publication.

An original and, in retrospect, provocative contribution to the argument was made in 1936 by Philippe de Félice.[2] He was unable to identify Soma but he described the plant in terms that seem remarkably up-to-date: he would probably have called Soma an 'hallucinogen', had the word existed. He deduced from the record that *Sarcostemma brevistigma* and similar plants had served to replace Soma toward the end of the Vedic period, when the Indo-Iranians in emigration were forced to use new drugs in place of Soma. Alcoholic drinks such as *súrā,* he thought, might also have come to be used. Then he continues:

> Instead of indulging in suppositions that no document supports, ought we not rather to ask ourselves whether, to arrive at the drink that plunged them into ecstasy, the ancient Indo-Iranians did not have recourse, like so many other primitive peoples, to some plant whose toxic properties they had discovered? This is what the examination of the texts seems to make clear. The liquor about which they speak is always drawn from a plant. This plant grows on the mountains which, as time passes, seem to become more and more

1. '*unvermindertes Wohlbefinden und fast geruchlose Exkremente.*'

2. Philippe de Félice: *Poisons Sacrés, Ivresses Divines: Essai sur quelques formes inférieures de la mystique,* Paris, 1936, pp. 265-266.

distant, more and more inaccessible.[1] What serves to produce the mystical potion is neither the leaves nor the fruit, but always the stems. The juice is either red or clear yellow. It must be filtered or decanted, to eliminate certain elements that are disturbing or that perhaps risk rendering it too toxic. Sometimes it has a bad taste or even smacks of carrion; thus it is certainly not for pleasure that one drinks it. The inebriation that it provokes can present grave dangers: the spirit wanders, the drink can lead even to madness. It happens sometimes that the inebriation is accompanied by organic disturbances, which are in reality symptoms of an acute intoxication.

Men know and fear the baleful effects of the drug, and, though he was a god, Indra himself did not escape them, since one day the Soma came forth from every opening in his body. This emeto-cathartic effect is confirmed in an old book of Hindu medicine. . . .

One may conclude from all this, it seems to us, that from the most remote antiquity the Indo-Iranians, when they were still dwelling together in their original home, possessed a special beverage reserved exclusively for the ceremonies of their religion and drawn from a toxic plant. The information that we possess about this plant unfortunately does not permit us to determine the species; but it is enough for us to classify it as among the toxic plants the use of which is widespread, for reasons of a mystical nature, in all parts of the world.

De Félice's reference to the emeto-cathartic effect of Soma is to the epithet *vamanī* in the Āyurvedic description and to the *Brāhmaṇa* story in which Soma injured Indra and flowed from his mouth and 'all the openings of his vital airs, and from his urinary tract'.[2]

Meanwhile the more conventional line of Vedic studies continued. L. van Itallie published an article in Dutch investigating the *Sarcostemma acidum* stalks in the light of the Soma plant; he described acids and alkaloids, carbohydrates and phytosterins, tannic acids and glycosides, but drew no new conclusions.[3] Johannes Hertel pointed out that there

1. The Indo-Aryans, spreading out east and south from the Indus Valley, were finding themselves increasingly remote from the Hindu Kush and the Himalayas. W.D.O'F.

2. *Śatapatha Brāhmaṇa* 1.6.3.1-7.

3. L. van Itallie, '*Soma-Haoma*, de heilige plant der Indiërs en der Perzen,' *Natuurwetenschappelijk tijdschrift* No. 19 (1937), pp. 1 and 9-11.

were many different kinds of Soma which scientific botany was unable to distinguish.[1] Henrik Nyberg referred to Soma-Haoma as an intoxicating drink made from a plant no longer identifiable, but he contributed obliquely to the *bhang* theory with his suggestion that Zoroaster was a shaman who drugged himself with hemp.[2] Nyberg himself, however, distinguished clearly between Haoma and hemp as plants, explaining Zoroaster's aversion to Haoma in the light of the belief that Haoma-drinking induced a state of intoxication (*Rausch*) rather than ecstasy.[3] Joges Candra Roy then developed Mukherjee's *bhang* hypothesis, adding to the original argument the facts that Soma is actually called *bhang* in the ṚgVeda;[4] that Soma is called a creeper nowhere in the ṚgVeda, but rather an herb (*óṣadhi*), a term which could apply to *bhang*; and finally that the Soma sacrifice 'was a feast and the drink added hilarity; *bhang* has been in use on similar occasions.'[5] Delli Roman Regni repeated the argument that Soma was not a fermented liquor but rather a non-alcoholic 'syrup-like thing,'[6]

1. Johannes Hertel: 'Das Indogermanische Neujahrsopfer im Veda,' in *Mitteilungen der Sächsischen Akademie der Wissenschaften*, Vol. 90, 1938, p. 83.

2. Henrik Samnel Nyberg: *Die Religionen des Alten Irans*, Leipzig, 1938, pp. 177, 190, 290 and 341. He cites several Avestan references to hemp, among them *Yašt* 13.124, *Vendidad* 15.14, *Vendidad* 19.20, *Yasna* 44.20, and *Vendidad* 19.41. W. B. Henning rebutted this argument in his *Zoroaster: Politician or Witch-Doctor?*, Oxford, 1951; he considered Nyberg's theory one to be rejected 'without further consideration . . . if one reflects on the effects of hemp, the physical, mental and moral deterioration it brings,' and he maintained that the Avestan citations, 'if correctly interpreted, can at the most serve to show that they cultivated hemp, possibly for the purpose for which hemp is cultivated all the world over — *i.e.*, to obtain its fibre.'

3. Nyberg, *op. cit.*, p. 288.

4. ṚgVeda IX 61[13]. *Bhañgá* here seems to be an epithet meaning 'intoxicating' (from *bhañj*, to break, *i.e.*, to disrupt the senses), but the reading occurs only once and is uncertain. Macdonell suggests that the word *bhañgá* first applied to Soma, meaning 'intoxicating,' and then 'came to designate hemp' (*Vedic Index*, Volume II, p. 93). Otto Schrader suggests that *bhang* 'was originally prized for the intoxicating effects of its decoctions' (*Prehistoric Antiquities of the Aryan Peoples*, translated from the German and published in London, 1890, p. 299), but after the one questionable Vedic occurrence, *bhañga* appears in classical Sanskrit several times only to designate the plant, with no reference to any narcotic effects. The 8th century A. D. *Śārṅgadhara Saṃhitā* (Bombay 1888; 1.4.19) is the first extant source that considers *bhañgā* a drug; it likens it to 'the saliva of a snake,' *i.e.*, opium (*phenaṃ cāhisamudbhavam*) and the commentator adds that the effect is like that of inebriating liquor (*madyaviṣayavat*). The *Bhāvaprakāśa* (Chowkhamba edition, 1.1) refers to *bhañgā* (*mātulāni*) as intoxicating and causing hallucination (*moha*) and slow speech (*mandavāk*); and the *Dhanvantarīya-nighaṇṭu* describes *bhañgā* as intoxicating, bitter, stimulating talk, inducing sleep, and producing hallucinations.

5. Joges Candra Roy: 'The Soma Plant,' *Indian Historical Quarterly*, No. 15, June 1939, pp. 197-207.

6. Delli Roman Regni: 'The control of liquor in Ancient India,' *New Review*, Calcutta, November 1940, p. 382.

and C. Kunham Raja maintained that, according to both the Vedas and the Avesta, Soma produced happiness (*máda*) while *súrā* produced evil intoxication (*durmáda*) and that Soma was a creeper with leaves, no longer available. Soma, he reasoned, had not time to ferment and if it had been an alcoholic drink the Indians would have substituted for it another alcoholic drink when it became unavailable, instead of the known non-alcoholic substitutes that they in fact used.[1] Yet the wine hypothesis – dead and killed again so often – reappeared in Ernst Herzfeld's *Zoroaster and his World*,[2] where he remarked that in 1931 he had received a letter from a New York gentleman who believed that a plant growing only in Persepolis – the *Salvia persepolitana* – was the Haoma, but that nothing had ever come of this communication. Herzfeld maintained that Soma must be a fermentation from grapes; that *aṃśú* means 'shoot, tendril, or bunch of grapes'; that the god Homa is the Aryan Dionysos; and that 'to thus define *homa* means to explain how wine [cultivated all over Iran before the advent of the Aryans and after] could remain unknown to the Avesta, and how the cultivation of *homa* [common in the Avesta] could disappear in Iran long before the Arab conquest. The solution is evident: *homa* is vine, wine.'[3] He adds that Haoma could not possibly be *bhang* because 'the use of hashish in Zoroaster's time is an imagination. The mysterious *homa* is wine, a reality . . . Nothing is known of the use of hemp as a narcotic prior to the Arsacid period.'[4]

1. C. Kunham Raja: 'Was Soma an intoxicating drink of the People?', *Adyar Library Bulletin*, No. 10, 1946, pp. 90-105.
2. Ernst Herzfeld: *Zoroaster and his World*, Princeton, 1947, Volume II, pp. 543 ff.
3. Bracketed material appears elsewhere in Herzfeld.
4. The first certain European reference to the use of hemp as a drug in India is in García d'Orta's record of a conversation with the Sultan Badar in Goa in 1563. The Sultan confessed that whenever in the night he had a desire to visit Portugal and Turkey and Arabia, 'all he had to do was to eat a little *bangue* . . .' In 1676 Henrich van Rheede (van Draakenstein), the governor of Malabar, published his *Hortus Indicus Malabaricus* in Amsterdam (written in Latin and Dutch), in which he called attention to a plant named *bangi* that was smoked like tobacco and caused inebriation (Vol. 10, p. 67). George Everhard Rumph in 1755 described at length, in Latin, the effect of the *Cannabis sativa* (also known as *ginji*) upon the natives, likening the plant to the Homeric νηπενθές: '. . . For the inhabitants of these regions, not content with the natural virtue of wine, which, however, they do not possess in great quantity, or with the other types of wine tree of that place, on the ground that in their opinion it elates one for a short time only, have devised from this plant such things as are able at a moment to remove anxieties of the heart, grief, or fear of dangers; yet this cannot be, except by a violent commotion of the senses and clouding of the intellect, which they themselves call

Still R. N. Chopra held to the Sarcostemma theory,[1] mentioning also the possibility of *Periploca aphylla*;[2] earlier, Chopra had also done extensive work with the Ephedras.[3] Karl Geldner in his translation of the ṚgVeda maintained that Soma could only be a species of Ephedra, probably *Ephedra intermedia* or *E. pachyclada*, the fruit of which is red and eaten by children and the stem of which is dried and taken in water as a treatment for fever.[4] Though he himself admitted that the juice was described as red, he nevertheless reiterated the old tradition that the 'milk' of the plant must be white, citing as support for this the verses that he had translated as Soma juice *mixed* with milk (IX $91^{2\text{-}3}$-'*nicht ganz klar*') or Soma delighting us with the 'milk' (*i.e.*, the expressed fluid) of the stem (IX 107^{12}).

Chinnaswami Sastri published, in 1953, an article demonstrating that Soma was not wine, reviewing the *Brāhmaṇa* substitutes, and concluding that the Soma juice was neither an intoxicant nor a stimulant.[5] The orthodoxy of Sri Chinnaswami's position may be indicated by the fact that his article is written in Sanskrit. Reinhold F. G. Müller pointed out the references to Soma in the Hindu medical books and concluded that fermented drinks, *súrā*, and brandy could have been used as substitutes, but he remarked that evidence of the process of distillation in India before Islam had not been proven.[6] Mircea Eliade expresses the view that ritual intoxication by means of hemp, opium,

drunkenness but we call inanity, which is usually followed by mania or a stupid condition.' (Rumphius, *Herbarium Amboinense*, Amsterdam, 1755; volume 5, p. 208). By 1868 *bhang* was so well known that Lord Neaves honored it with this bit of doggerel:

> The hemp — with which we used to hang
> Our prison pets, yon felon gang —
> In Eastern climes produces Bang,
> Esteemed a drug divine.

1. R. N. Chopra, S. C. Nayar, I. C. Chopra: *Glossary of Indian Medicinal Plants*, 1956.
2. R. N. Chopra *et al.*: *Poisonous Plants of India*, 1949, Vol. I, p. 683.
3. R. N. Chopra, S. Krishna, and T. P. Ghose, 'Indian Ephedras, their Chemistry and Pharmacology,' *Indian Journal of Medical Research*, No. 19, 1931-2, pp. 177-219; and *cf.* S. Krishna and T. P. Ghose, 'Indian Ephedras and their Extraction,' *Journal of the Indian Society of Chemistry*, No. 48 T, 1929, p. 67.
4. Karl Geldner: *Der Rig-Veda*, published in Harvard Oriental Series, Volumes 33-35, 1951, Vol. III, p. 2. Part I of an earlier version of Geldner's translation had been published in Leipzig in 1923, but Geldner died before the completion of Part III (which included the Soma hymns).
5. Chinnaswami Sastri: 'Soma-svarūpavimarśa,' in *Our Heritage*, No. 1, Calcutta, 1953, 80-86.
6. Reinhold F. G. Müller: 'Soma in der Altindischen Heilkunde,' *Asiatica, Festschrift für Friedrich Weller*, Leipzig, 1954, p. 428-441.

and other plant drugs is characteristic of a decadent period of shamanism, and that such means were only reluctantly admitted into the sphere of classical Yoga. He remarks, however, upon Patañjali's reference to herbs as a source of meditative powers and on the ṚgVedic description of the sage drinking poison.[1] He points to the use of opium and hashish in ecstatic and orgiastic sects in India, but he does not comment on Soma's possible status as such a drug, nor indeed upon the nature of the ecstasy induced by Soma, though he treats of the Vedas – and of ecstasy – at great length.[2]

A. L. Basham says that Soma is not what the Parsis now call Haoma (for the latter has no inebriating qualities), nor is it alcoholic (for it was consumed on the same day that it was pressed), and he goes on: 'The effects of Soma, with "vivid hallucinations" and the sense of expanding to enormous dimensions, are rather like those attributed to such drugs as hashish. Soma may well have been hemp... from which modern Indians produce a narcotic drink called *bhang*.'[3] Several Indian works were then published which investigated certain previously ignored plants as possible Soma plants: Nadkarni and Nadkarni identified the 'moon-creeper' with *Pœderia fœtida*, a plant used for the treatment of rheumatism;[4] P. V. Sharma called *Crinum latifolium*, a plant whose leaves and roots are emetic and purgative, the *som-vel*.[5]

V. G. Rahurkar arose to combat the alcohol theory, insisting that the Vedic references to *ṛjīṣá* and *tiróahnya*[6] show that Soma did not have time to ferment; he supported Oldenberg's mead hypothesis and concluded: 'Soma juice, thus, seems to be an orgiastic . . . non-alcoholic syrup-like . . . enervating drink. It was not even a fermented

1. Patañjali (*Yogasūtram*, Kāśī Sanskrit Series No. 85, Benares, 1931; verse 4.1) states that *auṣadhi* (herbs) and *samādhi* (meditation) are methods of obtaining *siddhi* (perfection through yoga).

2. Mircea Eliade: *Le Yoga, Immortalité et Liberté*, Paris, 1954; republished by Bollingen, 1958, in English translation as *Yoga: Immortality and Freedom*, p. 338.

3. A. L. Basham: *The Wonder That Was India*, London, 1954, pp. 235-6.

4. K. M. Nadkarni and A. K. Nadkarni, *Indian Materia Medica*, Bombay, 1954.

5. P. V. Sharma, *Dravya Gun Vigyan*, Benares, 1956.

6. *ṛjīsá* (which occurs only once, ṚgVeda I 32[6]) is thought by some to refer to the part of the Soma stalk left over after the juice has been expressed; Grassmann takes it to mean 'rushing forward freely', and, as it applies to Indra rather than Soma, this is more likely. *tiróahnya*, 'having lasted through yesterday', is an adjective applied to Soma.

liquor.'[1] B. H. Kapadia says that Soma's 'fruit' is red and fleshy and liked by children;[2] that it seems to have been a creeper (he cites *atasá* and *vána*);[3] and that it is an inflexible bush with dense, upright, leafless stalks. Still, Pentii Aalto states that Soma fermented for one to nine days; 'the alcohol percentage cannot have been high. Perhaps the juice of the plant contained narcotic ingredients.'[4]

The possible nature of these 'narcotic ingredients' was the subject of an article by Karl Hummel, who gave Vedic citations to establish that the Soma plant must be mountainous, yielding copious sap, and golden red in color.[5] Noting the opinions of Regel and Sir Aurel Stein, Hummel maintained that rhubarb best satisfied the requirements, for rhubarb is known to grow in the mountains of inner Asia, to yield a copious green-gold sap which turns reddish brown after standing for a while, and, of course, to be bright red, or, in some species, golden. Moreover, rhubarb is known to contain the drug Emodin; the problem of the supposed sweetness of the Soma juice, a quality absent from rhubarb juice, is easily solved by the assumption that the Indo-Aryans mixed Soma with honey, as well as with barley and milk. And as for intoxicating properties, also absent from the rhubarb juice, Hummel maintained that these were sufficiently supplied by the mere *sight* of the glorious red stalks in the eyes of 'the naive people.' A more conventional source of intoxication – *i.e.*, fermentation – was cited by N. A. Qazilbash to support his choice for Soma – *Ephedra pachyclada*.[6] Qazilbash maintained that the absence of latex in the Ephedra did not disqualify it, for only the later Sanskrit writers, and never the ṚgVeda, attributed the presence of latex to the Soma plant. *Ephedra*

1. V. G. Rahurkar: 'Was Soma a spirituous liquor?', *Oriental Thought*, No. 2, 1955, pp. 131-149.
2. B. H. Kapadia: *A Critical Interpretation and Investigation of Epithets of Soma*, Mahavidyalaya, Vallabha, 1959, pp. 4 and 35.
3. *atasá* means 'bush, undergrowth, or shrub'; *vána* means 'tree'. Their relationship with Soma in the ṚgVeda is inconclusive.
4. Pentii Aalto: 'Madyam apeyam,' in *Johannes Nobel Commemoration Volume*, Jñānamuktāvalī International Academy of Indian Culture, New Delhi, 1959, pp. 17-37.
5. Dr. Karl Hummel, Tübingen: 'Aus welcher Pflanze stellten die arischen Inder den Somatrank her?', *Mitteilungen der Deutschen Pharmazeutischen Gesellschaft und der Pharmazeutischen Gesellschaft der DDR*, April 1959, pp. 57-61.
6. N. A. Qazilbash: 'Ephedra of the Rigveda,' *The Pharmaceutical Journal*, London, November 1960, pp. 497-501.

pachyclada grows abundantly all along the Hindu Kush and Suleman ranges and yields a number of alkaloids, including L-ephedrine, Ephedra, and Pseudo-ephedrine, which act similarly to the hormone adrenalin and are used (in the form of crushed green twigs of *Ephedra pachyclada*) in Khyber and parts of Afghanistan as aphrodisiacs; it was Qazilbash's belief that at the time of the ṚgVeda the Ephedra plant was allowed to ferment, yielding a liquor that contained alcohol and ephedra alkaloids; 'the liquor, therefore, was intoxicating and possessed invigorating and stimulating effects . . . [and] . . . aphrodisiacal effects.'[1]

G. M. Patil remarks upon Soma's unparalleled intoxicating and invigorating nature. 'This intoxicant made [the Vedic people] talkative and inspired them to fight. It made them forget their mental and physical agonies, and therefore, it was a wonderful herb . . .'[2] Jan Gonda reiterated the hydromel theory,[3] and Alain Daniélou at first supported the theory of Soma as the creeper *Sarcostemma brevistigma*,[4] but later he seems to have amended his views in favor of Indian hemp ('*le chanvre Indien*') as the plant from which the ancient Soma drink was made.[5] Yet the growing dissatisfaction with conventional theories of Soma and increasing familiarity with drug-induced religious experiences have led many modern scholars to venture onto new territory; Leopold Fischer suggests that the state of mind evinced by the Soma texts 'comes much closer to alkaloid drug experiences than to alcoholic intoxication.'[6]

It is ironic that one of the earliest Vedic beliefs about Soma – that it was brought by a bird[7] – appears as a scientific criterion in two of the

1. R. Hegnauer, in his *Chemotaxonomie der Pflanzen* (Basel, 1962, Vol. I, pp. 451-460), discusses in detail the toxic effects of various Ephedras, identifying several of them with the Soma plant on the basis of Qazilbash's work.
2. G. M. Patil: 'Soma the Vedic Deity,' *Oriental Thought*, No. 4, 1960, pp. 69-79.
3. Jan Gonda: *Die Religionen Indiens*, Stuttgart, 1960, Vol. I, p. 64.
4. Alain Daniélou: *Le Polythéisme Hindou*, Paris, 1960, p. 111.
5. Alain Daniélou: *L'Erotisme Divinisé*, Paris, 1962, p. 53.
6. Agehananda Bharati (Leopold Fischer): *The Tantric Tradition*, London, 1965, p. 287.
7. *Vide supra*, p. 96. The belief that the Soma was brought from heaven by a falcon appears frequently throughout the ṚgVeda and other Indo-European sources. *cf.* A. Kuhn, *Mythologische Studien, I: Die Herabkunft des Feuers und des Göttertranks*, Gütersloh, 1886. Also *cf.* ṚgVeda I 80^{2}, III 43^{7}, IV 18^{13}, IV 26^{6}, V 45^{9}, IX 68^{6}, IX 77^{2}, IX 86^{24}, X 11^{4}, X 99^{8}, X 144^{4}, *etc.*

most recent studies of the Soma plant. Varro E. Tyler discards *Periploca aphylla* as a possibility because it 'occurs only at low altitudes in the mountains, contains a gummy latex not utilizable as a beverage, and lacks fleshy fruits attractive to birds . . . [Soma's] fruits are eaten by birds which disperse the seeds in the mountains, thereby propagating the plant.'[1] He also discards the Ephedra theory because 'it is very difficult to express much juice from these xeromorphic plants,' though he adds that *Ephedra pachyclada* when boiled in milk is used as an aphrodisiac and that 'the ash of the plant is mixed with tobacco to produce an intoxicating mixture which is applied to the gums.' Bringing similar objections against the rhubarb plant, he concludes: 'Either the ancient hymns of the *Rigveda* and the *Avesta* are gross exaggerations of fact or there grows in the vast mountain ranges of north-west India a plant whose CNS-stimulating properties, so well-known to the old inhabitants, still remain hidden from modern man.'[2]

A far more thorough article was published in the same year by J. G. Srivastava, but he too held fast to the importance of the agency of the bird, and to another long-disputed criterion: the ṚgVedic verse that some have misinterpreted as attributing to the Soma plant a thousand boughs.[3] He supports the latter implication with a verse from the *Rājanighaṇṭu* which according to him describes the climber as having 'several stems from the root-stock',[4] and with a verse from the *Suśruta Saṃhitā* which gives the Soma plant 'a tuberous root'.[5] Granting however that the ṚgVeda itself never attributes to the plant the qualities of a climber or milky latex, Srivastava goes on to mention several other plants with which the Soma has been identified, including the *Centella asiatica*, *Cocculus hirsutus* (used as a laxative and a cure for venereal

1. Varro E. Tyler, Jr.: 'The Physiological Properties and Chemical Constituents of Some Habit-Forming Plants: Soma-Homa, Divine Plant of the Ancient Aryans', *Lloydia*, Vol. 29, No. 4, December 1966, p. 284.
2. *Ibid.*, p. 285.
3. ṚgVeda IX 5[10]. *Sahásravalśam*, the adjective in question, modifies *vánaspátim*, which refers to the tree of sacrifice, not to the Soma plant.
4. J. G. Srivastava, 'The Soma Plant,' *Quarterly Journal of Crude Drug Research*, No. 6 (1966), 1, p. 811. This seems to be a translation of the phrase cited above and translated as 'having great clusters' (*vide supra*, p. 99).
5. Srivastava does not give the Sanskrit for this particular phrase but it may be a translation of *aṃśuman* (*vide supra*, p. 99).

diseases), *Fraxinus floribundus*, *Psoralea corylifolia* (used to cure leprosy), *Cæsalpinia crista*, and *Thespesia lampas* – most of these cited by the *Āyurvedic Kosh*. He disqualifies the genus Sarcostemma and the *Periploca aphylla* because birds do not disperse their fruits; he disqualifies *Vitis vinifera* because it does not have the 'twigs and stalks' that 'the *Rig-veda* clearly states' to have been used; and finally he settles back upon the Ephedra, whose 'seeds are covered by red, succulent, edible bracts, and are dispersed by birds,' and whose medical properties – which he describes at length – are commensurate with the fame of the Soma plant.[1]

Most recently an article by J. P. Kooger appeared to summarize briefly the major Soma theories, including Wasson's *Amanita muscaria* thesis, based on the note presented by Wasson in 1966 at the Peabody Museum of Natural History at Yale University. Referring to Wasson's theory as 'sensational' (*opzienbarende*) and likely to deal a great blow to extant theories, he nevertheless concluded that the mystery surrounding the identity of the Soma plant was by no means solved.[2]

Some years ago, R. C. Zaehner revived the rhubarb suggestion by referring to Haoma in these rather qualified terms: 'The Haoma plant (from its description as yellow and glowing probably something very like our rhubarb, which is found in the Iranian mountains to this day)...'[3] More recently, he mentioned to me the possibility that Haoma might be the wild chicory, which grows in the mountains of Persia, though he considers this suggestion merely a shot in the dark. But in actual fact, what else have all the other theories been but just that? Some ingenious, some thoughtful, some obviously silly, some plausible, some vague, some stubbornly wrong-headed, some Procrustean, some groping toward the truth – but all shots in the dark.

Not knowing what plant the poets of the ṚgVeda had in mind, modern scholars have often jumped to the conclusion that the hymns are vague and obscure in speaking of Soma. The *Brāhmaṇas*, dealing as they do with involved chains of substitutes, add to the confusion in almost

1. Srivastava, *op. cit.*, p. 816.
2. J. P. Kooger, 'Het Raadsel van de Heilige Soma-Plant der Indo-Iraniers,' *Pharmaceutisch Tijdschrift voor België*, 44th year, No. 7, 1967, pp. 137-143.
3. R. C. Zaehner: *The Dawn and Twilight of Zoroastrianism*, London, 1961, p. 88.

geometric progression; the few Avestan parallels are rendered more or less useless by the overlay of purely Iranian elements; and by the time the Europeans enter the scene, with their fixed ideas and various axes to grind, the situation approaches bedlam. Handicapped by a rudimentary knowledge of the vernacular and ancient languages of India and by inadequate communication in the academic world, scholars covered the same ground over and over again. Time and time again the same ideas appear, are disproved, and reappear as if they were proven theories; scholars draw upon the work of their colleagues and occasionally upon newly discovered primary materials, but there is remarkably little attention to the ṚgVeda itself. (Hillebrandt, Roth, and Geldner are notable exceptions.) Even within these limits, there seems to have been little contact between botanists and Vedists, Indian scholars and Europeans.

Fairly convincing evidence that Soma was not an alcoholic beverage was established quite early, yet Europeans continued to identify it with various forms of alcohol, and Indians continued to put pen to paper in order to assure the world that wine – which is of course anathema to an orthodox Hindu – was not Soma. The word for intoxication is ambiguous both in Sanskrit (*máda*) and in European languages; it denotes drunkenness or inebriation resulting from alcohol, but it may also apply to a mental state *similar* to that produced by alcohol. Thus the statement that Soma was 'intoxicating' as it appears in various discussions and in the ṚgVeda itself does not really exclude *any* plant capable of producing a state of exhilaration, including narcotic or psychotropic plants. The further vagueness of such terms as 'liquor' and 'strong spirits' blurred the distinction between fermentation and distillation, as does the uncertain connotation of the term *súrā*. All of this served to mask the inappropriateness of the identification of Soma with alcohol.

Often scholars tended to confuse the question of the identity of the plant with the nature of the process by which the drink was made from it, overlooking the fact that the beer theory, the mead theory, and the Sarcostemma theory are complementary rather than opposed, while only such theories as those postulating wine or *bhang* exclude all others. And, on the other hand, this failure to distinguish substance from method led several scholars to attempt a combination of various theories that are in fact incompatible.

It was difficult to resist the temptation to identify the Vedic plant with the plant actually used by the descendants of the authors of the Vedas, no matter how many facts argued against this identification; and in fact one is inclined to believe that there must be *some* relationship between the original and the substitutes, some quality in the substitute which resembled a quality of Soma enough for it to have been chosen in the absence of Soma, but the question remains as to which quality – taste or colour or effect or shape – this might conceivably be.

B. H. Kapadia remarked, 'Many Latin names are given for this plant like Ephedra, *etc.*, but we do not know exactly about it,' and this could surely stand as the epitaph for the greater part of the research done in those halcyon days of science – the nineteenth century, particularly the nineteenth century in Germany – when one still felt that by giving a categorizing Latin name to an unknown quantity one had somehow settled something. To argue whether Soma was *Sarcostemma brevistigma*, or *Ephedra pachyclada*, or *Periploca aphylla* was to assure oneself that the Soma plant had been found and that there only remained a few messy details to be cleared up; this led to smugness and a general disinclination to delve further that might not have existed had one been forced to call the plant 'milkweed' or 'some sort of rue'. It is at first striking that *bhang* was not considered as a possibility until 1921, but it is more understandable when one takes into consideration the greater attention that the psycho-physiological effects of drug-taking have received in recent times, especially in contrast with the universal disapprobation with which they were formerly associated.[1] Only in the last few generations have the anthropologists, botanists, and pharmacologists of the West entered fully into the problems presented by psychotropic plants and their role in the history of human cultures. The use of hashish in the Middle East has, of course,

1. In this context, it is interesting to note that by 1884 Regel had rejected *bhang* (*Cannabis indica*) as a possibility for the Soma plant. *Vide supra*, p. 114.

long been known, but until twenty years ago only as a curiosity. The discovery of mescaline by the modern world is almost a century old, and for some years has provoked widespread attention.

Aldous Huxley, one of the leading writers on this subject in recent times, gave the name 'Soma' to an unspecified marvellous drug in his novel *Brave New World*, in 1932. In his last novel, *Island* (1962), he depicted a Utopia that is clearly Indian throughout – Sanskrit is the language of the cult, Śiva is worshipped, and Yoga is essential to the philosophy of the islanders. And the drug upon which the cult of the Island is based is an hallucinogenic mushroom. That the mushroom is yellow and traditionally collected high in the mountains might suggest that Huxley had *Amanita muscaria* in mind – even that he was thinking of the Soma of the ṚgVeda, although he says expressly that the 'moksha-medicine' (as it is called) is *not* one of 'those lovely red toadstools that gnomes used to sit on.'[1] Later in 'Culture and the Individual'[2] Huxley, discussing the genesis of *Island*, says that he had been thinking of 'a substance akin to psilocybin', the active agent in the divine mushrooms of Mexico. The Wassons played a major part in the re-discovery of the Mexican psilocybin cult and Wasson himself had discussed his Mexican mushrooms and the Soma problem with Huxley in the late 1950's.

Certainly as soon as one rids oneself of the assumption that anything 'intoxicating' must be alcoholic, an hallucinogen of some kind seems the likely candidate for Soma, far more likely than millet or Afghan grapes or rhubarb or any other of the many plants that have been suggested. Few Vedic scholars knew any botany and some of them may not have realized that they were dealing with a problem primarily botanical. The botanists on the other hand could not read the ṚgVeda, by far the most important source about Soma, and so they permitted themselves to enter upon speculations that often seem ludicrous in the light of the Vedic hymns. But on behalf of both Vedists and botanists it is only fair to recall that for the most part the

1. Aldous Huxley: *Island*, Penguin ed., 1964, p. 140.
2. An essay published in *The Book of Grass*, an anthology edited by George Andrews and Simon Vinkenoog, Grove Press Inc., New York, 1967, pp. 192-201. The passage cited is on p. 200.

Soma question had for them merely a peripheral interest. Though the identification of Soma remained a desideratum of Indian studies, no outstanding figure applied himself directly and fully to the solution of this enigma. The historians of religion seem likewise to have given it only glancing attention.

Philippe de Félice in 1936 offered a significant description of Soma but, as he was not a member of the academic establishment, little notice was taken of it. Much the same is true of Huxley's speculative writings on the subject. If professional scholars attached small importance to the theories of these 'outsiders', it must be said that they offered no satisfactory alternative. Although the effort to identify the Soma plant has produced one of the most spirited and imaginative chapters in Vedic studies, it has also resulted in considerably more confusion than clarification. Wasson's novel solution of this old question revivifies a body of speculation that has become increasingly sterile and repetitive, and throws important problems of Indo-European and even Eurasian cultural history into a new perspective. This is indeed a welcome contribution, and it is to be hoped that its implications will be exploited in wide-ranging debate and fresh syntheses.

PART THREE

NORTHERN EURASIA AND THE FLY-AGARIC

I

EXPLORERS, TRAVELERS, AND ANTHROPOLOGISTS

IN the EXHIBITS we have assembled all the basic sources that we could find on the use of the fly-agaric for inebriating purposes in Siberia. They include a miscellany of travelers (a few of them Innocents Abroad), explorers and scientists on governmental mission, and anthropologists; also some linguists who discuss a fascinating word cluster that relates to our theme. We have added three secondary sources for particular reasons.

The earliest of our authors [1] is a poor Polish lad, not endowed with much education, who kept a diary of his stay in Siberia as a prisoner of the Tsar. An entry in that diary records the fact that in 1658 he saw the Ob-Ugrian Ostyaks getting drunk on the fly-agaric. Our latest paper is by two Soviet scientists [42] living in Vladivostok who in the fall of 1966 contributed an epitaph to the ancient practice:

> The minor nationalities of Siberia and the Far East now do not use any psychoactive drugs.... After the October Socialist Revolution and the establishment of Soviet Power these not numerous peoples have embarked on a new way of historical development. Formerly backward nationalities on a borderland of Russia, with the help of the Russian people and Soviet Power, they soon reached the socialist phase of social development.

The Soviet Union has never allowed foreigners to visit these 'minor nationalities'.

The testimony of our writers is of course worth only as much as the writers are worth. A few of them – notably the English-speaking contingent [14, 15, 20] and the one Frenchman [7] – lack understanding. They are superficial and disdainful, and most of their information seems to come from local Russian informants of dubious reliability, or from earlier writers whom we have also quoted, or else read and discarded as worthless. (*Vide, e.g.*, Kennan [14] citing Oliver Goldsmith, who in turn paraphrased von Strahlenberg [3].) The ignorance about mushrooms of these writers in English and French seems to have been complete. We reprint what they had to say because they are often quoted as sources but they are worthless.

When Erman [11] informs us that Toyon's wife transplanted a number of baby fly-agarics from the forest to her garden where he saw them some days later flourishing in their big scarlet caps, the mycologist raises his eyebrows. When he says that a native pays a reindeer for one fly-agaric in the off-season, we accept this, as other witnesses testify to the high value placed on the marvelous plant by those who used it.

Some of the Siberian tribesmen live in filth and indulge in practices that shock and revolt those Europeans whose sense of scientific detachment is not phenomenally developed. If in addition the observer is a mycophobe, taught from childhood to turn away with a shudder from a 'toadstool', his impulses will surely overpower his objectivity. Europeans have long held the fly-agaric to be lethal. The testimony of our writers on this will certainly be found surprising. Many of them and among the most trustworthy expressly acquit the fly-agaric of causing death; in fact, they testify that, properly dried, it has no bad effects. When a few of the writers mention deaths from the fly-agaric each of those deaths should be weighed carefully. Is there a single witness (some of whom spent not months but years and even a decade among the natives) who saw a man die from fly-agaric eating, or who was present in a settlement at the time of such a death? I think not. Does the reported death come from a native informant or from a fellow European, probably a mycophobe? If the latter, is it only hearsay in the foreign community, perhaps the very same death that was reported by Krasheninnikov [4] in his book published in the middle of the 18th century, a death that he himself reported as hearsay? If the source is native, is it of the kind – highly significant – that is accepted as veridical, according to Kai Donner [28], in the upper Yenisei among the Ket and the Selkup; *viz.*, that only shamans and those who are to become shamans can eat the fly-agaric with impunity; all others will surely die. (Lehtisalo [24] records similar shamanic beliefs.) Donner is not discussing vital statistics: he translates us to a different realm. We are in the presence of fossil survivals of remote beliefs, a sanction for violating a tabu, and it is precisely the sanction (at the time a projection of a surmise) that led my wife and me decades ago to suspect that our own remote ancestors had worshipped a mushroom and that a heavy tabu, surviving to this very day, had been laid on eating it. Death will come if the layman presumes to eat the forbidden fruit, the Fruit of Knowledge, the Divine Mushroom of Immortality that the Taoists talk about and that the poets of the ṚgVeda celebrated. The fear of this 'death' has lived on as an emotional residue, long after the shaman and his religion have faded from memory, and here is the explanation for the mycophobia that has prevailed throughout northern Europe, in the Germanic and Celtic worlds, in particular for the macropsia with which the north Europeans – and not least the mycologists and toxicologists – have viewed the poisonous properties of the fly-agaric.

There is a consensus among our writers that the natives of Siberia consumed the fly-agaric only after drying it. Von Strahlenberg [3] says that water is poured on it, then the water and mushrooms are boiled and the liquor is drunk. He does not mention dried mushrooms, but as he is speaking of winter they must have been dried. He and von Langsdorf [10] are the only ones who mention cooking the fly-agaric. Krasheninnikov [4] says that the dried mushrooms are steeped in the must of an Epilobium, which later writers have identified as the Russian *kiprei, Epilobium angustifolium.* (The 'must' is *not* fermented.) Krasheninnikov is the only source to mention Epilobium, but there seems no reason to doubt his word. It is natural, given the scattered population in Siberia and the poor communications, that customs vary from settlement to settlement, and even within the same settlement, from shaman to shaman. Steller [5] says that 'the fly-agarics are dried, then eaten in large pieces without chewing, being washed down with cold water.' Georgi [6], on the other hand, is responsible for the statement that in the Narym region the natives either eat one fresh mushroom or drink a decoction of three; but he is suspect as a witness. According to von Langsdorf [10], the Kamchadal prefer to leave the mushrooms to dry in the ground, exposed to the natural air and the sun. Small ones with many warts, he says, are the best and the most powerful. They are swallowed whole, being swallowed with the juice of bilberries (*Vaccinium uliginosum*), either one large mushroom or two small. Erman [11] and von Maydell [12] mention that they are taken dry, von Maydell adding that they are smoked and shrivelled up. Enderli's [19] description is classical on this point:

> At the man's order, the woman dug into an old leather sack, in which all sorts of things were heaped one on top of another, and brought out a small package wrapped in dirty leather, from which she took a few old and dry fly-agarics. She then sat down next to the two men and began chewing the mushrooms thoroughly. After chewing, she took the mushroom out of her mouth and rolled it between her hands to the shape of a little sausage. The reason for this is that the mushroom has a highly unpleasant and nauseating taste, so that even a man who intends to eat it always gives it to someone else to chew and then swallows the little sausage whole, like a pill. When the mushroom sausage was ready, one of the men immediately swallowed it greedily by shoving it deep into his throat with his indescribably filthy fingers (for the Koryaks never wash in all their lives).

The Koryak told Sljunin [18] that the fresh fly-agarics are highly poisonous and hence they do not eat them. They are dried, either in the sun or over the fire, and then consumed with fresh water to wash them down. Karjalainen [26], speaking of the shaman's ways among the Irtysh Ostyak, says that they swallow whole either three or seven mushroom caps, which may be fresh or dried: the Irtysh usage runs counter to the trend among our witnesses in that the shaman may take fresh mushrooms. Kannisto [33] speaking of the Vogul cultures reports that the mushrooms are dry when they are used. Lehtisalo [24], discussing the Yurak shamans (northern Samoyeds), tells us that the mushrooms must be fully grown and dry. Jochelson [21] also confirms that they are taken dry. There is no aspect of the fly-agaric on which there is more testimony than this, and the witnesses are almost unanimous. Particularly impressive is the quotation that we give in [30] from an heroic hymn of the Vogul people, where the hero, the 'two-belted one', addressing his wife, says, 'Woman, bring me in my three sun-dried fly-agarics!'

THREE MAPS

The data for our maps are drawn from the following sources:

A. the ethnic distribution from the map accompanying *The Peoples of Siberia,* M. G. Levin and L. P. Potapov, Editors, originally published in Moscow and Leningrad in 1956 under the title *Narody Sibiri*; translated into English under the editorship of Stephen Dunn and published by the University of Chicago Press, 1964;

B. for the distribution of the genus Betula we have relied chiefly on *Trees & Shrubs of the USSR,* S. Ja. Sokołov, ed., Ac. of Sciences, USSR, Moscow and Leningrad, 1951, vol. II, text fig. 72, p. 267; supplementing this source by data from the Gray Herbarium;

C. for the distribution of the genus Pinus we rely on *Geographic Distribution of the Pines of the World,* by Wm. B. Crichfield and Elbert L. Little, Jr., Forest Service of the U. S. Department of Agriculture, Washington, D. C., February 1966. The distribution of firs and other conifers is generally within the confines set by the pines.

A

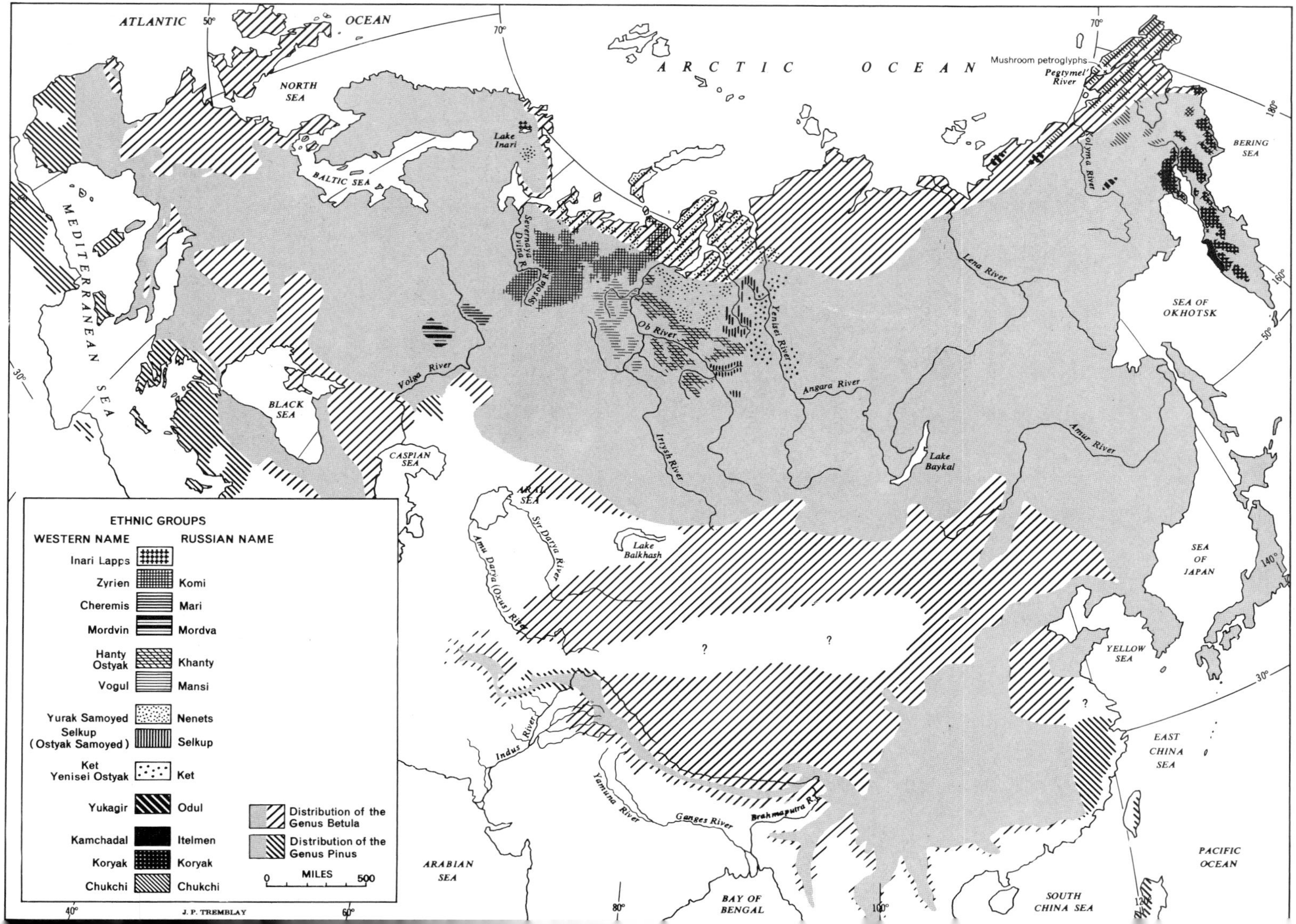
ATLANTIC OCEAN
NORTH SEA
BALTIC SEA
MEDITERRANEAN SEA
BLACK SEA
CASPIAN SEA
ARAL SEA
ARCTIC OCEAN
Mushroom petroglyphs
Pegtymel' River
Kolyma River
BERING SEA
SEA OF OKHOTSK
SEA OF JAPAN
YELLOW SEA
EAST CHINA SEA
PACIFIC OCEAN
SOUTH CHINA SEA
BAY OF BENGAL
ARABIAN SEA
Lake Inari
Severnaya Dvina R.
Sysola R.
Volga River
Ob River
Yenisei River
Angara River
Irtysh River
Lena River
Amur River
Lake Baykal
Lake Balkhash
Syr Darya River
Amu Darya (Oxus) River
Indus River
Yamuna River
Ganges River
Brahmaputra R.
50°
70°
180°
160°
140°
120°
100°
80°
60°
40°
30°
ETHNIC GROUPS
WESTERN NAME
RUSSIAN NAME
Inari Lapps
Zyrien
Komi
Cheremis
Mari
Mordvin
Mordva
Hanty Ostyak
Khanty
Vogul
Mansi
Yurak Samoyed
Nenets
Selkup (Ostyak Samoyed)
Selkup
Ket Yenisei Ostyak
Ket
Yukagir
Odul
Kamchadal
Itelmen
Koryak
Koryak
Chukchi
Chukchi
Distribution of the Genus Betula
Distribution of the Genus Pinus
0 MILES 500
J. P. TREMBLAY

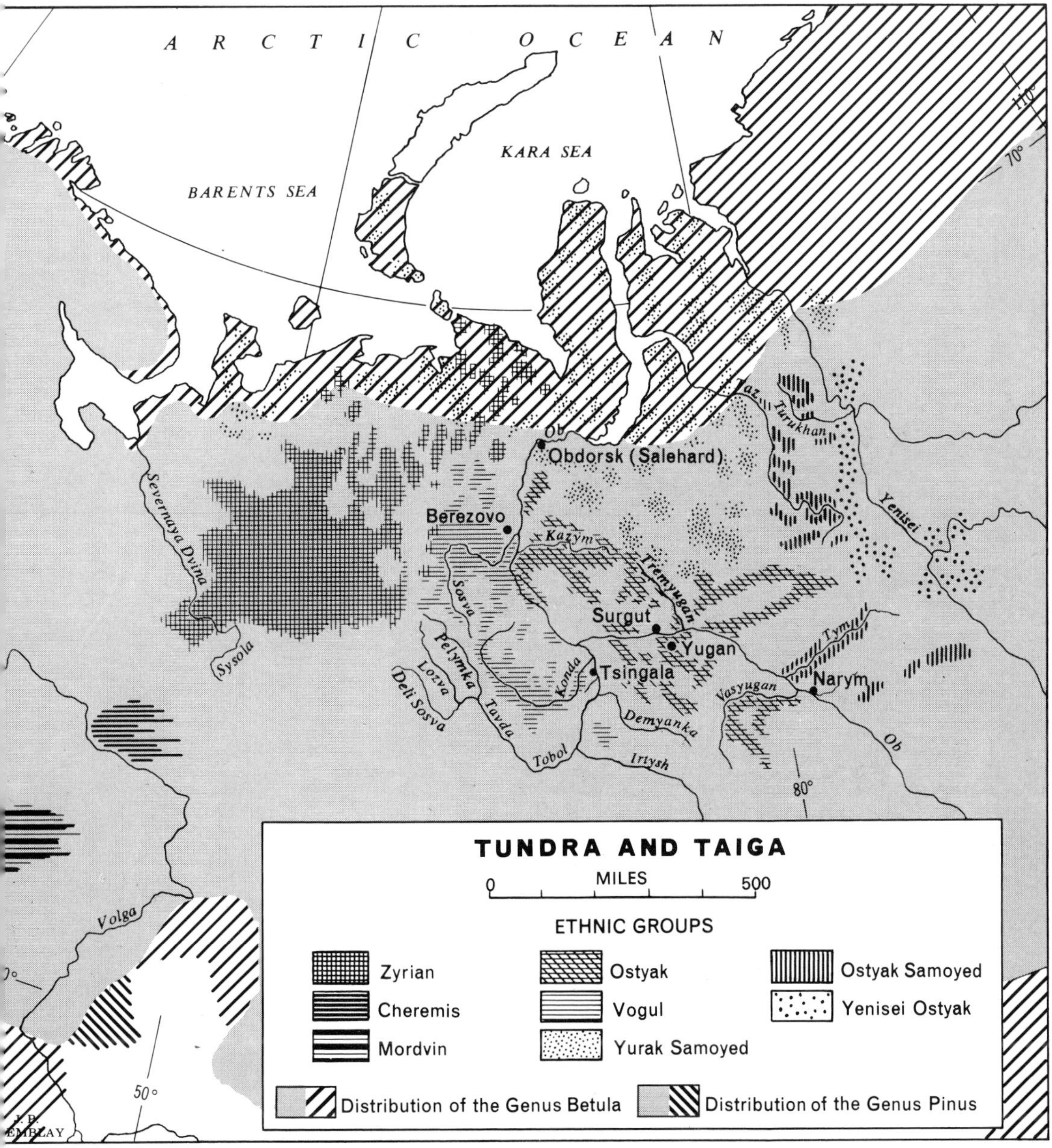

ARCTIC OCEAN
KARA SEA
BARENTS SEA
Ob
Obdorsk (Salehard)
Taz
Turukhan
Yenisei
Berezovo
Kazym
Tremyugan
Severnaya Dvina
Sosva
Surgut
Yugan
Tym
Tsingala
Narym
Pelymka
Konda
Lozva
Deli Sosva
Tavda
Vasyugan
Demyanka
Sysola
Tobol
Irtysh
Ob
80°
110°
70°
Volga
50°
TUNDRA AND TAIGA
0 MILES 500
ETHNIC GROUPS
Zyrian
Cheremis
Mordvin
Ostyak
Vogul
Yurak Samoyed
Ostyak Samoyed
Yenisei Ostyak
Distribution of the Genus Betula
Distribution of the Genus Pinus

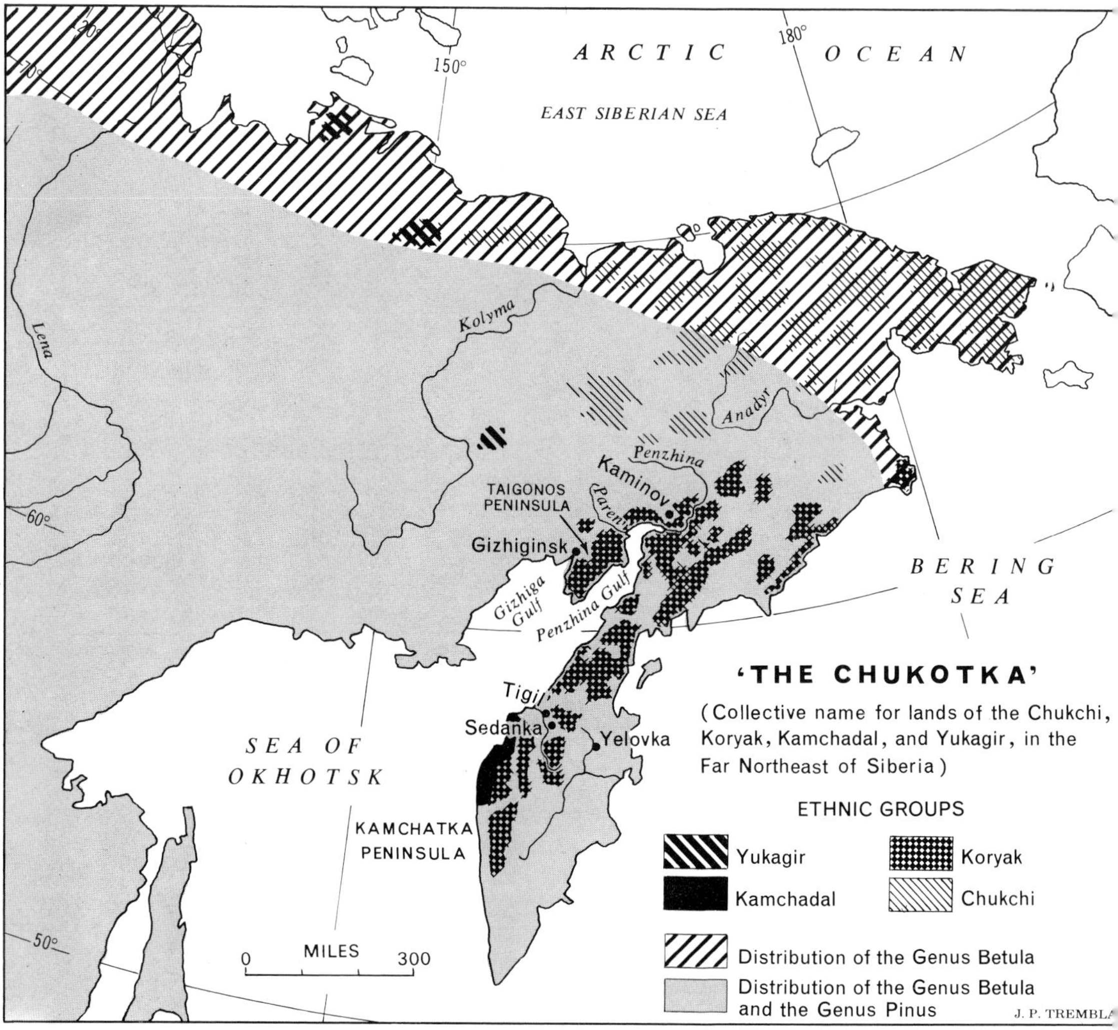
ARCTIC OCEAN
EAST SIBERIAN SEA
150°
180°
Kolyma
Lena
Anadyr
Penzhina
Kaminov
Paren'
TAIGONOS PENINSULA
Gizhiginsk
Gizhiga Gulf
Penzhina Gulf
BERING SEA
60°
50°
Tigil'
Sedanka
Yelovka
SEA OF OKHOTSK
KAMCHATKA PENINSULA
0 MILES 300
'THE CHUKOTKA'
(Collective name for lands of the Chukchi, Koryak, Kamchadal, and Yukagir, in the Far Northeast of Siberia)
ETHNIC GROUPS
Yukagir
Koryak
Kamchadal
Chukchi
Distribution of the Genus Betula
Distribution of the Genus Betula and the Genus Pinus

C

Thus it seems that drying is of the essence. The natives think that the fresh ones are dangerous, or at least not satisfactorily inebriating. Some say that eating many fresh mushrooms will kill you, but how do we know? I did not realise this when I went to Japan in the fall of 1965 and 1966, and with Japanese friends tried the fresh fly-agarics. The ṚgVeda had not prepared me for the drying. I had known of course that the Soma plants were mixed with water before being pounded with the pressing stones, but I had supposed that this was to freshen up the plants so that they would be capable of yielding juice when pressed. The desiccation, I thought, was an inevitable consequence of bringing the mushrooms from afar and keeping them on hand. There was nothing to tell me that desiccation was a *sine qua non* of the Soma rite. The reader may think that I should have familiarized myself with the Siberian practice before going to Japan. I agree. Imazeki, who by chance toasted his caps on one occasion before eating them, alone had satisfactory results, insistently declaring that this was nothing like alcohol, that this was far superior, in fact in a different world. Alone among us all, he has known *amṛta*, the ambrosia of the Immortals.

As for the harm that the fly-agarics might be expected to provoke, our witnesses give revealing testimony. Georgi [6] says that the natives

> feel much less head after this method of intoxication than is produced by spirituous liquors; nor is the use of it followed by any dangerous consequences.

Kopeć's experiences [9] with the mushrooms, for what they are worth, were benign and pleasant; he was not moved to violence. Von Langsdorf [10] says that although he made great efforts to find out something about the harmful or possibly deadly effects of the fly-agaric, he could obtain no satisfactory information on the subject.

> The Koryaks [he goes on] greatly prefer fly-agarics to the Russians' vodka and maintain that after eating fly-agarics a man never suffers from headaches or other ill effects. It is true that in extremely rare cases (of which no one could recall any specific example) persons who consumed an extraordinarily large quantity of the mushrooms are

> said to have died in convulsions, senseless and speechless, after six or eight days. However, it is not reported that moderate consumption ever produced any harmful after-effects. If, contrary to expectations, immoderate consumption of fly-agarics should nevertheless be followed by pressure on the stomach or some other disturbance, two or three spoonfuls of fat, blubber, butter, or oil are reputed to be an infallible remedy.

(Sljunin [18] gives an alternative antidote: '. . . a glass of vodka or diluted alcohol. A quarter of an hour after swallowing the vodka, the Koryak who is totally oblivious of his environment under the effect of the mushroom, completely regains consciousness.') Erman [11] quotes a native informant as saying that

> mushroom intoxication had a quite different effect from alcoholic drunkenness, since the former put the Kamchatka natives into a peaceful and gentle (*skromno* in Russian) mood, and they had seen how differently the Russians were affected by spirits.

Von Maydell [12], who passed the decade from 1861 to 1871 in Siberia, confirms the impression conveyed by his predecessors:

> . . . the mushroom produces only a feeling of great comfort, together with outward signs of happiness, satisfaction, and well-being. Thus far the use of the fly-agaric has not been found to lead to any harmful results, such as impaired health or reduced mental powers.

In a footnote he adds that he had been told of one fatal case, a Russian who died after eating rather large quantities of fresh mushrooms. He 'had been told': again we are in the realm of hearsay.

Von Dittmar [13] says substantially the same:

> Mukhomor eaters describe the narcosis as most beautiful and splendid. The most wonderful images, such as they never see in their lives otherwise, pass before their eyes and lull them into a state of the most intense enjoyment. Among the numerous persons whom I myself have seen intoxicated in this way, I cannot remember a single one who was raving or wild. Outwardly the effect was always thoroughly calming – I might almost say, comforting. For the most part the people sit smiling and friendly, mumbling quietly to themselves, and all their movements are slow and cautious.

According to Sljunin [18], the Koryak maintain that the continued consumption of the mushroom has no ill effect on the person's health. Enderli [19] reports that after a period of lethargy and monotonous singing, the mushroom eater is suddenly seized with a frenzy, raving and calling for drums, and then begins a deafening spell of singing, drumming, and running around within the confines of the yurt. He concludes with this sentence: '... immoderate consumption involves the danger of madness or death, but such cases occur rarely.' If he had seen or heard of an actual death, he is such an excellent reporter that he would certainly have told us about it. Jochelson and Bogoraz [21, 22] report no murderous outbreaks by fly-agaric eaters, and Jochelson says that to his question as to which they preferred, brandy or fly-agaric, many Koryak answered, 'Fly-agaric.' Jochelson added: 'Intoxication from the latter is considered more pleasurable, and the reaction less painful, than that following brandy.'

In these comments of various observers there is nothing that suggests the berserk-raging of the Vikings. Murderous ferocity marked the Viking seizures almost always, whereas murderous ferocity is conspicuously absent from our eye-witness accounts of fly-agaric eating in Siberia. In most of the reports the effect is soothing, *skromno*; sometimes there is a noisy inebriation. Both are familiar to us as expressions of alcoholic inebriation and both are harmless. The ardent advocate of a link between the fly-agaric and berserk-raging must content himself as best he can with the testimony of Krasheninnikov [4]: 'The Kamtschadales and the Koreki eat of it when they resolve to murder anybody.' This generalisation is hearsay: had he known of a particular episode, he would have reported it. Thus he tells us about four Russians who ate the fly-agaric, three of them in his entourage and one by hearsay. They were a menace only to themselves. This is the consensus of all the witnesses, and even here the threat to themselves is often hearsay, perhaps a reflex of the conventional attitude of Europeans when speaking of inebriation. Reguly and, after a lapse of some forty years, Munkácsi took down from native Vogul singers an Heroic Song concerning the Creation of the World. This song [30], drawn from the depths of the Vogul culture, is alone enough to demolish the berserk-raging notion of the Scandinavians. Our Hero had eaten three sun-dried fly-agarics and lay in a stupor when there bursts in upon him a messenger with news of the enemy's imminent invasion. The messenger urges the 'bemushroomed' Hero to throw off his inebriation and to come and lead the fight. The Hero replies that he has no strength and sends the messenger to rally his younger brothers. It is only on the second call that the Hero pulls himself out of the fly-agaric stupor and calls for his arms.

So far as we know, only two of our witnesses ate the mushroom: Kopeć [9], whose colourful narrative we publish for the first time in English, and Donner [28], who unfortunately and rather primly gives us no details other than that it is a powerful intoxicant. None drank of the urine. But a number of the Russians who settled in Siberia took to the fly-agaric, at least for a time. Many of our witnesses speak of this, and there is a startling sentence in Erman [11, p. 253] reading as follows:

> . . . the Russians of Klynchevsk, who according to the man from Yelovka pick whole packhorse loads of this valuable plant, prepare an extract by decocting it in water, and try to take away its extremely disgusting taste by mixing the extract with various berry juices.

What is surprising in this passage is the formula that the Russians hit on: extracting the juice with the help of water and adding various vehicles to make the drink palatable. In ancient times Soma was mixed with water and pounded with stones, then mixed with milk or curds or barley-water or honey; in Klynchevsk they mixed the extract with the juice of berries.

A fairly consistent picture of the fly-agaric syndrome emerges from reading the accounts of our witnesses. Krasheninnikov [4] says that in moderation it raises the spirits and makes one brisk, courageous, and cheerful, but if indulged in to excess, it leads to trembling, a merry or melancholic mood according to one's disposition, and macropsia: 'a small hole appearing to them as a great pit, and a spoonful of water as a lake.' Von Langsdorf [10] confirms the macropsia: 'If one wishes to step over a small stick or straw, one steps or jumps as though the obstacles were tree trunks.' So do von Maydell, Sljunin, Bogoraz, and Jochelson [12, 18, 22, 21]. There is often vomiting and at some stage sleep supervenes, during and after which one sees marvelous visions, *not* erotic according to von Maydell [12]; 'highly sensuous' according to Enderli [19]. Kopeć [9] seems to confirm Enderli. (We wonder how von Maydell knew the fly-agaric did not give erotic pleasure.) A number of our witnesses testify to the increase in strength that can be expected: von Langsdorf [10] met eye-witnesses who said that a man inebriated with the mushroom had been able to carry a 120-lb. sack of flour 10 miles, and Bogoraz [22] describes one under the influence of the fly-agaric:

> His agility increases, and he displays more physical strength than normally. Reindeer-hunters of the Middle Anadyr told me that before starting in canoes in pursuit of animals, they would chew agaric because that made them more nimble on the hunt. A native fellow-traveler of mine, after taking agaric, would lay aside his snowshoes and walk through the deep snow hour after hour by the side of his dogs for the mere pleasure of exercise, and without any feeling of fatigue.

Von Langsdorf [10] declares that according to the statements of the natives, those who have taken the fly-agaric in moderation feel 'extraordinarily light on their feet and are then exceedingly skillful in bodily movement and physical exercise.' Erman [11] confirms this:

> There is no doubt . . . about a 'marvelous increase in physical strength', which the man praised as still another effect of the mushroom intoxication. 'In harvesting hay', he said, 'I can do the work of three men from morning to nightfall without any trouble, if I have eaten a mushroom.'

The Chukotka tribesmen personify the mushroom as little men or women. Krasheninnikov [4] was the first to call attention to this. 'It is observed, whenever they have eaten of this plant, they maintain that, whatever foolish things they did, they only obeyed the commands of the mushroom.' (We quote from the 18th century rendering by James Grieve.) Jochelson and Bogoraz [21, 22] ratify this. The former writes: 'The idea of the Koryak is that a person drugged with agaric fungi does what the spirits residing in them (*wapaq*) tell him to do;' and the latter: 'The spirits of the fly-agaric have an outward appearance similar to that of the actual mushrooms, and the agaric-eater feels impelled to imitate them.' The Koryak and Kamchadal tales that we reprint are shot through with this personification of the mushroom. Among the Ob-Ugrians Lehtisalo [24] observes the same phenomenon. The mushroom eater enters the realm of the little people, talks with them, learns from them what he wishes to know – the future, the outlook for a sick person, *etc.* Among the Ob-Ugrians the divine mushroom seems to have retained more of its sacred character. It plays a role in the myths of creation and, as we have seen, there are vestiges of stern tabus on its use by unqualified persons.

Of all the properties of the fly-agaric as it is used in Siberia, the one that has drawn the most attention is its effect on the urine of the person who eats it. The inebriating virtue of the mushroom passes into the urine, whence the custom in the Chukotka of drinking the urine. This is amply confirmed by all our best sources and there can be no doubt about its truth. No one knows to this day the chemical composition of the inebriant. No one can say whether the drug in the mushroom is identical to the drug in the urine. Perhaps alien elements in the mushroom, such as sometimes cause nausea, are filtered out in the urine. To von Langsdorf [10] alone goes the credit of having asked some of these questions:

> I was not able to ascertain whether the consumption of fly-agaric is followed by constipation or diarrhœa or by an increase or decrease in the urine.
>
> I was also unable to obtain any satisfactory answer when I asked whether the taste or smell of the urine had been changed; everyone was probably ashamed to admit that he had drunk his own or someone else's. Nevertheless it strikes me as not improbable that fly-agarics, like turpentine, asparagus, and other things, impart a special, possibly quite pleasant, smell and taste to the urine. By analogy it would be worth investigating whether other narcotic

substances, such as opium, *Digitalis purpurea*, cantharides, *etc*, also retain their properties in the urine.

Many of the best names on our list of sources discuss this practice of urine drinking: von Strahlenberg [3], Krasheninnikov [4], Steller [5], von Langsdorf [10], Erman [11], von Maydell [12], von Dittmar [13], Sljunin [18], Enderli [19], Jochelson [21], and Bogoraz [22]. But those who describe the Uralic and Ket practices are silent about the urine. It is to be assumed that these tribes do not know its virtue, or that, knowing it, they do not indulge in it. They are closest to Russia and perhaps have been under the Russian influence more strongly than the peoples of the far Northeast. Perhaps fly-agarics are so common in their forests that they have no need to resort to their own urine to supplement the supply. We can only regret that none of the writers about these peoples seem to have put the question to them. If we have interpreted ṚgVeda IX 74[4] correctly, the Indo-Aryans were drinking urine impregnated with the fly-agaric 3,500 years ago, and before then the Indo-Aryans had come south from a region adjacent to an Uralic people, with whom it is to be assumed that they shared their fly-agaric eating propensities, including perhaps the urine-drinking.

By assembling as many of the sources as I could find and laying them before the reader, his attention is called as it has never been before to the peculiar and perhaps significant role of the reindeer in the fly-agaric complex of Siberia. Reindeer have a passion for mushrooms and especially for the fly-agaric, on which they inebriate themselves. Reindeer have a passion for urine and especially human urine. (When the human urine is impregnated with fly-agaric, what a regal cate is there, to be served to a favoured reindeer!) Steller [5], who stayed behind in Siberia when Krasheninnikov [4] returned to Russia, was the first to call attention to this. Von Langsdorf [10] saw the importance of this and quotes Steller *in extenso* on it. Erman [11] bears independent witness to the facts in an astonishing paragraph. Bogoraz [22] stresses the passion of the reindeer for human urine, which is likely to make it dangerous to relieve oneself in the open when there are reindeer around. The curious account of the death of two reindeer given by Sarychev [8], slightly garbled though it must be, seems to relate to the same phenomenon. Before I had read Steller [5] and Erman [11], I made my suggestion on pp.75-76 that living as some of these tribesmen do in intimacy with the reindeer, almost in a symbiotic relationship with them, they may have found it easy to indulge in the drinking of the urine in imitation of the beasts. Here then would be the genesis of the urine-drinking that has astonished the West.

What is the relationship between the hallucinogenic mushrooms of Middle America and the fly-agaric complex of Siberia?

The mushrooms are utterly different. The fly-agaric is an Amanita. The sacred mushrooms in Mexico are far removed, belonging to the Psilocybe, Stropharia, and Conocybe genera, these genera being closely inter-related. The drugs in the Mexican mushrooms are psilocybin and psilocin. We do not know what the drug is in the fly-agaric, or perhaps the drugs. The Mexican mushrooms keep one awake for about four or five hours, and then one falls into a deep, dreamless slumber for a couple of hours; there is no hangover. Though one remains awake and experiences hallucinations, one has little or no desire to move about. There is no macropsia. At the very moment one can talk about the marvels that one sees, and exchange comments with one's neighbors, but there is no kinetic agent in the mushrooms. By contrast, in Siberia some tribesmen feel an urge to talk loudly or shout and sing, to dance, to run about, to perform feats of physical activity. In Siberia stress is laid on the visions one has during a profound sleep that, according to most of our sources, comes at the end of ten or twelve hours, and also during the waking hours that follow.

There is one point of similarity. In each case nausea and vomiting occasionally occur in some persons. When the mushrooms act as an emetic immediately or a few minutes after they are ingested, it is a remarkable fact that in both cases the vomiting has no effect on the later inebriation. I have experienced this myself in Mexico. Von Langsdorf [10] is our witness for Siberia:

> ... people who have eaten a large quantity of mushrooms often suffer an attack of vomiting. The rolled-up mushrooms previously

> swallowed whole are then vomited out in a swollen, large, and gelatinous form, but even though not a single mushroom remains in the stomach, the drunkenness and stupor nevertheless continue, and all the symptoms of fly-agaric eating are, in fact, intensified.

There is a striking similarity in the imaginative world of the Mexican Indians and the tribesmen of Siberia: both have created a community of dwarfs who take over. In Mexico the mushrooms command. They speak through the *curandero* or shaman. He is as though not present. The mushrooms answer the questions put to them about the sick patient, about the future, about the stolen money or the missing donkey. The mushrooms take the form of *duendes*, to use the Spanish term; dwarfs in English. Similarly the eater of fly-agarics comes under the command of the mushrooms, and they are personified as amanita girls or amanita men, the size of the fly amanita.

If I am asked whether there is any genetic relationship between the two areas in the light of the similar mushroom practices, I should say that to harbor such a thought would be hazardous. The Indians of the New World have shown themselves supreme in the arts of the herbalist, discovering properties in the plant kingdom from which we Europeans have learned much. The Indians were certain to discover the divine mushrooms (as they think them to be) and it was natural, given their cultural background, that they should personify them as *duendes* and think they were speaking through the shaman's words. The same thing was true with the Siberian tribesmen. Both cults are, in my opinion, thousands of years old and autonomous in origin.

II

A FAR-REACHING SIBERIAN WORD-CLUSTER

A remarkable pattern of linguistic evidence marks the fungal vocabulary of many Siberian tribes and the European peoples. Specialists in the Uralic family of languages have greatly contributed to the Uralic aspect of this problem, but no first class scholar has dealt with the linguistic and cultural aspects of the entire pattern of fungal words that are scattered throughout northern Eurasia from the Iberian peninsula to Bering Strait. I can do no more than assemble some of the evidence and pose the questions that call for answers.

I. THE ANTIQUITY OF FLY-AGARIC INEBRIATION

In the Vogul language all words relating to drunkenness are derived from the word for fly-agaric, *paɒχ*, and its innumerable variants according to the dialect. This means that Vogul speakers, when they talk of getting drunk, say that the man is 'bemushroomed'. But it is important to note that the Vogul speaker is not aware of the etymology of the word: he uses it without thinking of the fly-agaric, whether the man was 'bemushroomed' on alcohol or fly-agaric. This is similar to our use of the word 'drunk'. If we say the shaman is getting 'drunk' on fly-agaric, it would not occur to us that he does not 'drink' the fly-agaric. If the Vogul speaker says that that foreigner was 'bemushroomed' yesterday on vodka, it never occurs to him that one drinks alcohol, instead of eating a mushroom. Just as our basic word in English is a secondary meaning derived from a beverage that one drinks, so in Vogul the basic word is derived from the fly-agaric. The fly-agaric was the original inebriant and probably the only one.

Vogul is an Ugrian language. In Zyrian (= Komi), a Finnic language, there are a number of words, *pagal-*, *pagav-*, *etc.*, meaning to lose consciousness as from alcohol. Uotila [34, 38] tells us that they are derived from **pag-*, and are cognate with the Vogul *paɒχ*. The Zyrians do not use the fly-agaric today, but these words would indicate that their ancestors did so.

It seems certain that these words go back to a time when fermented drinks (not to speak of their distillate, alcohol) were either not known at all or were unimportant, to a time when the fly-agaric was *the* inebriant. It seems that they go back to a time before the Ugrian and Finnic languages became differentiated, centuries before Christ.

But our story does not end there. Castrén [24a] in the last century reported the word *faŋkáˀam*, 'to be drunk', in the Tavgi language, belonging to the northern Samoyed group. If this attestation is sound, and Uralic scholars treat it seriously, then the beginnings go back far indeed. For between the Finno-Ugrian languages and the Samoyed languages there exists precisely the same p∾t shift that we find distinguishing the Latin and Germanic languages; *e.g.*, Latin *pater*∾ English 'father'. The Tavgi word, manifesting this basic shift, cannot therefore be a borrowing from Vogul or Zyrian. It goes back to a common ancestor, before the Uralic peoples divided into the Samoyed and the proto-Finno-Ugrian, certainly thousands of years before Christ. We cannot say when the fly-agaric was first used in the northern reaches of Eurasia. We can say, if Castrén is to be relied on, that it was being used when the ancestral tongue of the Uralic peoples split up. In any case we feel safe in saying, on the evidence supplied by the Uralic languages, that the fly-agaric was being invoked as a divine inebriant before the Aryans left their ancestral home and long before the ṚgVeda was composed.

2. THE DISTRIBUTION OF FLY-AGARIC INEBRIATION

So far as I know, Franz Boas initiated the accumulation of evidence on this subject when he pointed out [31] that the three languages of the Chukchi group, in the far Northeast of Siberia, used for 'mushroom' words whose common root was *poŋ*. He made no mention of such a root in the Ob Valley, and he was probably not aware of it. His book appeared in 1922 but the materials for it had been accumulated from the turn of the century.

The Finnish linguist Artturi Kannisto [33] was gathering data among the Vogul in the first decade of this century. He recorded the names *pa:ɒχ* and *pə:ɒχ* for the 'fly-agaric' used by shamans to achieve ecstasy. His papers were published only in 1958, long after his death. Munkácsi, the Magyar scholar, working in part with collections made in the second quarter of the last century by Antal Reguly, drew attention [32] to the Ob-Ugrian and Volga Finnic cluster of fungal words having identical origin, though in Ob-Ugrian they meant a specific mushroom, the fly-agaric, whereas in Mordvin and Cheremis they were generic, 'mushroom'. He suggested an Iranian word, *banha*, as cognate with or the etymon of the Uralic words. At least one of his colleagues, Lehtisalo [24a], expressed doubt about this, and now that we know more concerning the Iranian word, Munkácsi's suggestion seems to be ruled out. Apparently Lehtisalo in 1928 [24a] was the first to bring in the Samoyed languages: it was he who drew attention to Castrén's discovery that in the Tavgi tongue (in the North Samoyed group) *faɒká?am* means 'to be inebriated' and that it must be cognate with the other words. Uotila in 1930 [34] then added the Southern Samoyed language, Selkup, seeing in *pöɒer* and *pəɒgar*, the 'drum' and another musical instrument used by shamans, derivatives of the root *poɒ*. He detects a connection between our cluster and the Zyrian *pagal-*, *etc.*, 'to lose consciousness'. Bouda [35] in 1941 confirmed Boas's findings for the Chukchi group and he for the first time linked the word for 'mushroom' in the far Northeast with the Uralic words that we have been discussing. Steinitz in 1944 [36] thought that the link with the Zyrian *pagal-*, *pagyr*, meaning 'to lose consciousness', *etc.*, was unclear. In a notable paper Balázs [38] in 1963 summed up the Uralic evidence. He is inclined to accept Uotila's judgment on the Zyrian words, rather than Steinitz's hesitant position. He is sceptical about Munkácsi's view concerning an Iranian origin for this cluster, leaving however the last word to Iranian scholars. Apparently he was unaware of Walter B. Henning's pronouncement, published in 1951, on the Iranian word *bang*. (*Vide* Eliade [41])

We show the distribution of these tribes in the recent past in our maps A, B, and C. They have changed their location slowly over the past centuries, and are now disappearing. The Yukagir, for example, were once an important people; now they are confined to three shrink-

CHART OF URALIC LANGUAGES AND OTHERS

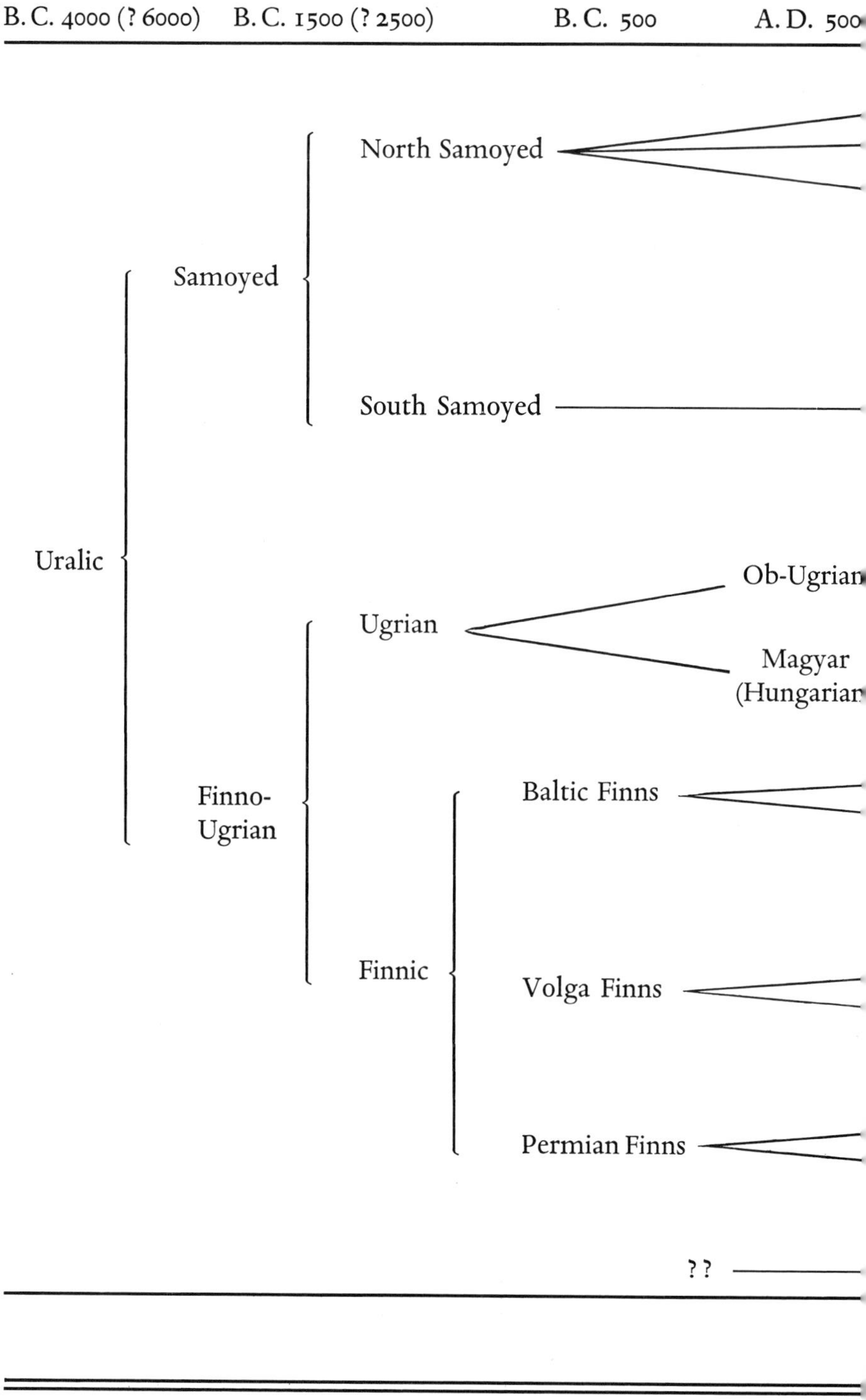

CHART OF URALIC LANGUAGES AND OTHERS

Contemporary	*Poŋ* Cluster
Yurak (Nentsy)	
Yenisei Samoyed (Entsy)	
Tavgi (Gnanasang)	*faŋkáˀam*: 'to be inebriated'
Selkup (and others, all extinct)	*pöŋer* : 'drum' *pəŋgar* : special Selkup musical instrument
Vogul (Mansi)	*paŋχ, piŋka*: 'fly-agaric', 'inebriation'
Ostyak (Khanty)	*poŋχ, paŋχ, paŋga*: 'fly-agaric', 'mushroom'
Magyar	bolond gomba: 'mad mushroom'
Finnish	——
Esthonians (and others)	——
Mordvin	*paŋga, paŋgo*: 'mushroom'
Cheremis (Mari)	*poŋgo, paŋgə*: 'mushroom'
Zyrian (Komi)	*pagalny*: 'to lose consciousness' etc. < **pag*: (root)
Votyak	
Lapp	
Ket (and others, all extinct)	*haŋgo*: 'fly-agaric'
Yukagir	
Chukchi Koryak Kamchadal	**poŋ*: (root, meaning 'mushroom')

ing areas. The Finno-Ugrian peoples, on the strength of linguistic evidence as interpreted by ethno-botanists and zoologists, seem to have had their original home in the bend of the Volga River, where the Mordvin and Cheremis tribes now are. The Ob-Ugrians and the linguistic ancestors of the Magyars presumably migrated to the east, the Magyars furthest to the East. Later the Magyars under Turco-Khasar domination migrated West to the Pannonian plain where they now are. They succeeded in surviving the Turkish domination and their own language emerged as dominant.

In the chart facing this page we show the linguistic families and the distribution of the *pov* cluster. We have even added a time scale but hasten to add that it is speculative, intended to give some idea of the linguistic history that we are dealing with.

A number of conclusions seem to emerge from our evidence. The rôle of the fly-agaric has been shrinking for centuries. Until a few generations ago it was deeply rooted in the Ob-Ugrian and Samoyed cultures, the words related to *pov* having given to these peoples many derivatives for 'inebriation', the musical instruments of the shaman, *etc.* In the tradition reported by Itkonen [23] among the Inari Lapps that they were once familiar with the fly-agaric as an inebriant, we get some idea of the wider range that this practice enjoyed. It is to be inferred from the distribution of the *pov* root in the Chukotka that these peoples were once adjacent to the Ob-Ugrians, far to the west, and that they were pushed into their present location by the infiltration of Tungus and Yakut from somewhere to the south. We must not think of this movement as resulting from wars of conquest to which precise dates might be given. It is much more likely to have been a relatively peaceful occupation of inhospitable wastes punctuated by occasional clashes with the sparse inhabitants. The Tungus (with the Lamut) and the Yakut have been where they are now for as far back as we have records. Perhaps a score of centuries have passed since the peoples of the Chukotka were living adjacent to the Ob-Ugrians.

The fly-agaric was accessible to man in the forest belt ages before he knew the art of distillation. For the gathering it was available to him probably before he possessed facilities for storing berries and the juice thereof, and before he had mastered the technique of fermenting liquors. It is hard to see why the shaman should not have been resorting to the fly-agaric for inebriation and ecstasy since the receding ice-cap of the last ice-age permitted him and the birch to exist together in the northern reaches of Eurasia.

3. THE 'POŊ' CLUSTER AMONG THE INDO-EUROPEANS

Now we shift the scene to the Indo-European world and Europe. In 1901 Holger Pedersen, then a young man and destined to become one of the leading figures in the comparative study of the Indo-European languages, published a lengthy paper in Polish [39] in which he found that the Proto-slavic **goHba*, the Old Church Slavonic *gǫba*, the Old High German *swamb*, the two Greek variants *sphóngos* and *spóngê*, and the Latin *fungus* were cognates. The late V. Machek, distinguished Slavic philologist, accepted this correlation in his *Etymologický Slovník* (Prague, 1957) of the Czech language. So does Jan Otrębski, the Polish philologist, in a paper that he published in 1939 in Vilno entitled *Indogermanische Forschungen*, where he gives scores of examples of the identical metathesis. Among Slavists the weight of evidence is in favor of this interpretation; only Berneker held that, though Pedersen's case was tempting, it was 'unclear'. He did not elaborate. Boisacq, following Berneker, deduced that the Pedersen thesis should be discarded.

Professor Roman Jakobson, solicited for his opinion on this matter, said:

> The etymology of Holger Pedersen, the great Danish specialist in the comparative study of Indo-European languages, seems to me and to many other linguists, *e.g.*, the distinguished Czech etymologist V. Machek, as the only convincing attempt to interpret the fungal name of the European languages. Not one single serious argument has been brought against Pedersen's 'attractive' explanation, as Berneker defines it, and not one single defensible hypothesis has been brought to replace this one.[1]

1. *Vide Botanical Museum Leaflet*, Harvard University, Vol. 19, No. 7 (1961), p. 150, ftn.

Was not the *poŋ* cluster of the Uralic peoples borrowed, perhaps as far back as Uralic times, from the neighboring Indo-Europeans? If the thesis of this book is right, the Aryans were using the fly-agaric in their religious rites before they left their homeland. The Indo-Iranians do not possess a word of the *poŋ* cluster, because under tabu influences they had replaced it by Soma or Haoma, and the original word was lost.[1]

But even if this supposition is right, there remains the further question as to what the original meaning of *poŋ* was, and whether it designated the fly-agaric when it was borrowed. We may take it for granted that specific meanings precede generic definitions: *i.e.*, names for individuals or species precede the names for classes of things. But this is not much help, for mushroom names easily change their meaning from specific to general, and from the name of one species to another, and meanings often shift when words go wandering from one people to another.

For the cultural evolution of man, the shamanic use of the fly-agaric may have been vital, greatly broadening the range of his experience, making known to him horizons beyond any that he knew in real life, in short sparking his imagination. But there was another spark, even more vital for his very survival. We all know the two ways by which most men generated fire in early times – by percussion, or by rubbing two wooden members together. But not enough stress has been laid on the primary tinder to catch the spark, a tinder so inflammable that it bursts into flames at once. The best tinder for this purpose in northern Eurasia has always been the dried *Fomes fomentarius*. This is a shelf fungus most commonly found on birch. At Maglemose, in diggings that date from soon after the last ice age, *Fomes fomentarius* has been found close by the hearth stones. At Star Carr in Yorkshire, in diggings of the same culture but perhaps somewhat earlier, *Fomes fomentarius*, sometimes still attached to the birch

1. R. L. Turner in his *Comp. Dict. of the Indo-Aryan Languages*, Entry 7643, assembles a group of cognate words in contemporary Indian vernaculars derived from Sanskrit that stem back to the hypothetical Sanskrit **paggala-*, 'mad', 'madness'. He suggests the possibility of a link to Sanskrit *paṅgú*, 'lame', 'crippled in the legs', of which the variant forms suggest to Turner a non-Aryan origin. May these be remote descendants of our *paŋχ* cluster?

host, have been found next to iron pyrites and the hearth stones. These archæological diggings date back almost to the last ice age, which began to recede *ca.* 12,000 B.C. As the ice cap receded and man pushed his way north, the control of fire defined his range of diffusion, and this touchwood or 'punk' (which in the style of this book we should spell *pəŋk*), this *amadou* or *yesca* or *Zunderschwamm* or *trut* or *zhagra* or *tapló* was what assured him of warmth and a cooked meal. Perhaps the men of Maglemose or Star Carr, whoever they were, as they busied themselves about those ancient hearths, were already calling their tinder 'punk'. The discovery of *Fomes fomentarius* and the simple methods of preparing punk for use had marked a long step forward in man's material progress and the comforts of human existence.

★

NOTE 1. Is *mo-ku*, written either 蘑菇 or 蘑菰, a member of the *poŋ* cluster? It is the ordinary market-place word for edible mushrooms throughout China. The word is not found in classical Chinese. The earliest citation that I can find is in the *Rules of Cooking* 飲膳正要, written in 1330 and published in 1456. These were compiled by Hu Ssu Hui 忽思慧, the senior chef of the Mongol Emperor and himself also a Mongol. (The Dynasty in 1330 was Mongol.) In Mongolian the same word in various dialectal forms is also used for mushroom. Philologists have not been sure which culture borrowed the word from the other. It seems to me that the circumstances indicate a Chinese borrowing from Mongolian. The Chinese characters possess different meanings, but in sound they lent themselves to express the Mongolian word. In Pekin *mo-ku* means specifically *Tricholoma mongolicum*, a delicious mushroom highly prized all over China, which was and still is imported in large quantities from Mongolia. It is also known as *k'ou mo* 口蘑, which means for the Chinese 'brought through the Kalgan gate'; in other words, through the Great Wall from Mongolia. With the passing of time the sense of *mo-ku* has become general, especially outside of Pekin.

If then the word is Mongolian, will the Mongolian experts tell us whether *mo-ku* could come from *poŋ*? (*s*)*pongos* > **bongos* > **mon-go*(*s*) > *mogu*.

NOTE 2. In 1963 I was visiting New Zealand and on August 6 was in Rotorua, in the center of the northern island, a Maori community. I was chatting with an elder of the Maori people, Keta Ehau, 74 years old at the time. Naturally the subject came around to mushrooms and he volunteered a story that seemed to him remarkable, and to me even more so. He had been in Siberia with two Canadians and a South African during the first world war, in the Government service, and he found that the natives there were still using touchwood, as the Maoris do in New Zealand. The natives where he was called it *paŋke* or *paŋkə*. This had made an impression on him because in Rotorua the fungus that serves as touchwood is called *paŋe*, without the *k*. But of course vacillation between *ŋ* and *ŋg* or *ŋk* is frequent. This fungus grows on the *rātā* tree, *Metrosideros robusta*, a member of the myrtle family. At Ruatoria, on the coast to the east of Rotorua, touchwood is made from a fungus growing on the *tawai* tree, Nothofagus spp. There the fungus so used is called *pú:tawa*, and *paŋe* is unknown.

How strange to find a word that might be a member of the *poŋ* family among the Maori of Rotorua! I looked the word up in Williams's *A Dictionary of the Maori Language*, of which many editions have appeared since the first in 1844. *Pú:tawa* was in them all. In the fifth edition, 1917, I found '*Pangē, pangī*: tinder, touchwood, made from spongy fungus.' In a copy of the fourth edition, 1892, that lies in the Trumbull Library, Wellington, there are the editor's notes made in anticipation of the fifth edition, and among them is this entry written by hand, probably around 1914.

III

EUROPE AND THE FLY-AGARIC

1

If I have established my case that Soma was the fly-agaric – that the *amṛta* of the Aryans was until only yesterday the divine inebriant still currently consumed by the shamans over vast reaches of Siberia, then at once the initial steps by which my wife Valentina Pavlovna and I started out on our inquiries more than forty years ago take on relevance and a cutting edge. For we did not begin by looking for Soma. Decades were to pass before Soma drew my attention. We started out by accumulating purely European ethno-mycological data, chiefly philological and folkloric, and those European data led us twenty years later to make a bold, many would say a wild, surmise: the striking pattern of our evidence would be understandable if we postulated a period when a mushroom had played a rôle in the religious life of our own remote ancestors, perhaps some 6,000 years ago, millennia before they could read and write, when the last ice-age was still yielding the frozen wastes to the pioneer food gatherers. We did not know which mushroom nor why, but it must have been hedged about with all the sanctions that attend sacred things in primitive societies. Judging by what we considered vestigial survivals in our own folkways, it must have been instinct with *mana*, an object of awe, of terror, of adoration.

Our later discoveries in Siberia and mine more recently among the Indo-Iranians were an immediate sequel to those early hesitant stumbling steps that we were taking in the 1930's, and they lend credence to our 'wild surmise' about our own European ancestors, for it is unlikely that a foolish misinterpretation of evidence would lead us to these rich finds. It is therefore in order to re-examine our early evidence, constituting as it does an exploration into the pre- and proto-history of our own European stock.

My wife and I embarked on this our intellectual foray late in August 1927. A little episode started us on our way. Valentina Pavlovna was Russian, a Muscovite by birth. I was of Anglo-Saxon ancestry. We had been married less than a year and we were now off on our first holiday, at Big Indian in the Catskills. On that first day, as the sun was declining in the West, we set out for a stroll, the forest on our left, a clearing on our right. Though we had known each other for years, it happened that we had never discussed mushrooms together. All of a sudden she darted from my side. With cries of ecstasy she flew to the forest glade, where she had discovered mushrooms of various kinds carpeting the ground. Since Russia she had seen nothing like it. Left planted on the mountain trail, I called to her to take care, to come back. They were toadstools she was gathering, poisonous, putrid, disgusting. She only laughed the more: I can hear her now. She knelt in poses of adoration. She spoke to them with endearing Russian diminutives. She gathered the toadstools in a kind of pinafore that she was wearing, and brought them to our lodge. Some she strung on threads to hang up and dry for winter use. Others she served that night, either with the soup or the meat, according to their kind. I refused to touch them. . . . This episode, a small thing in itself affecting only a peripheral aspect of our busy lives, led us to make inquiries, and we found that the northern Slavs know their mushrooms, having learned them at their mother's knee; theirs is no book knowledge. They love these fungal growths with a passion that, viewed with detachment, seemed to me a little exaggerated. But we Anglo-Saxons reject them viscerally, with revulsion, without deigning to make their acquaintance, and our attitude is even more exaggerated than the Slavs'. Little by little my wife and I built up extensive files concerning this modest corner of human behaviour, not only about the Slavs and Anglo-Saxons but about all the peoples of Europe, even to the Basques, the Frisians, the Lapps, and the Albanians.

Years passed. We had reached the 1940's before we pronounced our 'wild surmise', and we then gave utterance to it only to each other, since we were afraid of appearing ridiculous to our friends, perhaps even a trifle touched. Our evidence was airy and insubstantial, but it possessed a poetic consistency that carried conviction with us. We resolved to cast our net further afield and to explore the tribal cultures of Siberia. What was our amazement when we found, right on the doorstep of Europe, a mushroom – the fly-agaric – occupying the center of the stage in the shamanism of many northern tribes. We were hitting pay dirt and for long we thought we had reached the end of our road.

In 1952 our attention was diverted to Mexico, where we learned that there was a mushroom cult to be explored and studied *in situ*, both historically through the centuries or even the millennia, and also as a living anthropological practice in many Amerindian cultures of Oaxaca, Puebla, and Vera Cruz. The mushrooms used in Mexico were not the fly-agaric of Siberia and we could not discover any umbilical cord linking the Mexican and Siberian cults, but we enjoyed ample opportunity to dissect the modalities of a divine mushroom inebriant. The ten rainy seasons – 1953 to 1962 – spent in the remote mountains of Mexico[1] were a rewarding experience but they were only a diversion from our Eurasian preoccupations. In the course of these years Valentina Pavlovna's fatal illness manifested itself: she died on the last day of 1958, in the evening. Meanwhile we had rushed our *Mushrooms Russia & History* into print in May 1957. In it we expressed our 'wild surmise' hesitantly, by implication rather than directly, and not a single reviewer caught it. We were still unwilling to sponsor openly the notion of a divine mushroom among our own ancestors. Only one critic, a first-class mycologist, hinted at the point when he said that we had not succeeded in establishing our theory as to the origin of 'fly-agaric'.[2] The scientific mind is prone to measure evidence by austere scientific standards, by calibrating quantitative phenomena. But the myths and verbal origins of pre-literate communities are sheer poetry, to be understood only by poets and those with a gift for the play of imagery. There are values in those societies that do not lend themselves to quantitative calibration. This is where our critic went wrong. We were certain that the 'fly' of divine possession was the fly of the fly-agaric, and this fly has now led me to Soma.

1. These expeditions remain to be written up and I plan to do this in the coming years.
2. R. W. G. Dennis: *Kew Bulletin*, No. 3, 1957 (1958); pp. 392-395.

I do not recall when the Soma possibility first drew my attention: it was certainly after our first book went to press. My wife had never heard of the Soma mystery. My brother Tom and I, we had been told about Soma by our father in the first decade of this century, but in my case the question had lain dormant in the depths of sleeping memory until the '50's. From 1955 on I was in intermittent correspondence with Aldous Huxley, and often when he visited New York he would come down to Wall Street and have lunch with me. Perhaps it was he who revived my interest in this strange historical enigma. One day he and I were discussing the hallucinogens,[1] and I remember tossing out the fanciful suggestion that Soma might prove to be the fly-agaric, and describing to him the red and yellow phases of this remarkable mushroom, and its rôle in Siberia, with which I think he was already acquainted. I knew nothing about Soma at that time, and to aspire to the Soma secret was to be reaching for the moon. When *The Island* appeared some years later, I was surprised to discover that Huxley had set his story in an Indian setting, with Śiva being worshipped and yoga being practiced, and the drug that is the focus of the cult in the story is a yellow mushroom, surely the yellow fly-agaric: Huxley says expressly that it is the yellow mushroom rather than the red one on which gnomes sit! In his story he was coming close to the truth: he possessed the poet's intuition. Later, in 'Culture and the Individual' Huxley explained[2] that he had had in mind the mushrooms that yield psilocybin, the Mexican mushrooms that my wife and I had played a part in rediscovering and making known to the world. . . . I remember that in these same years, in the late

1. At this particular Huxley luncheon Stephan F. de Borhegyi was also present. . . . 'Hallucinogen' and 'hallucinogenic' were words coined by a group of physicians preoccupied with these mysterious drugs – Abram Hoffer, Humphrey Osmond, and John Smythies in America, and Donald Johnson in England. To Johnson goes priority. In 1953 he brought out a pamphlet, *The Hallucinogenic Drugs*, published by Christopher Johnson in London. But he says he picked up the word from the others, who however did not get into print until January 1954, in an article on 'Schizophrenia: A New Approach', in *The Journal of Mental Health*, London, Vol. C. The word quickly took hold and now trips off everyone's tongue as though it had been used for generations. The uninitiated layman is apt to think of the hallucinogens as merely a new kind of alcohol but by devising a fresh name the radical difference is established from the start. For those who know their effects, 'hallucinogens' may seem inadequate but it fits so long as one remembers that the hallucinations affect all the senses and also the emotions.

2. *Vide supra*, p. 146.

'50's, I also discussed the Soma problem with Mr. John P. C. Train, of New York.

After my conversations with Huxley and about the time when *The Island* appeared, in 1962, but before I had read it, in July, I engaged Wendy Doniger to write a *précis* of the Soma question, and she submitted her report on February 16, 1963. It was 33 pages of single-spaced typescript. In it she called my special attention to ṚgVeda IX 74^4, where the priests urinate Soma. This had astonished her and left her nonplussed: little did she suspect what it would mean for me, with my Siberian background. Her report lay dormant for months until I finally retired from my bank at the end of June of that year and translated myself to the Orient for a stay of some years.

2

I shall begin by saying where in Europe's past I have *not* found the cult of a sacred mushroom.

1. Mushrooms do not figure in the various witchcraft epidemics that raged in Europe in the late mediæval and renaissance times. The evidence here is voluminous and circumstantial, extending from Spain to Hungary and from France to Scotland and Sweden. That a rôle for mushrooms is never mentioned seems to me conclusive.

2. I have found no mushrooms in the records that we possess of the shadowy Druids. Our sources are meager. If we knew more, mushrooms might figure, but the evidence now is negative.

3. In Viking times, from the 8th to the 10th centuries, there was a special category of the Viking warriors known as the Berserks. Big powerful men, they fought in the forefront of the battle with a wild fury. They would 'go berserk' and this berserk-raging made them famous and feared. Today the belief prevails in certain Scandinavian circles that this berserk-raging was provoked by the fly-agaric. In Sweden and Norway even text-books and encyclopædias assert this as a fact. Samuel Ödman, a Swede, first propounded [43] the idea in

1784, deriving the notion from the accounts of travellers in Siberia earlier in the century. He cites Georgi [6] and Steller [5], and there is every reason to assume that he knew von Strahlenberg [3] and Krasheninnikov [4] also. A century later, in 1886, Fredrik Christian Schübeler, a Norwegian, expressed the same view. In Exhibits [43] through [46] we give in translation the principal statements in favor of the fly-agaric, for the reader to pass on their merit. The opposition has not been without able advocates, notably Fredrik Grøn, a specialist in the medical history of Norway, and Magnus Olsen, the authority on Norse traditions and literature.

Certain it is, in my opinion, that the fly-agaric was *not* used by the Berserks, and no time should be lost in expunging this yarn from the reference books. My reasons are two-fold.

First. The fly-agaric is never mentioned in the Sagas or Eddas. Of the fungal world only punk or touchwood (*knösk* and dialectal variants thereof) appears in them. No mysterious or unidentified plant plays a part in the berserk-raging. The early historians Saxo Grammaticus and Olaus Magnus made no mention of any such agent. There is no record of an oral tradition antedating the 19th century of such a practice. Ödman said expressly that he was basing his view on Georgi's [6] and Steller's [5] account of the Siberian shamanic usage.

The advocates of the fly-agaric as the cause of berserk-raging are constrained to place excessive weight on a shadowy episode alleged to have taken place in 1814. The story goes that in a brief war between Norway and Sweden the Swedish soldiers of the Värmland regiment were seen by their officer to be seized by a raging madness, foaming at the mouth. On inquiry the officer is said to have learned that the soldiers had eaten of the fly-agaric, to whip up their courage to a fighting pitch. But this episode has not been substantiated. On November 1, 1918, a Swedish physiologist named Carl Th. Mörner read a paper on the higher fungi before a learned society in Upsala. In the oral discussion that followed the paper a meteorologist, H. Hildebrandsson, disclosed for the first time the story of the Värmland regiment. Later, when Mörner published his paper, he cited what

Hildebrandsson had said, and this is our only evidence for what had happened. But Mörner did not seize the opportunity, at that time possibly available, to ascertain Hildebrandsson's source and to confirm his story with additional details. No one has ever heard of the episode other than from Hildebrandsson's account told offhand, in the discussion that followed a lecture, more than a century after the event and half a century ago. Bo Holmstedt, Professor of Toxicology at the Karolinska Institute of Stockholm, has lately made rigorous efforts to verify it in Värmland or elsewhere, without success. We must remember that Professor Hildebrandsson was speaking outside the field of his special competence and was merely contributing to an oral discussion.

Second. The symptoms of fly-agaric inebriation are the opposite of berserk-raging, and the Norwegians and Swedes who imagine that they are the same would do well to read Exhibits [1] through [38]. The effect of the fly-agaric is soothing, comforting, quieting, tranquillizing. At one stage there is a feeling of physical exhilaration, but in our case-histories there is not a single report of wild bellicosity. In an heroic hymn [30] of the Vogul a myth is told: the Hero has consumed three sun-dried fly-agarics and is lying in a stupor. A messenger rushes in, announces the imminent approach from the North of the fearful Mocking-bird Host with the red rump, and calls on the Two-Belted One to go forth and fight. But though the enemy is at the very gate, our Hero says he cannot stir because of his inebriation, and sends off the messenger to seek out his two younger brothers. The peril grows desperate and the messenger, returning, implores the Hero to fight. This time he *throws off* his inebriation, sallies forth, and slays the enemy right and left. The testimony of the myth only confirms what the travelers tell us over and over again. Not one of them describes a syndrome corresponding in the remotest degree to berserk-raging.

4. At the session of the Société Mycologique de France held on October 6, 1910, there was presented to the attendance a photograph of a Romanesque fresco from a disaffected chapel that had belonged

to the Abbaye de Plaincourault in the center of France. It was later the subject of a note published on pp. 31-33, Vol. XXVII, of the *Bulletin* of the Société. The fresco, crude and faded, is of the familiar temptation scene in the Garden of Eden. The gentlemen who presented the fresco to the Société Mycologique made the sensational statement that, instead of the customary Tree, the artist had given us the fly-agaric. A serpent was entwined around a gigantic fly-agaric and was engaged in a colloquy with Eve.

The interpretation put on the fresco by the mycologists has made an impression on their colleagues, particularly in England. Thus John Ramsbottom endorses the fungal message in *Mushrooms & Toadstools*[1] and in *A Handbook of the Larger British Fungi*;[2] also R. T. and F. W. Rolfe, in *The Romance of the Fungus World*;[3] *The Illustrated London News*, Nov. 21, 1953; finally, Frank H. Brightman more recently in *The Oxford Book of Flowerless Plants*, in 1966.[4] My wife and I visited the Plaincourault chapel on August 2, 1952. It is in the Berry, between Ingrandes and Mérigny, facing the Val de l'Anglais, hard by the Château of Plaincourault. The Chapel bears the date 1291 and the fresco must come down from that time or thereabouts. On April 2, 1959, Mme Michelle Bory, of the Muséum National d'Histoire Naturelle, visited the chapel at my request, and made the copy of the fresco that we offer our readers in PLATE XXI.

The mycologists would have done well to consult art historians. Here is an extract from a letter that Erwin Panofsky wrote me in 1952:

> ... the plant in this fresco has nothing whatever to do with mushrooms ... and the similarity with *Amanita muscaria* is purely fortuitous. The Plaincourault fresco is only one example – and, since the style is provincial, a particularly deceptive one – of a conventionalized tree type, prevalent in Romanesque and early Gothic art, which art historians actually refer to as a 'mushroom tree' or in German, *Pilzbaum*. It comes about by the gradual schematization of the impressionistically rendered Italian pine tree in

1. Collins, London, 1953, pl. 1b, p. 34; p. 46.
2. British Museum, London, 1949, p. 26.
3. Chapman & Hall, London, 1925, p. 291.
4. p. 112.

> Roman and Early Christian painting, and there are hundreds of instances exemplifying this development – unknown of course to mycologists. . . . What the mycologists have overlooked is that the mediæval artists hardly ever worked from nature but from classical prototypes which in the course of repeated copying became quite unrecognizable.

Professor Panofsky gave expression to what I have found is the unanimous view of those competent in Romanesque art. For more than half a century the mycologists have refrained from consulting the art world on a matter relating to art. Art historians of course do not read books about mushrooms. Here is a good example of the failure of communications between disciplines.

The misinterpretation both of the Plaincourault fresco and of berserk-raging must be traced to the recent dissemination in Europe of reports of the Siberian use of the fly-agaric. I think the commentators have made an error in timing: the span of the past is longer than they have allowed for, and the events that they seek to confirm took place before recorded history began.

3

Traditionally the European peoples vary like night and day in their attitude toward wild mushrooms. There are two areas that are on excellent terms with them. The northern Slavs and Lithuanians, and the Mediterranean littoral from Majorca and Catalonia to Provence and including apparently the whole of the *langue d'oc* area of France: these are the areas where wild mushrooms are considered friends, where children gather them for fun before they can read and write, where no adult feels the need of a mushroom-manual, where immense quantities of mushrooms are prepared for the table in innumerable ways, and where accidents are unknown. The gentle art of mushroom-knowing is a universal accomplishment. Mushrooms are a conversation piece among men and women. Novelists introduce them into their narratives, poets into their verses; and they recur in proverbs and ditties. Moreover – and here is the telling thing – all the references are friendly, favorable, wholesome.

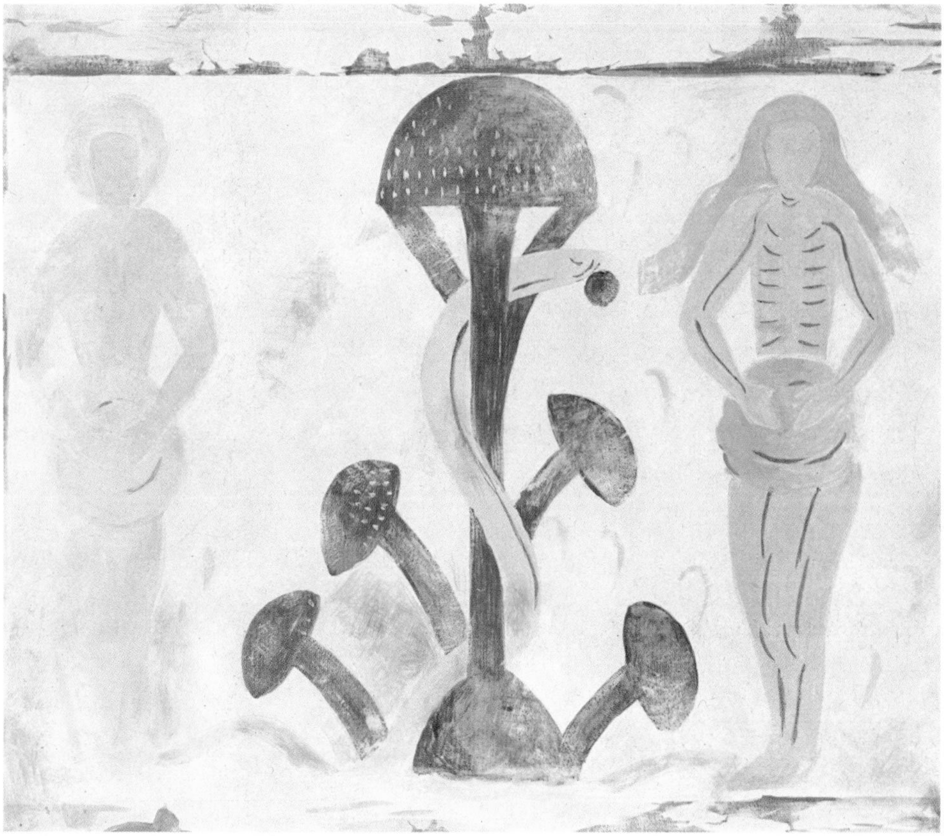

Plate xxi · Fresco of Plaincourault. Abbaye de Plaincourault, Mérigny, Indre, facing the Val de l'Anglais, in the Berry. (Copied April 2, 1959, by Mme Michelle Bory, staff member of the Laboratoire de Cryptogamie, Muséum National d'Histoire Naturelle, Paris)

On the other hand the Germanic and Celtic peoples are infected with a virulent mycophobia, coming down from pre-history. In recent generations there has been some improvement as the traveled and educated classes have begun to spread a different gospel, and as groups of zealous mushroom amateurs have begun to leaven the mass of the population. The criterion by which to judge this aspect of a people's culture is the pronouncements of the older writers and of the untutored country folk. The educated element does not offer a pure strain of the native ways. In *Mushrooms Russia & History* we presented an anthology of many such expressions and I have assembled other quotations since that book appeared. Here I shall cite only a few.

The English people to begin with had no name for a mushroom that would permit them to eat it: 'toadstool' was our chief word and one does not eat a toadstool. The Court under French influence in the 15th century introduced *mousseron*, which became 'mushroom', and from then on the English, or at least a few of the more enlightened ones, could consume one or two species. But even the French were unenthusiastic. The *Grete Herball* of 1526, a translation from the French, voices the hostility of both peoples:

> ... Fungi ben mussherons ... There be two maners of them; one maner is deedly and sleeth them that eateth of them and be called tode stooles, and the other dooth not. They that be not deedly haue a grosse gleymy [slimy] moysture that is dysobedyent to nature and dygestyon, and be peryllous and dredfull to eate & therefore it is good to eschew them.

So mushrooms are of two classes, those that are deadly and those that had best be eschewed. In Diderot's *Encyclopédie* the Enlightenment had not yet spread to mycophagy:

> But whatever dressing one gives to [mushrooms], to whatever sauce our Apiciuses put them, they are really good but to be sent back to the dung heap where they are born.[1]

1. Article on *Champignons* by Louis de Jacourt, 1753: Mais quelqu'apprêt qu'on leur donne, à quelque sauce que nos Apicius les puissent mettre, ils ne sont bons réelement qu'à être renvoyés sur le fumier où ils naissent.

For Keats, in *Endymion* a 'fungous brood' sends up, 'sickly and pale, Chill mushrooms coloured like a corpse's cheek.' Tennyson makes one of his heroines, Lynette, turn away from Gareth as though she smelled 'a foul-flesh'd agaric', deeming it 'carrion of some woodland thing'. Spenser in *The Shepheardes Calendar* identified the 'grieslie Todestool' with winter, and so does Shelley, thus also doing violence to nature, in *The Sensitive Plant*:

> And agarics and fungi, with mildew and mould,
> Started like mist from the wet ground cold;
> Pale, fleshy, as if the decaying dead
> With a spirit of growth had been animated!
>
> Their moss rotted off them, flake by flake,
> Till the thick stalk stuck like a murderer's stake,
> Where rags of loose flesh yet tremble on high,
> Inspecting the winds that wander by.

Montaigne, that giant of the Renaissance, pouring forth the rich contents of his mind and feelings in his Essays, ignores the fungal world. Here is a son of Périgord who never mentions truffles. Rabelais presents to his readers an obnoxious character called Lent-observer who has a *potiron*, mushroom, for a chin, and whose excrement consists of morels and toadstools. There is never a kind word for mushrooms in Rabelais, this native of Chinon in the heart of France. So far as I know, neither Chaucer nor Milton mentioned them, and Shakespere barely.

For the modern poets mushrooms are unchanged. Emily Dickinson repeats the old refrain:

> Had nature any outcast face,
> Could she a son contemn,
> Had nature an Iscariot,
> That mushroom – it is him.

Or D. H. Lawrence in *How Beastly the Bourgeois Is*:

> How beastly the bourgeois is
> especially the male of the species –

Nicely groomed, like a mushroom
standing there so sleek and erect and eyeable –
and like a fungus, living on the remains of bygone life,
sucking his life out of the dead leaves of greater life than his own.

And even so, he's stale, he's been there too long,
Touch him, and you'll find he's all gone inside
just like an old mushroom, all wormy inside, and hollow
under a smooth skin and an upright appearance.

Full of seething, wormy, hollow feelings
rather nasty –
How beastly the bourgeois is!

Standing in their thousands, these appearances, in damp England
what a pity they can't all be kicked over
like sickening toadstools, and left to melt back, swiftly
into the soil of England.

The poets invoke mushrooms only when they seek a loathsome figure of speech. Of course mushrooms decay, but why pick on mushrooms? Everything that lives will rot. The poets never see the infinitely subtle, fresh colouration, quivering with life, of the mushroom world, varying from species to species and from individual to individual; the delicate softness of their texture, their shapes, graceful, grotesque; the aroma of each species different from all others, conveying by its scent its own proper signature.

Wild fungi are an emotional trip-hammer for mycophile and mycophobe alike, and in the poets with their heightened sensibilities the contrast in the response to fungi is sharpest. Professor Roman Jakobson was spending the summer of 1919 in Pushkino, near Moscow, with the poet Vladimir Majakovskij, who would go out almost daily into the forest to look for mushrooms. He would usually return with a large basket-full of them. He knew them all and where to look for every kind. He told his companion that mushroom gathering offered the ideal accompaniment for the composition of poetry, and in the course of that summer he composed the best parts of his epic, *150 Million*, while engaged in this pastime. During the previous season, in

1918, he had conceived his play *Mystery Bouffe* in the woods among the mushrooms.

In the beginning the Germanic world was steeped in darkest mycophobia. Certainly nowhere in European literature is there a more perfect expression of loathing for mushrooms – all the more eloquent because taken for granted – than in the writings of Saxo Grammaticus, the Danish historian who flourished about A.D. 1200. He was telling of a military campaign waged in Sweden by Hadding the Dane in the 10th century, and how the Danes ran out of provisions, and were driven to the last extremities of hunger. Here in Book I: vii: 7 of his *Saxonis Gesta Danorum* we discover the low opinion in which the Danes of olden times held wild mushrooms:

> ... After the spring thaw, Hadding returned to Sweden and there spent five years in warfare. By reason of this lengthy campaign, his soldiers, having consumed all their provisions, were reduced virtually to starvation, and resorted to forest mushrooms to satisfy their hunger. Finally under pressure of extreme necessity they ate their horses, and in the end they satisfied themselves with the carcasses of dogs. Even worse, they did not scruple to eat human limbs.

Now that the passing centuries have dimmed for us the personal sufferings of Hadding's host, we may permit ourselves to be amused by the graduated stages of their desperation as reflected in their diet, and our thoughts turn to what soldiers of other origins would have done in a like pass. Had they been Celts, they would surely have eaten horses, dogs, and each other before turning to the foul fungi of the forest. If they had been Slavs of the North, they would have been feasting on those noble mushrooms from the outset of their long campaign, and, fortified by the delectable fare, would have engaged the enemy like lions, and most certainly turned the tide of war. Until General Bernadotte, a son of Pau in the Pyrenees who became King of Sweden, spoke well of ceps, neither Lapps nor Swedes ate mushrooms. We know this because Linnæus tells us so. In the section on the fungi in his *Flora Lapponica*, he observes that in Sweden only foreigners consider mushrooms fit for eating, nor does he except himself from the general rule. What a pity that the great Linnæus was a mycophobe! It is said that when he was naming the famous *Lactarius deliciosus*, he thought he was naming a Lactarius of the Mediterranean that had been described to him as excellent eating: he thought the specimen before him was the same *because it smelled as though it ought to taste good.*

4

Members of a community observing a tabu are far from sensing that it is a tabu. Their obedience to the tabu is in the natural order of things. It lies along the grain of the wood. As I am writing for the English-speaking world, I fear my readers will put aside my book, saying: 'The poor idiot just doesn't know you may get poisoned from mushrooms. Come, what'll it be, a highball or a Martini?' The breathtaking aspect of it is the unanimity of the witnesses. Those from mycophilic peoples are invariably mycophile; those from mycophobic races are invariably mycophobe. The only exceptions are those who have traveled and read widely, and because they have traveled, at least intellectually, they are not really exceptions. In the 1920's and 1930's this subject was a frequent conversation piece in gatherings frequented by White Russians, but it was only table-talk. My wife and I thought it deserved better than that. Today everyone is aware that deep-seated emotional attitudes acquired in early life are of profound importance. It seems to me that when such traits betoken the attitudes of whole tribes or peoples, and when those traits have remained unaltered throughout recorded history, and expecially when they differ sharply from one people to a neighboring people, then you are face to face with a phenomenon of profound cultural importance, whose primal cause is to be discovered only in the well-springs of cultural history. In this instance we are exploring from the inside (not through cave paintings or mute archæological artifacts) one aspect of the religious life of our ancestors in protohistory.

'Toadstool' is an astonishing folk-word. For centuries it has incapsulated the inspissated loathing and fear of the English-speaking people for wild mushrooms. For the Englishman, as commonly used it means any mushroom he does not know and therefore distrusts, which means all or almost all wild mushrooms. One of the two most important words in the fungal vocabulary of Europe, it has nevertheless lost almost everywhere its application to a particular species. For the sinister mark of the toad is not confined to the English fungal vocabulary. You will find it in Norwegian and Danish, though not in Swedish; in Low German, Dutch, and Frisian; in Breton, Welsh, and Irish. It cannot be translated into standard French, and the other Romance languages know it not. Nor does it survive in standard High German, though it lingers on in High German dialects. Thus the citadel of the 'toadstool' is in the ring of peoples who dwell around the shores of the North and Irish Seas, a gigantic and evil fairy-ring, as it were, embracing the surviving Celts, many of the Germanic peoples, provincial France (where the 'toad' figure may have come down from the Gauls), and the Spanish Basque country of Guipuzcoa and Biscay. The Bretons, let it be remembered, emigrated from Britain to their present home across the Channel in the fifth and sixth centuries after Christ, and are thus remote heirs, folkwise, of old Britain.

Not all of these peoples use the figure of the toad's stool. The Norwegians and Danes speak of the toad's hat; the Low Germans, of the frog's stool; the Dutch say toad's stool; and the Frisians refer to an old fungus as a toad's hide. The Irish term is the frog's pouch; the Welsh, toad's cheese; the Bretons, toad's cap, but by the addition of a single initial sibilant, their term becomes toad's stool, and this is a recognized variant in their language. Here are the words in these tongues: in Norwegian and Danish, *paddehat*; in Low German, *poggenstohl*; in Dutch, *paddestoel*; in Frisian, *poddehûd*; in Irish, *bolg losgainn*, with *bolg* meaning pouch; in Welsh, *caws llyffant*, with *caws* meaning cheese; in Breton, *kabell tousec*, and also *skabell tousec*. The Pennsylvania Dutch speak a dialect of High German that comes down from the language of the Palatinate in the 18th century, and in Pennsylvania Dutch we find both toad's stool and toad's foot: *grottestuhl* and *grottefuss*. We know that toad's bread, *pain de crapault*, was used for wild fungi in 16th century France, and this same expression has been reported in modern times in the Calvados region of Normandy. All these names hinging on the toad seem indefinite in their application, and all of them are pejorative. But there are two contiguous or almost contiguous areas that give the term specific meaning. The fly-agaric is called *crapaudin* in many parts of France and in the form *grapaoudin* this word has been reported as far south as the Hérault, on the Mediterranean. In the Basque of Guipuzcoa and Biscay the fly-agaric is the *amoroto*, the precise equivalent of *crapaudin*, 'the toad-like thing'. Often the peripheral cultures of the world preserve archaic traits and meanings better than the throughways of trade and communication, and when we find that the toad is linked to the fly-agaric in Basque and in French provincial usage, our attention is alerted.

All these words, in varying degrees, exhale a bad odor. They designate wild fungi that the speaker considers, rightly or wrongly, inedible and dangerous. The English toadstool, freighted with evil, is typical of the class. In the dialects of England there are numerous variants, and these are interesting because they echo the figures of speech that are current in our list of foreign words. Thus we find toadcheese or taddecheese, toad's bread, toad's cap or toadskep, and toad's meat. For the toad itself there is an ancient variant, pad or paddock, which gives us paddock-stool and puddock-stool. This 'pad' is the same word for toad that the Dutch and Frisians, the Norwegians and Danes, use. This is the witches' word in the opening scene of *Macbeth*:

> Padock calls anon: faire is foule, and foule is faire,
> Hover through the fogge and filthie ayre.

Here in our argument we interrupt its course for a necessary diversion.

Today civilized men have a kindly feeling for the toad. Lewis Carroll and Kenneth Grahame have planted the seeds of their benign influence in the minds of successive generations of well brought up English-speaking children. The Victorians were inclined to foster sympathy for the whole animal world. (Was this because the industrial revolution released increasing numbers of men from slavery to the soil, from intimate conflict with cantankerous nature?) As for the toad, there has been an additional influence: men of science have undertaken to show that it is the farmer's friend.

Far different was the repute of the toad in times past. There was no other member of the animal kingdom that inspired such revulsion and fear. Chaucer spoke of the 'foule tode', and Spenser of the loathly and venomous toad. 'A pad in the straw' was what our ancestors said when they meant 'a nigger in the woodpile'. (Now that this last phrase is banned in polite society and perhaps vanishing, why not revive the earlier expression?) Shakespere reveled in the toad as a potent term of abuse. In *Richard III* the toad is a recurring theme, as is fitting for a play about a king described as:

> That bottel'd Spider, that foule bunch back'd Toad.

Among all of Shakespere's many references to the toad, there is not one that is neutral, much less friendly. Edgar in *King Lear* denounces Edmund as 'a most toad-spotted traitor'; and the witches in *Macbeth*, when they concoct their hellish brew, give to the toad pride of place in the cauldron:

> Toad, that under cold stone
> Days and nights hast thirty-one
> Swelter'd venom sleeping got,
> Boil thou first i' the charmed pot.

Not only were toads venomous; to the mediæval mind they were also a symbol of lechery, as were warts and moles, with which toads were supposed to be covered.

The evil repute of the toad is not yet dead. There are English circles where 'Toad!' flung in anger would be a fighting insult now. The derivative 'toady' brings to mind the sycophantic and hypocritical squat of the creature, with its upturned watchful eyes. The bad name of the toad survives among untutored countryfolk in England and the United States, where farmers cling to the belief that the spittle of toads is poisonous, and that warts will grow on the skin where a toad has touched. French peasants down to recent times, and perhaps even now, put toads to death by methods shocking for their cruelty, methods that reveal an ingenuity in torture ordinarily reserved by man for his fellow-man.[1]

The unpleasant abuse heaped on the toad, as well as the serpent, seems to have been a fruit of Christianity. In Old French *le bot* was a name for Satan, resorted to as an evasive term, a word derived from a Germanic root, meaning the club-footed one, or the splay-footed, or the limping one. (Among Satan's traditional attributes was a bad foot causing him to limp.) That same word *bot* was a designation also for the toad and the toadstool, constituting thus a sinister trinity linked together in verbal identity. In the Carpathians and the Ukraine the toad theme recurs in the mushroom vocabulary in conjunction with the 'mad-mushroom', which as we shall see can be traced to the fly-agaric, *povχ*, *pavχ*, of the Ostyak. Surprisingly, in China the common name for the fly-agaric is 蛤蟆菌, *ha-ma chün*, 'toad-mushroom';[2] in that country suffering from deforestation the fly-agaric today is found chiefly in Manchuria, along the Amur River. But the foul reputation of the toad in Western Europe is absent, significantly, in Russia and China.

One asks oneself why the early Churchmen in the West convicted the toad of heinous crimes. Was it because the toad occupied an honoured place in the Pagan pantheon? So it seems. In a remarkable paper[3] Marija Gimbutas has shown how paganism lingered on in Lithuania long after it had disappeared elsewhere, until the last century, and one can study the surviving practices there. The Lithuanian peasants in the conservative areas continued to regard toads and snakes gently, encouraging them to live in their homes and considering their welcome presence a happy augury. Only a hundred years ago these peasants were still making wooden grave markers carved in the shape of stylized birds and of toad's hind legs (Fig. 6). Professor Gimbutas shows her readers prehistoric pottery from central Europe with the toad motif incised on the clay (Fig. 7). The toad

1. *Vide Le Folklore de France*, by Paul Sébillot, vol. III, *La Faune et la Flore*, Paris, 1906; pp. 280 ff.
2. 劉波 Liu Po: 蘑菇及其栽培 Mo-ku chí ch'i tsai-p'ei [Mushrooms and Their Cultivation], 科學出版社 K'o-hsüeh Ch'u-pan Shê [Scientific Publication Association], Peking, 1964; pp. 21, 88.
3. *Ancient Symbolism in Lithuanian Folk Art*, Memoirs of the American Folklore Society, Vol. 49, 1958.

seems to have been a beneficent deity, a chthonic spirit compact with earth force and sexuality; the snake likewise. But Christianity changed all that, and not for the better.

How strange it is that the most spectacular, the most potent, mushroom lacks a name in the English language. A people priding ourselves on our love of nature has not bestowed a name on this regal plant

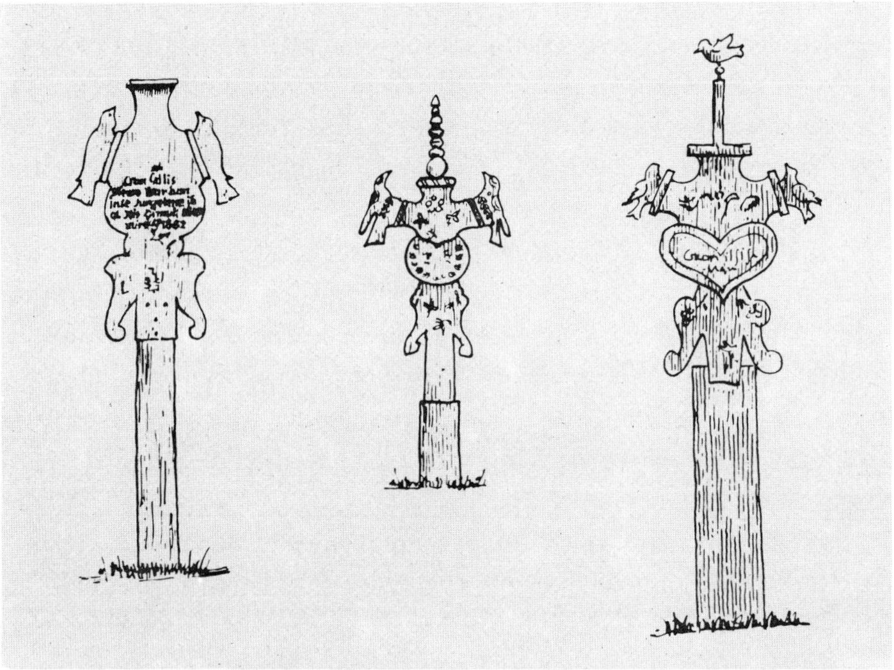

FIG. 6. Wooden grave markers carved in bird and toad's leg motifs from the 19th century cemeteries in Lithuania Minor. (After Marija Gimbutas: *Ancient Symbolism in Lithuanian Folk Art*, p. 32, Mem. of Amer. Folklore Society, Philadelphia, Vol. 49, 1958).

bedecking our woods in the fall of the year. (For 'fly-agaric' of course is not a name. I use it in this book as a term of convenience. It has no circulation among the genuine country-folk of the English-speaking world. A post-Linnæan invention, 'fly-agaric' has led its exsanguinated existence mostly between the pages of pallid mushroom manuals.) On the other hand, the most important fungal word in English, 'toadstool', has no specific meaning; though on the Continent, in those regions where it retains a specific association, the link with the fly-agaric is unmistakable. Our earliest citations for the word are of course after the Christian fathers had introduced writing into northern Europe, and already it had lost its moorings. I suggest that the 'toadstool' was originally the fly-agaric in the Celtic world; that the 'toadstool' in its shamanic rôle had aroused such awe and fear and adoration that it came under a powerful tabu, perhaps like the Vogul tabu where the shamans and their apprentices alone could eat of it[1] and others did so only under pain of death; that people hesitated to pronounce the very name of this mushroom, so that in time it became nameless and the name it formerly carried hovered thereafter ambi-

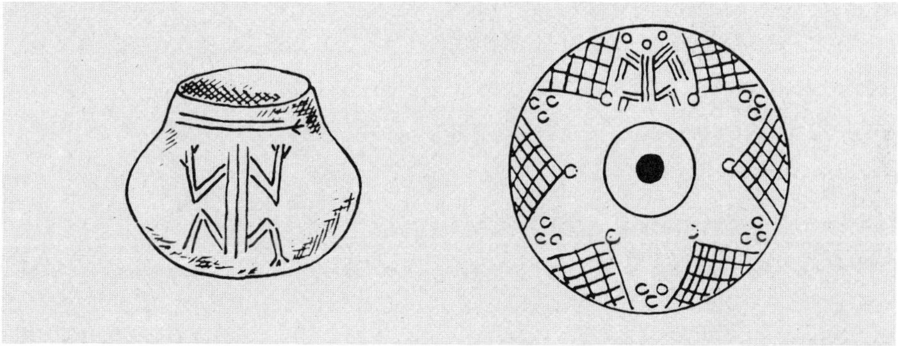

FIG. 7. Toad motif on prehistoric pottery. Left: Neolithic pot with incised toad form. Second half of 3rd millennium B. C. Danubian I culture in Czechoslovakia. Right: Figure of toad on bottom of Early Iron Age pot from Central Germany. (After Marija Gimbutas: *Ancient Symbolism in Lithuanian Folk Art*, p. 35, Mem. of Amer. Folklore Society, Philadelphia, Vol. 49, 1958).

guously over the whole fungal tribe so that all the mushroom world fell under the same floating tabu. This tabu was a pagan injunction belonging to the Celtic world. The shamanic use of the fly-agaric disappeared in time, perhaps long before the Christian dispensation. But in any case the fly-agaric could expect no quarter from the missionaries, for whom toad and toadstool were alike the Enemy. (We remind the reader of St. Augustine's censure of mushroom-eating by the Manichæans supported much later by the excoriations of St. François de Sales and of Jeremy Taylor, p. 71.) Today we are dealing with a deep-seated emotional attitude born in a tabu long forgotten,

1. *Vide supra*, pp. 152-153; also Kai Donner [28] p. 286, and Lehtisalo [24] p. 280.

a tabu on a Sacred Element, the fly-agaric, a tabu overlaid by and mixed up with the venom of the Christian Church's anathema.

The truly lethal mushrooms – *Amanita phalloides, A. verna, A. virosa* – have played virtually no role in Eurasian history. They are of importance only to the rare individual who eats one of them and dies from it, and to his kin. (They have occasionally served the assassin's purpose.) In many of the languages of Europe the peasants have no name for them: they pass them by and ignore them. But everyone knows the notorious reputation of the fly-agaric and shudders at the thought of eating it. Thus in our own day a sanction having its origin in a purely religious tabu thousands of years old is better known and more effective than the lethal properties of the deadly species.

5

In the spring and again in the summer of 1967 I visited Dr. János Gulya, in Budapest, to consult with him in the field where he is the master, the Finno-Ugrian languages, and especially the Ugrian cluster – Magyar, Ostyak, and Vogul. Out of his knowledge of these, to us, remote languages he drew to my attention a usage in Ostyak that may have relevance for our inquiry into the fly-agaric. The word *tul-pavχ*[1] occurs in two Heroic Songs, in one as part of the hero's name, and in the second, repeatedly in the course of the narrative. In both cases the word signifies the fly-agaric, '*tul-*' meaning 'fool' and 'foolish'; *pavχ* means of course 'fly-agaric'. This serves to nail down the meaning of expressions that have circulated in past centuries in many parts of Europe, but whose specific sense has long been up in the air.

In Magyar there is a phrase, a conversational cliché, *bolond gomba*, 'fool-mushroom', circulating especially in rural areas, as when one asks of a person behaving foolishly, 'Have you eaten of the fool's mushroom?', or when one rejects a proposition by saying, 'Do you think I have eaten a fool-mushroom, that I should do such a thing?' Or

1. The citations are in *Osztják Hősénekek* [Ostyak Heroic Songs], Reguly A. és Pápay J. hagyatéka [*The Legacy of Anton Reguly and J. Pápay*]. 1. kötet. [Vol. 1.] Reguly Könyvtár 1. [Reguly Library 1.] Edited by Miklós Zsirai, Budapest, 1944. The first citation is from the first heroic song, pp. 2-165, on pp. 2-3. The second citation is from the second song, pp. 392, 398, 400. The first song was from Obdorsk, now Salehard. The second was from the northern Sosva river basin.

again, 'He is laughing as though he had eaten fool's mushrooms.' In Hungary the 'wise-woman', *javas asszony*, is said to use this same mushroom in love philtres, and the angry lover sends the philtre on to the object of his passion. At our request considerable effort has been made in Hungary to find out whether peasants in any region of that country identify the *bolond gomba* with a particular species, but with no success. This is why Dr. Gulya's discovery of a linguistic link with the Ob-Ugrian peoples is gratifying. The link bridges the centuries and securely fixes a knowledge in former times of the properties of the fly-agaric in the Pannonian Plain among the Hungarian people.

The fly-agaric's peculiar virtue was widely known. In Yugoslavia one still says, *Najeo se ljutih gljiva*, 'He has eaten enough of the fool-mushroom'. In Vienna one may hear, *Er hat verrückte Schwammerl gegessen*, 'He has eaten the mad mushrooms,' and all Germans recognize the meaning of *Narrenschwamm*, 'fool-mushroom'. John Parkinson in his *Theatricum Botanicum* (1640) speaks of the 'foolish mushroom', but he is leaning on his Continental sources. The 17th century Polish poet Wacław Potocki warns his readers in *The Unweeded Garden* against a kind of mushroom called *szmer*, lest it render the reader foolish (*szaleć*), 'as from opium'. He was clearly passing on hearsay, since no mushroom could be confused with opium. A Slovak informant from near the Tatra Mountains tells us that rejected mushrooms are variously called by his people *žabací huby*, toad-mushrooms, *hadáci huby* and *hadúnke huby*, both meaning viper-mushrooms and *šalené huby*, mad-mushrooms. The toad and serpent, prehistoric deities chased away by Christianity, are thus associated with the inebriating mushrooms. To the east of Slovakia, in the Ukraine, the natives today call any wild mushroom that they reject *zhabjachyj hryb* – the toad-like mushroom.

The Hungarians invaded Europe and settled in the Pannonian plain late in history, at the end of the 9th century. Their 'mad mushroom', *bolond gomba*, establishes a link with the fly-agaric of the Ob-Ugrians, but not with European pre-history. Whether the verbal traces of the 'mad-mushroom' that we have found are all derived from Ugrian sources, *via* the Magyar peoples, we cannot say. It seems probable that a knowledge of the properties of this extraordinary fungus survived independently. How otherwise are we to explain the name of the fly-agaric in Catalan – *oriol foll,* the 'mad-oriole', the 'oriole' being *Amanita cæsarea* and the fly-agaric thus being the 'fool's *A. cæsarea*'; or the *mujolo folo* around Toulouse, or the *coucourlo fouolo* in the Aveyron, or the *ovolo matto* in the Trentino, all meaning 'mad mushroom'? In the dialect of Fribourg, Switzerland, the fly-agaric is the 'devil's hat', *tsapì de diâblhou.* Eugène Rolland reports *bò* as a designation for any gilled fungus in the Haute Saône and *botet* in the Loire, both presumably meaning originally the Satanic mushroom, by way of *bot,* 'toad' and Satan in Old French. In the Aude two words designate the fly-agaric, *mijoulo folho,* the 'mad-mushroom', and *tuo-mouscos,* the 'fly-killer'; they are synonyms.

6

This discussion of the fungal vocabulary of Europe has finally brought us to the second of the two important mushroomic words of the European languages – *Fliegenpilz* (or *Fliegenschwamm* in the older tongue), the 'fly mushroom' of the Germans and corresponding words in all the other Germanic languages, except English in which I find no trace of the 'fly' figure. The word in the Germanic languages is simply 'fly-mushroom'. In Russian and French the mushroom is popularly called the 'fly-killer'. In Russian *mukhomor* is the only name for this conspicuous and decorative mushroom. In French, as we have seen, *tue-mouche* competes with the more important *fausse oronge,* the 'false Caesar's amanita', and also with the *crapaudin* of the provinces.

In *Mushrooms Russia & History* we gave what for our time was a new explanation of the German 'fly-mushroom': the fly, in our opinion, was certainly the fly of madness, of divine possession. The association of madness with insect activity, and particularly with the fly, *sensu lato,* is exceedingly old and can be documented throughout Eurasia. In the course of our Siberian readings we found Jochelson [21, p. 267] reporting in the Chukotka that if the eater of the fly-agaric vomits, the people believe that the spirits of the demonic mushroom can be found in the vomit as 'worms', which then quickly vanish into the earth. He was speaking of the Koryak. Bogoraz in turn says of the Chukchi [22, p. 277] that the shamans think psychic disturbances are caused by insects, and they use insects in their treatment of mental disorders. The shaman catches an insect from off his drum, swallows it, spits it up, and applies it to the head of the victim. He then sucks the sick man's head. Sometimes the insect is imaginary, sometimes real. This widespread notion that insects cause madness was based, as I will venture to suggest, on a natural phenomenon: the herdsmen saw their reindeer and sheep driven into a frenzy when insects lodged in their nostrils and procreated there, and the animals did not recover until the new generation emerged from the nose.

Bugs, flies, moths, all kinds of larvæ – in short, the insect world – constituted for our ancestors until recent times an order of nature instinct with supernatural powers, mostly malevolent and always awesome. Their strange shapes and stranger behaviour, their incredible numbers and countless kinds, perhaps most of all their undeniable faculty of metamorphosis, may be at the root of this rôle that they played in the thinking of untutored mankind. The fly was divinely possessed and so was the *Fliegenpilz*. Already in the Old Testament the neighbors of the Israelites worshipped Beelzebub, whose name meant the Lord of Flies. In the Greek New Testament, where that heathen god does not appear, the same name was used as a synonym for the Prince of Demons. The Biblical term crops out in modern literature, as in line 1334 of Goethe's *Faust*: *Wenn man euch Fliegengott, Verderber, Lügner heißt* . . . 'When one calls you Fly-god, Destroyer, Liar . . .' In Nordic mythology the god Loki assumes the appearance of a 'fly' to enter the tightly closed apartment of the sleeping goddess Freya. He pricks her, and when she starts, deftly detaches her necklace and steals it. Whatever that 'fly' was, no one thought of it as a housefly, for the housefly does not bite.

In the Middle Ages and the Renaissance it was the convention to represent demons in the shape of flies. Hieronymus Bosch, the greatest of Europe's painters of the demonic world, presents us with a superb illustration of flies in their demonic rôle. We find it in the left hand panel of the Hay Wain, where the momentous events that took place in the Garden of Eden leading up to Man's First Fall are portrayed with moving beauty. We reproduce the upper portion of that panel in Plate XXII, wherein the observer sees that the angels expelled and tumbling down from Heaven assume the form of 'flies' – a diverse lot of winged demons, Beelzebub's host. In recent times the poet echoes this theme skeptically when, a few years ago, he uttered in *The Times Literary Supplement* the following sprightly lines:

Has Freud not hit the Devil on the snout?
  Is not Beelzebub destroyed by flit?
Are we important? Do we really sprout
  Immortal souls that priests may manumit?

H. S. Mackintosh, issue of Oct. 2, 1953.

Sometimes accidents of sound led to confusions in names. In English 'flea' and 'fly' are words of different origins, but the phonetic similarity caused them sometimes to be used one for the other. When in *Henry V* Falstaff lay dying, he saw a 'flea' stick upon the toper Bardolph's flaming nose, whereupon he said it was a 'blacke Soule burning in Hell'. What he saw was of course a fly, and the black soul in hell was, according to the beliefs of that day, its incorporeal counterpart, the Demon domiciled in every fly. Though in English the fly is not linked with the fly-agaric, the supernatural associations of the word survive. The Oxford Dictionary reports that 'fly' has always been a designation for a 'familiar spirit', and one still hears occasionally 'fly' as an adjective meaning nimble, dextrous, sharp, with the suggestion of an uncanny faculty in those directions.

One of the most interesting examples of the Satanic fly in European literature is to be found in that classic of the Danish stage, Ludvig Holberg's *Jeppe of the Hill*, first produced in 1722. Two physicians are conversing. One of them speaks of his Lordship, who has had a strange, ugly dream, which so excited him that he imagines himself a peasant. Whereupon the other physician recalls a remarkable case ten years back: '... a man who thought his head was full of flies. He could not rid himself of the delusion, until a most clever doctor cured him in the following

PLATE XXII · Hieronymus Bosch: the Hay-wain.
Upper part of left panel of triptych.
(*Courtesy of Prado Museum, Madrid*)

manner. He covered his patient's whole head with a plaster in which he had embedded masses of dead flies. [Query: were they the Spanish flies of the Pharmacopœia?] After a while he removed the plaster and showed the flies to the patient, who naturally believed that they had been drawn from his own head and therefore concluded that he was cured.' Here is a beautiful instance of the way a dramatist (or physician) puts to use an outworn belief that lingers on in the penumbra of man's consciousness: an outworn belief, but one that has survived until only yesterday, as we have just seen, in the Chukchi country.

In the Middle Ages delirium, drunkenness, and insanity were attributed to insects that were loose inside the head of the victim. This belief, strange for the modern mind, survives in many familiar locutions. A man has a bee in his bonnet, a fly (or bug) in his ear, or demonic bats (= 'bots') in his belfry. The Norwegians get flies into their heads or put flies into others' heads. To 'put a bee on someone' means to fix him willy nilly for a given purpose: in this locution the demonic intent is scarcely fossilized. It used to be said in French, when a man was becoming angry, that *la mouche lui monte à la tête,* a fly is climbing up into his head. Down to recent times *avaler les mouches* was a phrase for saying that someone had summoned up his courage, and the flies thus swallowed were of course demonic. Rabelais at the very end of Book IV makes the coward Panurge protest that, far from being afraid, he is braver than if he had eaten all the flies cooked in the pastries of Paris betwixt St. John's Day and All Saints'. There is a colloquial expression that circulates around Lyons and perhaps elsewhere: *Ne prends pas la mouche*: don't catch a fly, don't get excited. Of a man who is unbalanced one says, *Il a l'araignée dans le plafond,* he has a spider in his ceiling, *i.e.*, in his upper story. The Czechs use an identical phrase: *míti mouchu* (or *pavouka*) *na mozku,* to have a fly (or spider) on the brain. The Italian is apt to say, *Gli è saltata la mosca al naso,* the fly jumped to his nose, that is to say, he became upset. When the Dutch say, *Hij ziet ze vliegen,* he sees them flying, he has bats in the belfry, does not the turn of phrase refer to the 'flies' that he sees? In Russian they say of a man who is tipsy: *on c mukhoj,* so-and-so is 'with fly'. There is a gesture peculiar to the

Russians that we feel sure stems from this association of psychic possession with flying insects. The Russian when he suggests having an alcoholic drink is like as not to perform a fillip against his neck below the ear: this is so habitual that it is become a Pavlov reflex. Does not that familiar fillip invoke the demonic 'fly' entering the mind? According to the Icelandic-English lexicon of Cleasby-Vigfusson, under *fluga*, in Icelandic belief sorcerers would bewitch flies and send them to kill their enemies. This explains the modern Icelandic phrases, 'to swallow the fly' or 'to carry the fly', meaning 'to be the tool of another man in a wicked business'. In Basque folklore, sorcerers and other malevolent beings would work their will through demons that assumed the shape of flies, and sorcerers would carry the demonic creatures in a sheath such as anglers use for their flies today.

7

Everyone who knows the first thing about wild mushrooms knows why the fly-agaric is so called: it kills the flies that feed on it and until modern times it was used as a household insecticide on the Continent. This is what all the books say. What is more, there is a large part of Europe where the untutored rustics, the people who read no books, also accept the story as part of their legacy of folk knowledge. It belongs to that curious fund of 'facts' that people keep repeating to each other and believing, without verification or analysis, like the saying that all Russians are good linguists. The area of Europe where our folk belief prevails is extensive but not all-inclusive. It embraces the Slavic world, the Germanic world except the British Isles, the Vosges, where Franco-German bilingualism prevails, and one or two enclaves elsewhere in France. The ancient authors, though they have much to say about the fungi, never refer to a fly-killer, and in modern Italy among the country folk we believe the association with the fly is unknown, but our inquiry has not been exhaustive. It is unknown among the Basques, and apparently to all the rural population of the Iberian peninsula. If we judge by Eugène Rolland's evidence in his *Flore Populaire*, in France the name *tue-mouche*, 'kill-fly', is indigenous only in Alsace and the Aude, though thanks to the mushroom manuals, it is now familiar to a sprinkling of educated Frenchmen elsewhere. (Rolland reports that at Val-d'Ajol in the Vosges the natives hang the fly-agaric from the ceiling, where, he says, it draws flies to their death.) By contrast the German *Fliegenpilz* or 'fly-fungus' has its variants in all the German dialects, and also in Dutch and the Germanic languages of Scandinavia, but not in English.

Albertus Magnus, the Universal Doctor of the Church, supplies us with our earliest citations[1] attributing insecticidal properties to the fly-agaric; his writings date from the 13th century. Other learned clerks follow suit down the centuries. They differ in details disconcertingly. The distinguished Carolus Clusius, for example, whom some consider the father of mycology, describes the fly-killing fungi as a genus having five or six species, of which one is the fly-agaric.[2] Clusius was a widely travelled man and he places the use of the fly-agaric for fly-killing only in Frankfort-on-the-Main. The great Linnæus in *Flora Svecica* repeated the statement about killing flies, but instead of Frankfort he said it was a custom in Smolandia, a Swedish province where he had spent his childhood. This is not the only reference to the insecticidal agaric in Linnæus. In his famous *Skånske Resa* ('Journey through Scania') published in 1751, on page 430, he tells how a certain Swede in Upsala got rid of bedbugs from two of his rooms by the use of *Amanita muscaria*. He describes the remedy and then concludes with what we consider a most significant statement:

> One takes in the autumn fresh specimens of the fly amanita, pounds them with a pestle quite small in a jar, lets them stand well closed until they become slimy or like gruel. Then one takes a feather or brush and smears all the cracks and corners where they (the bedbugs) keep themselves, and this procedure is repeated several times at monthly intervals. The room stinks for two or three days, but then the smell disappears. These nasty creatures die of it as if the plague had come amidst them, and whole bug-families perish as if from the Black Death. Although this remedy is simple, it is surer than anything else hitherto invented, and with its aid several houses in Upsala have now become free of bugs.

1. *De Vegetabilibus*, Book II, Chap. 6:87; Book VI, Chap. 7:345.
2. *Rariorum Plantarum Historia*, 1601, Genus XII of the Pernicious Mushrooms.

In a note on the next page Linnæus adds that he has learned of this method for the first time from a Mr. Bern, Cashier of Interests in Upsala. The text is in Swedish.

For us the striking thing about this description is that it is a report on a novelty. Certain families of Upsala were putting the popular reputation of the fly-agaric to a test, and their first impression was enthusiastic. Why had they not been using it for centuries? What were their final conclusions, say five years later? The answers to these questions are not vouchsafed to us. Was not Europe in these last centuries finally reaching a standard of living that made people more impatient with insect pests, and as the older and original meaning of *Fliegenpilz* was being forgotten, were they not now misinterpreting the name, and experimenting here and there to try out the mushroom's insecticidal virtues, only to discover that they were of no practical value?

After Linnæus the references to the fly killing potency of the fly-agaric in mushroomic writings are innumerable. All the mycologists believe in it – with one dissenting voice. None puts it to a test – with one exception. The French mycologist Jean Baptiste Bulliard, in his *Histoire des Plantes Vénéneuses et Suspectes de la France*, which he finished in 1779, dares to strike a sceptical note. Speaking of the fly-agaric, he says:

> I have never noticed that it kills flies, as several authors assert. I have had specimens, raw and cooked, for long periods in my apartment. Flies light on them, and seem even to eat them, without bad effects. But I intend to repeat this experiment with certain new precautions.

We know not the results of the promised experiments, but we know that when Bulliard died in 1793, he held the view that a new scientific name should be bestowed on *Amanita muscaria*, and he suggested '*Agaricus pseudo-aurantiacus*', presumably because he considered the old name false. The mycological world has not deferred to his wishes, nor until the last few years even tested his premises. In the last fifteen years steps have been taken to remedy this. As long ago as 1953, on our suggestion, the mycologist F. E. Eckblad of the Botanical

Museum of Oslo, working under Professor Rolf Nordhagen, performed experiments leading him to the tentative conclusion that there was no evidence of insecticidal virtue in the fly-agaric. More recently Professor Roger Heim has supervised experiments by some of his assistants seeking the answer to the problems that we have raised,[1] and work has been in progress also in Switzerland, Japan, and England. It seems that the complex chemical make-up of the fly-agaric varies, perhaps materially so far as the fly-killing and the inebriating constituents are concerned, according to place and season of growth.

Albertus Magnus was relying on hearsay, and so indeed was Linnæus when he quoted a Mr. Bern about bedbugs. Most writers have placed the fly-killing at a distance from themselves in time or place, Clusius in Frankfort, Linnæus in Smolandia, John Ramsbottom in Poland, Bohemia, and Rumania. Mr. Ramsbottom[2] also says that the fly-agaric was 'formerly' used in England and Sweden for killing bedbugs, as though it had been an habitual practice in those places, thus justifying the occasional use of the name 'bug agaric'. But Linnæus was reporting a novelty in Sweden, and 'bug agaric' in English cannot be old, since Linnæus by his great authority imposed the name

1. The Eckblad results are in a letter that he wrote me dated Sept. 8, 1953, and that we summarized in *Mushrooms Russia & History*, p. 213.

More recently, the *Revue de Mycologie*, Roger Heim editor, has carried the following articles under the general heading of '*Un Problème à éclaircir: celui de la Tue-mouche*':

Tome xxx (1965), Fasc. 1-2, July 15, 1965:

L'Amanite tue-mouche, bien ou mal nommée? by Gabrielle Bazanté, pp. 116-121.

Etude de l'action de l'*Amanita muscaria* sur les mouches, by Monique Locquin, pp. 122-123.

Tome xxx (1965), Fasc. 4, 1966:

L'Amanite tue-mouche nord-américaine n'est pas la *Muscaria*, by Roger Heim, pp. 294-298. This is a *mise au point* of the various forms of *A. muscaria* that have been reported in the journals.

Tome xxxi, Fasc. 3, Nov. 30, 1966:

Continuation of Gabrielle Bazanté's paper, pp. 261-268.

Continuation of Monique Locquin-Linard's paper, pp. 269-276.

Tome xxxii (1967), Fasc. 5, July 1968:

Continuation of Monique Locquin-Linard's paper, pp. 428-437.

For the papers of the Eugster-Waser team in Zürich, the Bowden team in England, and the Takemoto team in Japan, *vide* bibliographies that accompany the papers by Conrad H. Eugster and Peter G. Waser published in *Ethnopharmacologic Search for Psychoactive Drugs*, a volume edited by Daniel H. Efron, Bo Holmstedt, and Nathan S. Kline, 1967, Public Health Service Publication No. 1645; being the 'Proceedings of a Symposium held in San Francisco, Calif., January 28-30, 1967.'

2. *Vide* his *Poisonous Fungi*, Penguin Books Ltd., 1945, p. 21.

'agaric' on the gilled fungi and 'bug' for 'bedbug' is a modernism in the English language. Only Bulliard put the fly-agaric to the test, and his findings were negative. But in support of Bulliard's scepticism we can add a story told to us by a Russian friend, Ekaterina Apollinarievna Bouteneff. Her nurse in childhood was an unlettered peasant woman from the region of Riazan. Our friend remembers having seen this good woman time and again put out a saucer with a crushed cap of the fly-agaric in it, a lump of sugar on top of the fungal mess. This was going to kill the flies, she would always say. But our friend always observed that the flies did not die. When she would ask her nurse why they did not die, the reply was always the same: 'They are sure to die later.'

In the past ten years much work has been done on the chemistry and pharmacology of the fly-agaric by Drs. Conrad H. Eugster and Peter G. Waser of the University of Zürich, Switzerland, and by Dr. T. Takemoto and his colleagues at Sendai University in Japan. An acid has been isolated from *Amanita muscaria* which the Japanese named 'ibotenic', and it possesses slight insecticidal properties. Under favorable conditions flies, imbibing the juice of the freshly cut specimens, fall into a stupor. In the stupor they are apt to succumb to their enemies, but if not, they recover after some hours or even days.

8

Twenty-five years ago we had gathered, Valentina Pavlovna and I, much of the information that has been offered to the reader in this chapter. 'Toadstool' and *Fliegenpilz* were folk-words coming down to us from our remotest ancestors, for whom they were freighted with supernatural meaning. To amplify our information concerning them we set out to explore the primitive cultures of Siberia, with the results that now lie before the reader. This in turn led me to work up and here present an interpretation of certain aspects of the ṚgVeda, and of the Soma that lies at the core of the ṚgVeda. Out of the depths of Asia we gain the vast perspective of an ancient cult, now finally disappearing, and what it must have meant for man's imagination and emotional life in the pre-literate phases of his past. How strange and stirring that the Soma of the Aryans should be linked to the 'toadstool' of our day in subtle ways that no one has suspected until now!

The primordial inebriant of northern Eurasia was the fly-agaric. This was the divine inebriant that inspired the astonishing lyrics of Maṇḍala IX of the ṚgVeda, and the Heroic Hymns of the Vogul in our own time. Through a different set of circumstances the same inebriant was responsible for laying a blight on the mushroom world throughout the English-speaking world. I suppose that the 'toad' figure of the 'toadstool' was peculiarly the property of the Celts, and the tabu must have been enforced by singularly effective religious sanctions. The high reverence that must have accompanied the tabu changed to intense revulsion when the divine inebriant became under the Christian dispensation demonic possession. These are only speculative guesses, of course, but the strength of the tabu even today is impressive. In the English-speaking world there are victims of allergy who are peculiarly sensitive to mushrooms. It is even said that among mycologists there are those who, poor souls, must refrain from eating the objects of their study, for fear of reactions. My wife, who was a physician specializing in allergies and with a large Russian practice, had never heard of a Slav who complained of sensitivity to mushrooms. There have been many distinguished mycologists in the English-speaking world, but would not incomparably more talent have flowed into this field if 'toadstools' had not been of its theme?

The attitude of the Germanic world has been somewhat different from the English. Whether Albertus Magnus believed the story of the fly-killing mushroom we cannot say. As time has gone on and the world has become increasingly fly-conscious it has become easier to believe that the fly-agaric is so named because it kills flies. Perhaps this alternative explanation served the Church's purpose, diverting attention from the awesome truth. Certainly in recent generations, when with increasing education the fly has come to be considered a pest, this watered-down meaning of the name has completely won the day for mothers, governesses, nursery maids (now called baby-sitters), and children.

There is a further distinction to be drawn between the *Fliegenpilz* and the 'toadstool'. No one eats the fly-agaric in Germany, of course, but the attitude toward it is not unfriendly. The red mushroom with white spots appears frequently on greeting cards to convey good wishes and a seasonal message of happy augury, precisely as the *ling chih* is used in China and throughout the Chinese orbit.

A noteworthy thing: this attribute of happy augury belonging to the *Fliegenpilz* is regarded by chimney-sweeps as peculiarly theirs. The chimney-sweep pursues an ancient craft that still lingers on in Central Europe. One occasionally sees its devotees hurrying through the streets of the ancient cities, dressed in their formal black garb, the leader wearing his top-hat, with ladders and brushes, the utensils of their trade. The fire of the hearth is their traditional preoccupation. Has the chimney-sweep made the fly-agaric his own for the same reason that the Vedic poet identified Soma with Agni?

EPILOGUE

THE TREE OF LIFE AND THE MARVELOUS HERB

EPILOGUE

THE TREE OF LIFE AND THE MARVELOUS HERB

In *Part Three* and the Siberian Exhibits I have laid before the reader evidence for the part played by the fly-agaric as a divine inebriant in the proto- and pre-history of Eurasia. A recapitulation of this evidence will bring out aspects that the more detailed account tended to obscure and will chart the course for further inquiries.

Our surprising discovery in Siberia is linguistic. Just as we of Indo-European stock say today that so-and-so is 'drunk', a word derived from the fermented beverage that we 'drink', and just as we say of the Siberian shaman that he gets 'drunk' on fly-agaric without giving thought to the semantic contradiction, so the corresponding word in the Ob-Ugrian and one (at least) of the Samoyed languages can be translated as 'bemushroomed', and where that word survives it is used for alcoholic inebriation with no awareness of philological anomaly. Moreover we can assert that that word goes back far into the past. The Finno-Ugrian languages and the Samoyed group together make up the Uralic family. There is a characteristic consonant shift between Samoyed and the other languages of the family: an 'f' in Samoyed turns up as 'p' in the others. As the reader will perceive when he reads [34], the Ob-Ugrian root *paŋx* appears as *fankáˀam* in Tavgi, a north Samoyed language, both of them connoting inebriation. (This happens to be precisely the same shift that occurs between the Latin and Germanic families, – *e.g.*, the 'p' in *pater* and the 'f' in 'father'.) The cluster of words that interests us shows this characteristic shift, and therefore it was not borrowed at some later time but must go back to common Uralic. According to the weight of scholarly opinion, Uralic ceased to be spoken *ca.* 6000 B. C., or according to some authorities as recently as 4000 B. C. At that remote period there was not yet writing in the world: the Sumerians seem to have been the first to devise a method for making speech visible, and this they did shortly before 3000 B. C.

The use of the fly-agaric as an inebriant therefore dates back to the period when common Uralic was last spoken, but this is the minimum age. There is no reason to suppose that the peculiar virtue of this miraculous herb went for long undiscovered after it became common in the birch and pine forests as these spread over the Siberian plains in pursuit of the retreating ice cap of the last glacial age, *ca.* 10,000 B.C. After all, the first inhabitants probing the northlands were food gatherers, and how could they fail to see this spectacular plant with its solar disk growing around the base of the noble birch? And given their mental equipment and physical appetites, how could they fail to discover and then to take advantage of its inebriating qualities? None of our writers, not even the able anthropologists Bogoraz and Jochelson (who as Russians were surely mycophiles), seem to have discerned the rôle that it must have played in the past of the north-Eurasian peoples. Perhaps the Soviet authorities, now that they are under less pressure from urgent problems, will be disposed to allow able, sympathetic observers to go among the tribesmen and learn what they still know about their former practices with the fly-agaric, the normal inebriant over that vast expanse of the earth's surface for thousands of years.

We must be thankful for the anthropological testimony that we possess but we must not exaggerate its importance. For a shamanic practice that has lasted six, or eight, or ten millennia our soundings reach back only three centuries, ripples on time's surface. Some of the observers were supercilious and none of them saw the implications of their observations. None of them seems to have been prepared in botany and none probed the questions that are compelling for us. The circumstances that brought them on the scene were at the same time bringing about the end of the beliefs and performances for which they were to be the sole witnesses. They were observing the fly-agaric cult only in its dying phase, when the area of its diffusion was being lopped away, when in some places the integrity of belief in it had been undermined, and when the tribes themselves were mostly in a pitiful state of physical and psychological disarray. There is ample evidence that the ethnic movements, often gradual and more or less peaceful, in the inhospitable tundra and taiga of Siberia have been continuous, and we are far from unraveling them. In recent centuries the peoples practicing the fly-agaric cult have been living in areas to the north of where they were in their heyday, some of them on or close to the Arctic Ocean. They have been displaced from their former homes by Altaïc tribes who do not, apparently, take the fly-agaric[1] but who have absorbed into their shamanic practices the corpus of beliefs that go with the fly-agaric, beliefs that seem to accompany the ecology of the forest belt, especially the reverence for the birch. For the historian of human cultures it is a matter of regret that the impact of the modern world is inevitably brutal, bulldozer-like, in its disregard and contempt for the beliefs and ways of life of primitive peoples whom our industrial civilization wrenches from their traditions and tries, usually without success, to bring into step with our contemporary ideas. This holds true for the communist as well as the capitalist world: witness the authors that we quote in [42].

In northern Europe there is circumstantial evidence that in former times, long before the advent of literacy, the fly-agaric held sway over our own ancestors. This evidence, suggestive but falling short of proof, was enough to launch us on our inquiry, and the collateral confirmation that we found in Siberia buttresses our initial supposition as to our own ancestors. The picture that begins to emerge of a united field in Eurasia where the fly-agaric evoked religious adoration is mightily reenforced by my discovery that the Aryans, hailing from northern Eurasia and settling in the second millennium before Christ on the Iranian plateau and in the Indus Valley, brought down with them as one of their gods the fly-agaric, incorporating it into their elaborate religion of basic Indo-European pattern. The fly-agaric appears to have given those who ate it (or drank its juice) a feeling of elation, of ecstasy, so powerful that they felt they were sharing, for

1. For this conclusion we rely chiefly on negative evidence. Anthropologists like S. M. Shirokogorov, who specialized in the Tungus and Yakut cultures, make no mention of the fly-agaric. Ivan A. Lopatin, also with extensive personal experience, has assured me in a personal communication that the Tungus shamans know nothing of the practice. *Vide* also Brekhman and Sem [42] p. 334.

the nonce, the life of the immortals. As we shall see, we think that the renown of this divine inebriant spread far beyond northern Eurasia and the Aryan world. We have already suggested that under the First Emperor of China, Shih-huang, toward the end of the third century before Christ, rumours about the marvelous herb erupted in the Imperial Court and led to the conception of the *ling chih*, the 'Divine Mushroom of Immortality', which the Taoists made peculiarly their own and which survives to this day throughout the orbit of Chinese culture.

There is I think an inference that we may draw: a plant with properties that could be plausibly named the Herb of Immortality responded to one of man's deepest desires in the early stages of his intellectual development. The superb fly-agaric gave him a glimpse of horizons beyond any that he knew in his harsh struggle for survival, of planes of existence far removed and above his daily round of besetting cares. It contributed to the shaping of his mythological world and his religious life.

Now that the hallucinogens are again becoming familiar to us all, perhaps vicariously we are vouchsafed a glimpse into the subjective life of peoples known to us heretofore only by the mute artifacts uncovered by the archæologists. To weigh the effects of those hallucinogens is a formidable task, today rendered doubly difficult (perhaps even impossible) by the emotions they inspire in our own community, not least among the students of religion. Some of these seem loath to admit even the possibility that the hallucinogens encouraged the birth of religion, and may have led to the genesis of the Holy Mysteries. For them the hallucinogens are the abomination of abominations. Moreover, the fixation of our Western world on alcohol, often a stultifying intoxicant and seldom an invigorating one, closes our minds to other inebriants, older perhaps for the race than our fermented drinks and in their effects utterly different.

In the face of the Siberian testimony and the Vedic hymns, I am at a loss to explain what I write down as the partial failure of my own experiments with the fly-agaric. True, these experiments have confirmed that its reputation as a lethal mushroom is only a superstition, a tabu handed down to us from our remote forebears. (I use 'tabu' not as a figure of speech but in its strictest anthropological sense.) But why did we not feel the elation that the writers and poets describe, comparable with what my companions and I experienced after eating the hallucinogenic mushrooms of Mexico? Is there a difference between the fly-agaric of continental Asia and the Japanese and European specimens? This is possible; we must remember that von Langsdorf [10, p. 240] thought the Siberian fly-agaric displayed a 'navel' lacking in the European ones, in short, to use the mycologists' word, that it was 'umbonate'. But it seems unlikely. There was the divergent experience that Rokuya Imazeki once enjoyed, on October 1, 1965, in Sugadaira. (*Vide* p. 75) All the fly-agarics that we have eaten were gathered in October. We find evidence[1] that at the end of the season their potency falls off. We must either greatly increase the dose or try specimens gathered in summer. The affirmative testimony about the fly-agaric in Siberia is compelling, not to speak of the astonishing lyrics addressed to Soma in the ṚgVeda.

In her recent little book[2] Miss Barnard has driven home brilliantly the need to seek the genesis of myths in natural phenomena. Let us see how this fits the case of the Tree of Life and the Wondrous Herb of Immortality.

In the Siberian and Altaïc cultures, wherever the birch grows it plays an exalted rôle, sometimes also the pine, more rarely the fir. The tall Siberian birch with its delicate dancing foliage and its dazzling white bark is a thing of ethereal beauty, and this alone is enough to give it a favored place in the affections of the Russians. But beyond the Urals it enlisted more than the affections of the tribesmen: it is the nodal point for their shamanism, for their beliefs about the supernatural. All or almost all of the serious writers about these cultures speak of the conspicuous place of the birch in their practices and

1. *Vide* Brekhman and Sem [42] p. 335; also [10] p. 247. Dr. Conrad Eugster informs me that chemical analysis shows more ibotenic acid in the fly-agarics gathered in mid-summer.
2. Mary Barnard: *The Mythmakers*, Ohio University Press, 1966.

thoughts.[1] Yet not one of them links that special place with the fly-agaric. Not one of them perceives why the birch is the Tree of Life.

The fly-agaric lives in mycorrhizal intimacy with the birch, especially the birch; sometimes with the pine, occasionally with the fir. Moreover, while *Fomes fomentarius* grows on several kinds of trees, it is popularly associated with the birch because the birch is the most common of its hosts. *Fomes fomentarius* is the shelf fungus, often reaching huge size, that has always supplied the north Eurasian tribesmen with punk or touchwood, the primary tinder that catches the spark from the fire-drill and bursts into flames. This also has a mystic rôle to play: among many primitive peoples the procreation of fire is analogous to the sexual act. In French 'punk' is *amadou*, a word that goes back to Latin *amare*, and in English a 'punk' until only a few centuries ago was the harlot who sparked her lover into flame. The parallel 'spunk' has to this day the scabrous meaning of 'semen', and 'spark', a different grade of the same word, carries various erotic meanings. We shall see that in Siberia the same associations hold good. In the northern latitudes it was only a ready fire that made life livable and punk (cognate as I have suggested with Ob-Ugrian *paŋx*, *poŋx*; with Chukchi *poŋ*) seems to have captured men's emotions. The birch, parent to both fly-agaric and punk, naturally held pride of place as the Tree of Life, providing in punk the key to fire for the body[2] and in the fly-agaric fire for the soul.

What must have been conspicuous facts of nature for the Siberian food-gatherers are almost completely ignored by the Europeans who have visited them. This is not surprising. Europe's intellectuals are largely recruited from the urban culture and are or soon become strangers to the countryside. Of all the writers about the fly-agaric in

1. Here are three secondary works that will introduce the reader to the vast bibliography of primary sources about the place held by the birch among the Siberian peoples:

a. *The Mythology of All Races*, John Arnott MacCulloch, Editor; Vol. IV, *Finno-Ugrian, Siberian*, by Uno Holmberg, Chap. V, 'The Tree of Life'. Archæological Institute of America, Boston, 1927.

b. Jean-Paul Roux: *Faune et Flore Sacrées dans les Sociétés Altaïques*. Adrien-Maisonneuve, Paris, 1966. *Vide* pp. 52-62 81, 89, 186, 359-361.

c. Mircea Eliade: *Shamanism: Archaic Techniques of Ecstasy*, Bollingen Series LXXVI, Pantheon Books, New York, 1964; pp. XIV, 70, 116-120, 244-246, 403. Translated from the French, *Le Chamanisme: Les Techniques Archaïques de l'Extase*, Payot, Paris, 1951; pp. 9, 78, 116-120, 245-246, 362.

2. *Vide* [4], the Editor's note on *zhagra*, a Russian word for 'punk', p. 238; also Jochelson [21], p. 269.

Siberia that we assemble in the Exhibits, only three mention its connection with the birch. Von Maydell in 1893 wrote of the fly-agaric that 'it is said to occur only among birch trees', and almost a century earlier, in 1809, von Langsdorf observed that 'isolated fly-agarics grow in Kamchatka, in birch forests and on dry plains'. (Probably the 'dry plains' of von Langsdorf were the habitat of the dwarf birch of the Arctic regions, *Betula nana*; for the fly-agaric will grow with any species of birch.) But even von Maydell did not perceive the implications of the accurate observation that he reported as having been made to him by others. Von Dittmar also associated the fly-agaric with birch forests but vaguely.[1] Even the Russians and Poles failed to point out the connection between the fly-agaric[2] and the place of the birch in Siberian folklore. Since 1885 mycologists have recognized the mycorrhizal relationships between certain species of mushrooms and certain species of trees, but this important advance in their science did nothing to broaden and deepen the knowledge of anthropologists because mycologists are prone to keep to themselves, and they often look down their noses on 'folklore' about mushrooms as a childish and irrelevant diversion from the grave questions of taxonomy and scientific nomenclature that preoccupy them; and most anthropologists, strangely, seldom study botany and never mycology. I say 'strangely' because plants fill a large part of the universe of the peoples we commonly call primitive.[3]

The birch is preeminently the tree of Siberian shamanism. This is so widely recognized that I need not argue the case and will only

1. *Vide* [12], p. 254; [10], p. 247, and [13], p. 256.

2. There is a traditional saying, exceedingly old, in Russian:

Po veleniju shchuch'emu,	On the pike's bidding,
Po prikazu mukhomorovu.	On the fly-agaric's orders.

The pike plays a potent rôle in Russian folklore. Here it is yoked with the fly-agaric. Is this couplet borrowed from a Finnic or Ob-Ugrian people? Do we hear in the fly-agaric's 'orders' a distant echo of the orders of the fly-agaric that are reported in Jochelson and Bogoraz? *Vide* [21], pp. 268 ff.; also [22], pp. 274 ff.

3. There are of course notable exceptions to this criticism of anthropologists; *e.g.*, the studies in ethno-botany carried out by Harold C. Conklin among the Ifugao on the island of Luzón, in the Philippines. *Vide* (1) *Studies in Philippine Anthropology; in honor of H. Otley Beyer*, edited by Mario D. Zamora, pp. 204-262; and (2) 'Ifugao Ethnobotany 1905-1965: the 1911 Beyer-Merrill Report in Perspective', *Economic Botany*, Vol. 21, No. 3, July-September 1967, pp. 243-272.

summarize it. We read, for example, that among the Buriat northwest of Lake Baïkal the inhabitants bow morning and evening to two birches that they have planted in front of their huts. We read that the birch with seven or eight or nine branches is favoured, these symbolizing the successive gradations in ascending to the ultimate heaven; and it is held that the trees' roots penetrate to the very depths of the earth. As though to symbolize the reach upwards and the reach downwards, an eagle (or a mythological bird that we conventionally call an eagle) surmounts the tree and a serpent dwells at its roots. Again we read that the shaman selects a stout birch, fells it, and places it in the center of the yurt that he is going to build for his performance. He cuts seven or eight or nine notches in it, representing the seven or eight or nine heavens through which he will ascend. Later in the course of his ecstatic performance he climbs this tree making use of the steps, and passes through the hole in the roof through which the smoke from the fire finds its way, going on his symbolic journey to the other world.[1]

Uno Holmberg in the *Mythology of All Races* summarizes the Siberian myths about the birch in his chapter on the Tree of Life. The spirit of the birch is a middle-aged woman who sometimes appears from the roots or the trunk of the tree in response to the prayers of her devotees. She emerges to the waist, her eyes are grave, she has flowing locks, her bosom is bare, and her breasts are swelling. She offers milk to the Youth who approaches her. He drinks and his strength grows a hundred-fold. This myth, which is repeated in myriad variations, clearly refers to the fly-agaric. But none of Holmberg's sources have called this to his attention. What are the 'breasts' but the 'udders' of the ṚgVeda, the swelling pileus of the full-grown fly-agaric? In another tale the tree yields a 'heavenly yellowish liquid.' What is this but the tawny yellow *pávamāna* of the ṚgVeda? Repeatedly we hear of the Food of Life, the Water of Life, the Lake of Milk that lies, ready to be tapped, near the roots of the Tree of Life. There where the Tree grows is the Navel of the Earth, the Axis of the

1. J.-P. Roux, *Faune et Flore Sacrées dans les Sociétés Altaïques*, Paris, 1966, pp. 54, 59. M. Eliade, *op. cit.*, pp. 9, 116-122; French edition, pp. XIV, 116-120.

World, the Cosmic Tree, the Pillar of the World. The imagery is rich in synonyms and doublets.

Mircea Eliade believes that this cosmological scheme probably has an oriental origin, or that its mythical features go back to a 'palæo-oriental' prototype (a toponymic designation that I have difficulty in assigning to a specific location; apparently it lies somewhere between the eastern Mediterranean and China), or (after Uno Harva) that the ensemble of initiation rites among the Siberian tribes comes down from a Mithraïc source.[1] It is the consensus of all who have written on the matter that the Siberians could not have fathered the myths and practices that they have made their own. Or, to put the thought more accurately, the very idea of such a possibility seems not to have been entertained by them.

On the contrary I now suggest that the source and focus of diffusion of all these myths and tales and figures of speech – all this poetic imagery – were the birch forests of Eurasia. The peoples who emigrated from the forest belt to the southern latitudes took with them vivid memories of the herb and the imagery. The renown of the Herb of Immortality and the Tree of Life spread also by word of mouth far and wide, and in the South where the birch and the fly-agaric were little more than cherished tales generations and a thousand miles removed from the source of inspiration, the concepts were still stirring the imaginations of poets, story-tellers, and sages. In these alien lands, far from the birch forests of Siberia, botanical substitutions were made for Herb and Tree. Here is where absurdities were introduced into the legends, where fabulous variations proliferated, where peoples who had never known the North such as the Semites were influenced by the ideas and in one way or another incorporated them into their religious traditions. The end-products of these extravaganzas have caused scholars much (and I think needless) trouble as they subjected them to sober exegesis and tried to reconcile them.

In the north to this day we find a notable consistency in the myths and poems of the Siberian people, having regard to the facts of nature. Their imagination never takes them more than one or two removes

1. Mircea Eliade, *op. cit.*, pp. XIV, 245-6, 60-70, 120. French edition, pp. 9, 247, 79, 121.

away from the life history of the fly-agaric, the birch (or pine or fir), and punk. The contradictions and wild embellishments begin only when the corpus of myths is translated to the exotic world of the Near East, Mesopotamia, Iran and India, and China; in short when the umbilical cord with the natural phenomena is broken. Here then is nature's triangle:

Birch

FLY-AGARIC Punk

The fly-agaric holds the place of honour in this Trinity: without it there would be nothing. Its beauty matches its magic powers. The birch is also indispensable. Some will find it astonishing that the Siberian peoples observed and understood, according to their lights, the mycorrhizal relationship, only rediscovered by mycologists in 1885. For the tribesmen the roots of the birch tapped the lake of the Waters of Life and filled to overflowing with tawny yellow milk the breasts of the fly-agaric. The noble stance of the superb birch befitted its rôle as host and divine guardian. The punk is the least of the Trinity, vital in the North but meaningless in the South where other methods for making fire were used. But we must not disregard it. Here for example is a legend about punk that survives in many recensions from Central Asia.

Speaking of the Uighur, a Mongolian tribe, Marco Polo tells us that

> They say of their Khan who first ruled over them that he was not of human origin, but was born of one of those excrescences on the bark of trees, and that we call *esca*. From him descended all the other Khans.[1]

In another version of the same myth we learn that two trees played a part in procreating the royal family of the Uighurs, a birch and an

1. The Italian text reads:

> Raccontano che il re che li resse per primo non era di origine umana, ma era nato da una di quelle escrescenze che la linfa produce sulla corteccia degli alberi, escrescenze che noi designiamo col nome di esca. — *Il Libro di Messer Marco Polo*, edited by L. F. Benedetto, Treves-Treccani, Milan and Rome, 1932, p. 73.

evergreen resembling a pine.[1] *Esca* is the Italian word for 'punk', in Siberia *Fomes fomentarius*.

J.-P. Roux, the latest writer on the rôle of the flora among the Altaïc peoples, raises the inevitable question. Speaking of the place that the birch holds in the shamanic *séances* and at the animal sacrifices, he observes:

> Nothing permits us to think that the tree is chosen because of its capabilities or its appearance or because it acquires by reason of the ceremony the nature of a venerated tree. The only point that merits our attention is perhaps without value, the effect of a simple coincidence: the shamanic tree is most often a birch, that is to say, as everyone knows, a tree whose bark is whitish. With people who accord so great an importance to the colour white, is not the choice of the birch motivated by its flashing bark?[2]

No one had pointed out to M. Roux that the birch is host to the fly-agaric and touchwood. True, the quality of whiteness has an almost magical meaning for the northern Eurasians. But the question presents itself whether this is not secondary, and whether whiteness enjoys its exalted status partly because it characterizes the host to the fly-agaric and punk. Or, to put it differently, the fly-agaric and punk are primary in the hold of the birch on the souls of the natives and it must follow as night the day that the whiteness of the birch is in most fitting and wonderful harmony with its supernal attributes.

The Siberian legends and myths as we possess them were recorded in recent generations. They never cite the link that ties the fly-agaric to the birch. Perhaps this is because for the Siberian tribesmen the connection was self-evident: any cretin would know as much. Students of the Siberian cultures, unaware of the thesis that I am developing, have only by chance, occasionally, asked a few of their Siberian informants a few of the relevant questions, and they have not followed through. The inquirer who goes among primitive people must know the questions to pose, must see the implications of the answers he receives, must probe slowly with utmost patience and tact, especially

1. J.-P. Roux, *op. cit.*, p. 359.
2. J.-P. Roux, *op. cit.*, p. 186.

where religious beliefs and practices are concerned. (The anthropologist has a thankless calling: no matter how thorough he is, his successors are certain to reproach him for not having put all the questions that later seem imperative.) Even so we could hardly ask for better than Holmberg's essay on the Tree of Life in Siberia, as it is presented in the *Mythology of All Races*. The fly-agaric and the Soma hymns of the ṚgVeda supply the key that unlocks the myths of his tribesmen.

The word for 'birch' in Sanskrit is *bhūrja*. Scholars have sometimes expressed mild surprise that the Aryans remembered, after their long migration, this Indo-European name, with cognates in almost all Indo-European tongues; in fact, the birch is, significantly, one of the few trees of which this can be said. The migration of the Aryans must have lasted for generations, even centuries. Yet when they first caught sight of the birch in the Hindu Kush or the Himalayas, we must assume that right away they exclaimed, 'What, the *bhūrja!*' Of course communications with the homeland may have been better in pre-literate Asia than we imagine: we may be victims of the bias of the literate world against periods in man's past about which we know almost nothing. But even if there was complete isolation, *bhūrja* as the Tree of Life held a place in their subjective life that they would not quickly forget. On the other hand *bhūrja* is not mentioned in the ṚgVeda, and Abel Bergaigne a century ago pointed out how trifling was the rôle of the Tree of Life in the Vedic hymns.[1] This should also not surprise us. In the mythological pattern that we are discussing the fly-agaric held the central position and the Tree of Life was secondary. The Indo-Aryans possessed the Marvelous Herb, which they bought from aborigines high in the mountains. With the fly-agaric in hand, what need had they of the birch? What was out of sight was for the moment out of mind. But elsewhere in the Near and Middle East the poets and sages had neither and from the renown of both their imaginations could embroider endless patterns.

On pp. 77-84 we gave three recensions – Indian, Iranian, Chinese – of the same tale going back to the Soma∾fly-agaric of the Aryans.

1. Abel Bergaigne: *La Religion Védique*, Vol. 1, p. 199.

These versions came down from the second half of the first millennium after Christ. Thus they were late, and we may ask ourselves how many at that time in the Southern latitudes had knowledge of the fly-agaric. In each of these versions a novel element was introduced: the notion that the leaves of the Herb of Immortality, if placed on corpses, would restore them to life. This absurd accretion was however of ancient provenience: we can trace it back a thousand years before the Vedic hymns were composed, to the Sumerian fragments of the Epic of Gilgamesh.[1] The Soma hymns are unique and precious for their textual integrity, but tales of the Herb of Immortality (that is, of the fly-agaric) long antedate the ṚgVeda.

When man first devised a method of inscribing words on clay, *ca.* 3000 B.C., he poured forth on his tablets, among other items, the ideas that seemed to him deserving of perpetuation, but ideas not necessarily indigenous, derived from sources not necessarily known to the learned men who were shaping the characters. There is a scholarly bias, as understandable as it is mistaken, to trace the origin of ideas according to the literacy of peoples, and sometimes to give to the Near Eastern and Mesopotamian cultures credit for conceptions that they were merely the first to record. This is, I suggest, the case with the Herb of Immortality and the Tree of Life, whose archetypes were brought down from the forests of Siberia in the fourth millennium before Christ or earlier. The Hittites and the Mitanni rulers were Indo-European invaders from the north who preceded the Aryans, and the Sumerians long preceded them, and there were doubtless others even earlier of whom we have no historical knowledge. We must avoid the temptation of supposing that the tribesmen of Siberia could not have possessed a rich world of the imagination simply because, not having mastered the art of writing, they are for us inarticulate. When the Sumerians wrote down the Epic of Gilga-

1. Geo Widengren; *The King and the Tree of Life in Ancient Near Eastern Religion* (King and Saviour IV), Uppsala Universitets Årsskrift 1951:4. Acta Universitatis Upsaliensis; p. 21. In the Near and Middle East graves of prehistoric cultures running back to *ca.* 6500 B. C. have been found in which the corpses were painted with ochre or cinnabar. It has been suggested to me that this practice may have had its origin in the crimson fly-agaric, the Herb of Immortality, and in the notion that the 'leaves' of the sacred *oṣadhi* if placed on a corpse would in some way assure it of Eternal Life.

mesh, we should not think of it as a fresh creation. It already belonged to the world of mythology and he is a rash scholar who today would say with assurance that that corpus of myths first saw the light of day in the Near East or Mesopotamia.

Was Uno Harva mistaken and did the Mithraïc beliefs and rites come down from the forests of what we now call Siberia? Let us look again at what is known of the Orphic mysteries, and reconsider the archetype of our own Holy Agapé. On what element did the original devotees commune, long before the Christian era? Certainly the overt vocabulary relating to the birch and the fly-agaric carried great prestige over millennia throughout the south and east of Asia: the Tree of Life, the Pillar of the World, the Cosmic Tree, the Axis of the World, the Tree of the Knowledge of Good and Evil – all these were variations stemming back to the birch and the fly-agaric of the northern forests. The Herb (or Plant) of Life, the Herb of Immortality, the fruit of the Tree of Life, the Divine Mushroom of Immortality – these are alternatives ultimately representing the fly-agaric, no matter how far removed the poet or sage or king might be from the real thing. In remote China we have seen the devotees of the Manichæan sect as late as the 12th century eating 'red mushrooms' in such quantity as to arouse the indignation of a pillar of the Chinese Establishment: is not this an echo of Siberian shamanism, not having passed direct from Siberia to China, but tortuously, through successive Middle Eastern religions, until we reach the last of Mani's followers, far from his Iranian home? The Water (or Milk) of Life and the Food of Life are doublets, the former being the *pávamāna* expressed from the latter, the resplendent Soma. If I am right, here is striking confirmation of the ideas advanced by Miss Barnard. When we seek the source of myths, we should *chercher*, not *la femme*, but *un phénomène de la nature*.

In the opening chapters of Genesis we are faced with the conflation, clumsily executed, of two recensions of the fable of the Garden of Eden. The Tree of Life and the Tree of the Knowledge of Good and Evil are both planted in the center of Paradise. They figure as two trees but they stem back to the same archetype. They are two names of one tree. The Fruit of the Tree is the fly-agaric harboured by the birch. The Serpent is the very same creature that we saw in Siberia dwelling in the roots of the Tree.

Of arresting interest is the attitude of the redactors of Genesis toward the Fruit of the Tree. Yahweh deliberately leads Adam and Eve into temptation by placing in front of them, in the very middle of the Garden, the Tree with its Fruit. But Yahweh was not satisfied: he takes special pains to explain to his creatures that theirs will be the gift of knowledge if, against his express wishes, they eat of it. The penalty for eating it (and for thereby commanding wisdom or education) is 'surely death'. He knew the beings he had created, with their questing intelligence. There could be no doubt about the issue. Yahweh must have been secretly proud of his children for having the courage to choose the path of high tragedy for themselves and their seed, rather than serve out their lifetimes as docile dunces. This is evidenced by his prompt remission of the death penalty. . . . It is clear that among community leaders the hallucinogens were already arousing passionate feelings: when the story was composed the authentic fly-agaric (or an alternative hallucinogen) must have been present, for the fable would not possess the sharp edge, the virulence, that it does if surrogates and placebos were already come into general use. The presence of the serpent is a happy necessity, for throughout Eurasia the serpent is intimately associated with the fungal nomenclature of the mushroom world, or with particular species of mushrooms, though in nature as it happens they have nothing to do with each other. Only in regions where snakes are unimportant, as in the British Isles, is the serpent replaced by the toad. The toad is then made heir to the curse visited on the serpent, and in turn the toad infects and infests the toadstool. The snake, the toad, and the toadstool are alike chthonic spirits.

If these perceptions are right, then the mycologists were right also, in a transcendental sense of which neither they nor the artist had an inkling, when they saw a serpent offering a mushroom to Eve in the Fresco of Plaincourault. And Ponce de León early in the 16th century was still seeking in Florida the pool of living water that he might have discovered in the Siberian taiga, the pool where Gilgamesh finally found his Herb of Immortality thousands of years earlier, only to lose it again to the Serpent who was more subtle than any beast of the field, the very same Serpent who engaged Eve in pleasant conversation, whose habitation is in the roots of the towering Siberian birch.

ACKNOWLEDGEMENTS

As everyone who knows me will realize, my dependence on others with specialized knowledge in different fields has been complete. In Sanskrit and Vedic I have enlisted the cooperation of Dr. Wendy Doniger O'Flaherty, and I have consulted Professor Daniel H. H. Ingalls of Harvard University, the late Professor Louis Renou of the Sorbonne, and Professor Georges Dumézil of the Collège de France. For my first steps in this field I am under obligations to Professor Bart van Nooten, now on the faculty of the University of California. In matters of general Indo-European linguistics my friend and revered mentor has been Professor Georg Morgenstierne of Oslo, who has guided my footsteps in the Indo-Iranian field, the Kafir and Dardic languages, and the older tongues of Scandinavia and the Northern Seas. Mrs. Inger Anne Lysebraate, mycologist of Oslo, has volunteered to help me in pointing up the possible references to mushrooms in the early literatures of Iceland and Scandinavia.

For my Chinese inquiries I am indebted to Wango Wêng 翁萬戈, of New York, to Chou C'hi-k'uên 周啟昆, formerly of the faculty of Hong Kong University, better known to his English-speaking friends as Steve Chou, now of Leeds University, to Professor Kao Yao-lin 高耀琳, formerly of Nanking University, now living in Hong Kong, and particularly to Kristofer M. Schipper, élève de l'Ecole Française d'Extrême Orient, who on the occasions of my repeated visits was living in Tainan, Taiwan. Mr. Wêng has written all of the Chinese characters that appear in my book. For guidance in Japanese fungal matters I have had the valuable assistance of Professor Yoshio Kobayashi of Ueno Museum, and of Rokuya Imazeki, co-author with Tsuguo Hongo of the standard manual in two volumes on Japanese mushrooms, an excellent field guide that the English-speaking world has yet to equal.

Dr. János Gulya of Budapest has generously given me days of his time reviewing with me many problems pertinent to my theme in the Uralic (especially the Ob-Ugrian) languages; and Tamás Radványi also, who has served us as a gifted and gracious interpreter. The Publishing Office of the Magyar Academy of Sciences has kindly consented to let me re-publish in English extracts from their *Glaubenswelt und Folklore der Sibirischen Völker*, edited by V. Diószegi and originally published in 1963. I wish also to record here my appreciation of the help given me in former years by the late Dr. Sándor Gönyey, formerly of the Ethnographical Museum of Budapest, and by Dr. Gabriel Bohus, of the Botanical Department of the Museum of Natural History, also of Budapest. Professor Robert Austerlitz managed to find time from the crushing pressure of administrative and teaching duties at Columbia University to guide my steps in matters of Siberian linguistics.

In the area of library research for efficient and intelligent service I thank Marcelle Lecomte Drakert of New York. Miss Mary Mahoney and Mrs. Evelyn Waters of Ridgefield, Connecticut, have laboured long and hard in the preparation of the manuscript for the printer.

The Human Relations Files of New Haven, Conn., were helpful to me in canvassing sources about the remote peoples of Asia, especially of Siberia. I am grateful to them for introducing me to Philip Lozinski; it was he who put me on the trail of St. Augustine and the Chinese Manichæans, and he has shown a continuing interest in my mushroom inquiries.

For a financial grant that aided my travels in the Far East I am beholden to the Bollingen Foundation of New York; and for my appointment as an Honorary Curator of Botany, to the Milwaukee Public Museum and its Director, Dr. Stephan F. de Borhegyi.

The Botanical Museum of Harvard University has graciously appointed me to its faculty as Research Fellow, and the New York Botanical Garden has made me Honorary Research Associate. These dignities have served to open doors for me in the Far East where I had been unknown. It has also been a privilege to enjoy the facilities of these two institutions and to have had access at all times to the guidance and counsel of Professor Paul C. Mangelsdorf, until lately director of the Museum, and of Dr. Richard Evans Schultes, its Curator of Economic Botany and now also the Executive Director of the Museum.

ACKNOWLEDGEMENTS

It is always a pleasure for me to express my continuing gratitude to Professor Roger Heim, Membre de l'Institut, for his help in furthering my inquiries. He and his excellent staff at the Laboratoire de Cryptogamie, of the Muséum National d'Histoire Naturelle, seem always delighted to work on my questions, little and big, and the reception that they accord me on my frequent visits to Paris moves me deeply. From the beginning Professor Heim has shown a particular interest in every aspect of my researches into the strange problem of the fly-agaric: he has initiated his own programme of experimental studies, which are being reported from time to time in the *Revue de Mycologie*. In pursuit of our mutual interests he and I have traveled together to the farthest reaches of the inhabited world.

Among those to whom I am beholden for collaboration I cannot refrain from citing Dr. Giovanni Mardersteig of Verona. Eminent typographer, scholar, and artist, he has not only shown endless patience in the printing of my manuscripts: on numerous occasions out of the store of his learning he has contributed to my argument, and he and his family by their kindnesses make my visits to Verona memorable events in my life. What a pleasure it is, when probing the unknown in the cultural history of our race, to be brought into touch with men like Professor Heim and Dr. Mardersteig.

I had hoped that this treatise on the Herb of Longevity would appear in 1966, on the occasion of the 70th birthday of Roman Jakobson, to whom I have dedicated it. But in 1968 it is by two years that much more appropriate. May the Tree of Knowledge, possessed as it is of a private line to the Herb, continue for many years to bestow abundant blessings on a favoured son!

R. Gordon Wasson

Danbury, Connecticut
June 1968

EXHIBITS

1.

THE FLY-AGARIC IN SIBERIA

PRELIMINARY NOTE

WE have assembled here the writings of those who have described the practice of eating the fly-agaric in Siberia. In Section A we include explorers, travelers, and anthropologists, as well as a few native folk tales and a Vogul hymn gathered and translated by the anthropologists and linguists. In Section B we have grouped together linguists – mostly Finnish and Hungarian – who give us evidence of a fly-agaric word pattern that presents us with a fascinating aspect of social history. Of the many writers about the peculiar mushroomic habits of the Siberian tribesmen who have passed on the information at second or third hand, we have chosen three for particular reasons to include in our Section C. One is Carl Hartwich, whose *Die menschlichen Genussmittel* set a landmark in the history of pharmacology; it was published in 1911. Another is Professor Mircea Eliade, who enjoys wide renown as a student of the history of religions and religious practices throughout the world. The third is a paper by two Soviet scientists.

Undoubtedly there are primary sources that have eluded us, perhaps important ones. We have searched diligently, but the sea of writings about Siberia is so vast, the sources so widely scattered, that some could easily have escaped our net.

We have arranged the selections roughly in their chronological order.

A. Explorers, Travelers, and Anthropologists

[1]

KAMIEŃSKI Dłuzyk, Adam. Dyarusz więzienia moskiewskiego, miast i miejsc. (A Diary of Muscovite Captivity, Towns and Settlements) In *Warta.* A collection of articles. Edited by the Rev. A. Maryański. Poznań, Poland. Published in 1874. pp. 378-388; the passage relating to mushrooms is on p. 382.

[This is the earliest report that we have found of fly-agaric eating among the tribesmen of Siberia. It is an entry in the journal of a Polish prisoner of war made in 1658. He is describing the habits of the Ob-Ugrian Ostyak of the Irtysh region in Western Siberia, tributary to the River Ob. – RGW]

... They neither sow nor plow; they live only on fish and fowl of which there is a great wealth there, namely swans, geese, ducks such as we don't have. These very Ostyaki go about in fish skins and they have footwear from the same and parkas from geese and swans. They dwell in camps on islands. They smoke various fish for the winter and tear off the fat from the fish into a birchbark vessel, two vessels full; and they drink it warm by the quart, to our great astonishment. And they make nets from nettles and some have shirts from nettles. They eat certain fungi in the shape of fly-agarics, and thus they get drunk worse than on vodka, and for them that's the very best banquet.

[2]

OGLOBLIN, N. 'The First Japanese in Russia in 1701-1705.' *Russkaya Starina.* October 1891, pp. 19-24.

[This article in the Russian review contains as Section V the story of 'Denbei', the first Japanese to arrive in Russia. He came from Osaka; and his vessel after leaving Yedo (Tokyo) in the last years of the 18th century was thrown off course by storms. The crew finally sighted Kamchatka and took refuge in the estuary of a small river. The natives, who are called in the text Kuril Islanders, took three of the Japanese prisoners, of whom two later died, Denbei being left. The Russian explorer Atlasov, passing that way, heard tell of the prisoner, and the latter was led to him. After further vicissitudes Atlasov took him to Moscow where he was presented to Peter the

Great, who insisted that he learn Russian and that he teach his own language – Japanese – to a few Russians. He gave an account of what had happened to him. Though he learned some Russian and could make himself understood in the language of Kamchatka, his account is filled with discrepancies attributable to errors in communication. 'Denbei' is the name of this Japanese merchant in the Russian transcription: his Japanese name is unknown to us. His testimony is almost, but not quite, worthless. We have included it for the sake of completeness and because it establishes the earliest date on record when the Russian Imperial Court was apprised of the Siberian practice of eating the fly-agaric. In telling of his misfortunes among the inhabitants of Kamchatka, he had this to say, according to the Russian text. – RGW]

They [the natives of Kamchatka] place their fish in pits, covering them on top with twigs and grass, and when the fish turn all mouldy, they put them into wooden troughs, add water, and heat this concoction with hot stones, also mixing in some fly-agarics. They drink this brew, and treat their guests with it, and get drunk on it. However he [Denbei] and his companions could not drink this concoction but ate roots and fish which were not yet too mouldy.

[3]

STRAHLENBERG, Filip Johann von. An Historico-Geographical Description of the North and Eastern Parts of Europe and Asia; But more particularly of Russia, Siberia, and Great Tartary; etc. . . . London. 1736. Second printing, 1738. Originally published in Stockholm in 1730. Translated also into French. p. 397 of the English edition.

[Von Strahlenberg was a Swedish Colonel who passed twelve years in Siberia as a prisoner of war. He was chiefly in Tobolsk where he assembled much accurate and valuable information about the peoples of Siberia. He is describing the practices of the Koryak tribe in the extreme northeast of Siberia. – RGW]

The Russians who trade with them [Koryak], carry thither a Kind of Mushrooms, called, in the Russian Tongue, Muchumor, which they exchange for Squirils, Fox, Hermin, Sable, and other Furs: Those who are rich among them, lay up large Provisions of these Mushrooms, for the Winter. When they make a Feast, they pour Water upon some of these Mushrooms, and boil them. They then drink the Liquor, which intoxicates them; The poorer

Sort, who cannot afford to lay in a Store of these Mushrooms, post themselves, on these Occasions, round the Huts of the Rich, and watch the Opportunity of the Guests coming down to make Water; And then hold a Wooden Bowl to receive the Urine, which they drink off greedily, as having still some Virtue of the Mushroom in it, and by this Way they also get Drunk. In Spring and Summer they catch a large Quantity of Fish, and digging Holes in the Ground, which they line with the Bark of Birch, they fill them with it, and cover the Holes over with Earth. As soon as they think the Fish is rotten and tender, they take out some of it, pour Water upon it, and boil it with red-hot Pebbles (as the Finnlandians do their Beer) and feed upon it, as the greatest Delicacy in the World. This Mess stinks so abominably, that the Russians who deal with them, and who are none of the most squeamish, are themselves not able to endure it. Of this Liquor they likewise drink so immoderately, that they will be quite intoxicated, or drunk with it.

[4]

KRASHENINNIKOV, Stepan. Opisaniye Zyemli Kamchatki. (Description of Kamchatka Land) St. Petersburg. 1755.

[Translations into English, French, and German appeared in the 18th century. We have had made a fresh translation from the Russian, finding that the 18th century translations corrupted the meaning and shortened the text. The edition that we used was edited under the auspices of the Soviet Academy of Sciences, with notes, and published in Moscow and Leningrad in 1949. – RGW]

Chapter 14

CONCERNING THE FEASTS AND GAMES OF KAMCHATKA

They hold feasts whenever the people of an *ostrog* [stockade] wish to entertain their neighbors, especially whenever there is a wedding or some successful trapping venture, and these are spent, for the most part, in overeating, dancing, and singing. On such occasions the guests are treated by their hosts to large goblets of *opanga* so generously that they have to vomit more than once.

Sometimes for their enjoyment they also use the *mukhomor*, the well-known mushroom that we ordinarily use for poisoning flies. [*Mukhomor* – fly-agaric] It is first soaked in must of *kiprei* [*Epilobium angustifolium*], which they drink, or else the dried mushrooms are rolled and swallowed whole, which method is very popular.

The first and usual sign by which one can recognize a man under the influence of the *mukhomor* is the shaking of the extremities, which will follow after an hour or less, after which the persons thus intoxicated have hallucinations, as if in a fever; they are subject to various visions, terrifying or felicitous, depending on differences in temperament; owing to which some jump, some dance, others cry and suffer great terrors, while some might deem a small crack to be as wide as a door, and a tub of water as deep as the sea. But this applies only to those who overindulge, while those who use a small quantity experience a feeling of extraordinary lightness, joy, courage, and a sense of energetic well-being, such as the Turks are said to experience when they have partaken of opium.

It is worth noting that all those who have eaten the *mukhomor* unanimously affirm that all their extravagant actions at the time are carried out on orders of the *mukhomor*, which secretly commands them. But all their actions are so harmful to them that, if there were no one to look out for them, not many of them would remain alive. Concerning excesses of the Kamchadals, which happen among them, I will make no mention, since I have not witnessed any personally and the Kamchadals do not like to talk about this; but then it could be, too, that it does not come to such extremes among them, either because they have become accustomed to the *mukhomor*, or because they do not use it to excess. However, in respect to the Cossacks who have eaten the above mushroom, I shall report some wild behavior, some of which I personally have witnessed, and some of which I heard from the perpetrators of those actions, or from other trustworthy persons.

An orderly of Lieutenant-Colonel Merlin who was with the investigation in Kamchatka was commanded by the *mukhomor* to strangle himself under a delusion that this would cause other people to admire him. This, in fact, might have come to pass, had not his friend restrained him.

Another of the local residents [Cossacks] had a vision of Hell and a terrifying fiery chasm into which he was to be cast; for which reason, following the orders of the *mukhomor*, he was forced to get down on his knees and confess all the sins that he could remember committing. His friends, of whom there were many in the common room where the intoxicated man was confessing, heard this with great amusement, while he seemed to believe that he was confessing his sins in the privacy of the sacrament to God alone. Because of this he was made the butt of much deliberate ridicule, since, among other things, he related some things which should have best remained unknown to others.

A certain soldier, they say, used to eat the *mukhomor* in moderate quantities whenever he had to go on a long journey, and thus was able to cover great distances without any fatigue; but, ultimately, having indulged to the point of delirium, he crushed his own testicles and died.

The son of a Cossack from Bolsheretsk, who was in my employ as an interpreter, had been made drunk on *mukhomor* without his knowledge, and attempted to cut his own abdomen on the *mukhomor's* orders, from which he was barely saved in the last minute, for his hand was restrained in the very act.

Kamchadals and settled Koryaks also eat *mukhomor* while planning to kill somebody. Incidentally, among the settled Koryaks the *mukhomor* is held in such high esteem that those who are intoxicated are not allowed to urinate on the ground but are furnished by others with a dish for this purpose, which urine they drink and also do wild things like those who have eaten the mushroom; for they get the *mukhomor* from Kamchadals, as it does not grow in their own country. Four mushrooms, or less, constitute a moderate use, but for a high degree of intoxication up to ten mushrooms are usually consumed.

Members of the female sex neither indulge in eating to excess nor partake in *mukhomor* consumption, owing to which all their amusements consist of talking, dancing, and singing . . .

[In the above account Krasheninnikov tells of four Europeans who ate the mushroom. Three were certainly exhibitionists, such as make a show of themselves in the West with alcohol. The third, who is said to have died, was reported as hearsay. Krasheninnikov was a careful author, and he concedes that such extravagant behaviour may not occur among the natives. He adds that the Kamchadal and the settled Koryak eat the fly-agaric 'while planning to kill somebody'. He gives no examples and as no subsequent writer, not even his own colleague Steller, repeats the statement, much less documents it, we may disbelieve this. Krasheninnikov wrote early and is of historical importance but we possess later informants immeasurably better equipped to tell us about the tribesmen. This single sentence, familiar to Ödman [*vide infra*, [43]] may be responsible for the belief that berserk-raging came out of Siberia. But Krasheninnikov nowhere makes mention of wild ferocious behaviour suggestive of berserk-raging. – RGW]

*

[In 1949 a definitive edition of Krasheninnikov was published by the Soviet Academy of Sciences, edited by Lev Semenovich Berg. In this edition Berg shows himself aware of the cultural significance of the fungi in the Kamchadal society in a note, p. 236, on a passage in Krasheninnikov where the latter speaks of the Kamchadal as 'omnivorous creatures, for they pass by neither *zhagra* nor *mukhomor*, though the former has no taste and does not satisfy hunger, and the latter is obviously harmful'. Berg's footnote on this passage follows: – RGW]

The Kamchadal ... will pass by neither *zhagra* nor *mukhomor*. *Zhagra*, according to the Dictionary of Dal′, is another name for punk, touchwood, tree fungus, Polyporaceæ. Referred to here is the white agaric (family Polyporaceæ), which grows on trees. The body of the fungus Fomes sp. was formerly used as tinder; some of them have medicinal application. Certain species of the genus Polyporus are edible. Concerning the use of the variety of fungi Polyporus that grows on birch trees (*der weisse Baum-Schwamm* in German) for alimentary purposes, Steller has this to report (p. 92): the Kamchadal knock them off birches with sticks, break them up with axes, and eat them frozen. S. Yu. Lipshitz and Yu. A. Liverovskiy (1937, p. 197) report that ashes of the fungus Polyporus (Polyporus sp.) are used in Kamchatka as snuff. Concerning the use which the Kamchadal make of the 'birch tinder' as a pain killer, Krasheninnikov reports on p. 443. Concerning fungi Polyporus of Kamchatka see: A. S. Bondartsev: *Fungi of the Family Polyporaceæ, Telephoreæ, and Hydneæ, Collected in Kamchatka by V. P. Savich.* (From the Expedition to Kamchatka of F. P. Ryabushinskiy. Botanical Series, Part 2, Moscow, 1914, pp. 525-534.) The most wide-spread fungus in Kamchatka is *Fomes igniarius* (L.), harmful to rock birch, white birch, alder, and aspen. *Fomes fomentarius* (L.) is found on both varieties of birch. There are many other species of this genus and also other genera. Of the genus Polyporus in Kamchatka, *Polyporus sulfureus* (Bull.) is found on the larch and *P. varius* Fries on the poplar.

Steller also reports (pp. 92-93) on the use of the *mukhomor* (fly-agaric) as an intoxicant by Kamchadal, Koryak, and Yukagir. See also L. S. Berg, *Discovery of Kamchatka ...*, Third Edition, 1946, pp. 163-164.

[*Zhagra* in Russian is a synonym of *trut*, the former being derived from a root 'to burn' and the latter from 'to rub'. Both mean primarily *Fomes fomentarius*, the shelf fungus that grows on many trees but that is primarily linked with the birch because the birch is its commonest host in the forest belt of Eurasia. Many species of fungus have been used as the primary tinder in the making of fire, but for 'touchwood' or 'punk' *Fomes fomentarius* has enjoyed primacy from the beginning. At Star Carr in Yorkshire quantities of *Fomes fomentarius* were found in the site of a settlement dating from

9,000 to 10,000 years ago, some still attached to birch logs, along with flint. The touchwood was adjacent to the hearthstones. (*Vide* J. G. D. Clark and others: *Excavations at Star Carr*, Cambridge University Press, 1954, pp. 17-18; also E. J. H. Corner: 'Report on the Fungus-Brackets from Star Carr, Seamer,' in the 'Preliminary Report on the Excavations at Star Carr, Seamer, Scarborough, Yorkshire', *The Prehistoric Society*, XVI, 1950, pp. 123-4.) The British Isles were still attached to the Continent by land, and the site was early Mesolithic, apparently a temporary camp of hunters, not long after the last ice age. The culture was the same as had been previously described at Maglemose (the 'Great Bog'), Denmark, where also *Fomes fomentarius* was found adjacent to the hearth stones. (*Vide* N. Fabritius Buchwald and Sigurd Hansen: 'Om Fund af Tøndersvamp fra Postglacialtiden i Danmark', *Danmarks geologiske Undersøgelse*, iv Række, Bd. 2, Nr. II, 1934.)

[The fungus that the Kamchadal knock off birches and then eat frozen is certainly different, probably *Polyporus betulinus*, a soft, white, rather spongy or rubbery growth on the birch, without much taste, which is sometimes eaten raw or cooked by mycophiles in Europe and the United States. – RGW]

[5]

STELLER, Georg Wilhelm. Beschreibung von dem Lande Kamtschatka, dessen Einwohnern, deren Sitten, Nahmen, Lebensart und verschiedenen Gewohnheiten. (Description of Kamchatka, its Inhabitants, their Customs, Names, Way of Life, and Different Habits) Leipzig. 1774. pp. 92-93.

[Steller, a member of the Krasheninnikov expedition, stayed on and later published a valuable account of his life in Kamchatka. – RGW]

Among the mushrooms the poisonous fly-agaric (in Russian *muchamoor*, in Italmen [Kamchadal] *ghugakop*) is highly valued. In the Russian settlements this habit has been lost for a long time. However, around the Tigil and towards the Koryak border it is very much alive. The fly-agarics are dried, then eaten in large pieces without chewing them, washing them down with cold water. After about half an hour the person becomes completely intoxicated and experiences extraordinary visions. The Koryak and Yukagir are even fonder of this mushroom. So eager are they to get it that they buy it from the Russians wherever and whenever possible. Those who cannot afford the fairly high price drink the urine of those who have eaten it, whereupon they become as intoxicated, if not more so. The urine seems to be more powerful than the mushroom, and its effect may last through the fourth or the fifth man. Despite the fact that I have personally made these observations in 1739, some people have contradicted my experiences. I have therefore taken great pain to establish the truthfulness of what has been recorded here. Reports from persons whose authority cannot be attacked have confirmed my findings. Thus a man from the lower gentry named Kutukov, having to guard the reindeer herd, has noticed that these animals have frequently eaten that mushroom, which they like very much. Whereupon they have behaved like drunken animals, and then have fallen into a deep slumber. When the Koryak encounter an intoxicated reindeer, they tie his legs until the mushroom has lost its strength and effect. Then they kill the reindeer. If they kill the animal while it is drunk or asleep and eat of its flesh, then everybody who has tasted it becomes intoxicated as if he had eaten the actual fly-agaric.

[6]

GEORGI, Johann Gottlieb. Russia: or, A Compleat Historical Account of all the Nations which Compose that Empire. London. 1780. First published in German in St. Petersburg, 1776-1780. pp. 189-190.

Their drink [of the Ostyak] is water, broth, and fish-soups, a great deal of milk, and brandy whenever they are rich enough to buy any. The Ostyaks are very fond of getting drunk; and, as they have but seldom the means of procuring strong liquors for that purpose, they get intoxicated by smoking a great quantity of strong tobacco, and by chewing a kind of mushroom called the fly mushroom. . . . Numbers of the Siberians have a way of intoxicating themselves by the use of mushrooms, especially the Ostyaks who dwell about Narym. To that end they either eat one of these mushrooms quite fresh, or perhaps drink the decoction of three of them. The effect shews itself immediately by sallies of wit and humour, which by slow degrees arises to such an extravagant height of gaiety, that they begin to sing, dance, jump about, and vociferate: they compose amorous sonnets, heroic verses, and hunting songs. This drunkenness has the peculiar quality of making them uncommonly strong; but no sooner is it over than they remember nothing that has passed. After twelve or sixteen hours of this enjoyment they fall asleep, and, on waking, find themselves very low-spirited from the extraordinary tension of the nerves: however, they feel much less head-ache after this method of intoxication than is produced by spirituous liquors; nor is the use of it followed by any dangerous consequences.

★

From the author's chapter on the 'The Yakoutes', Vol. 2, p. 394:

They never wash their hands or any part of their person. Their skin-sacks that hold the milk stink abominably, and communicate a horrid taste to their contents. In the summer season they drink so much koumiss, and smoke tobacco so constantly, that they are frequently drunk. When they intend to get fuddled in a decent manner, they endeavour to procure the Russian brandy; but, as they are but seldom able to obtain it, they supply this want by mushrooms of an inebriating quality, as the Ostyaks, and several other people of Siberia do.

[So far as I know, this is the only passage in the Siberian texts that attributes the eating of the fly-agaric to the Yakut, whose language belongs to the Altaïc family. Perhaps fringe settlements of Yakut had taken to the habit from their neighbors, *e.g.*, the Yukagir, who apparently in the 18th century were still eating the *mukhomor*. (Sten Bergman's statement [27] about the Lamut addiction to the fly-agaric falls into the same category of isolated testimony.) Or in Georgi's case he may simply be mistaken. His book with its wealth of detail about the peoples of Siberia holds the reader's attention but it is exasperating in that it is inadequately documented: the reader usually does not know what the author observed personally and what he took from others, nor who the others were nor whether they were reliable. The diaries of his travels in Siberia would be most interesting, but we do not know whether they survive. – RGW]

[7]

LESSEPS, Jean Baptiste Barthélemy, Baron de. Journal historique du voyage de M. de Lesseps. . . . Paris. 1790. Translated into German and English. The English edition appeared in 1790 in two volumes in London, entitled 'Travels in Kamchatka, during the Years 1787 1788'. The following is from the English edition, Vol. 2, pp. 90-91 and 104-5.

[How supercilious and superficial are this French aristocrat's remarks about the fly-agaric, which he seems not to have recognized as a mushroom common in his own country. – RGW]

Their passion for strong liquors, increased by the dearness of brandy, and the difficulty of procuring it on account of their extreme distance, has led them to invent a drink, equally potent, which they extract from a red mushroom, known in Russia as a strong poison by the name of *moukhamorr*.[1] They put it in a vessel with certain fruits, and it has scarcely time to clarify when their friends are invited to partake of it. A noble emulation inflames the guests, and there is a contest of who is best able to disburden the master of the house of his nectar. The entertainment lasts for one, two, or three days, till the beverage is exhausted. Frequently, that they may not fail of being tipsy, they eat the raw mushroom at the same time. It is astonishing that there are not more examples of the fatal effects of this intemperance. I have seen however some amateurs made seriously ill, and recovered with difficulty; but experience does not correct them, and upon the first occasion that offers, they return to their brutish practice. It is not from absolute sensuality, it is not from the pleasure of drinking a liquor, that by its flavour creates an irresistible craving for more; they seek merely in these orgies a state of oblivion, of stupefaction, of total brutishness, a cessation of existence, if I may so call it, which constitutes their only enjoyment, and supreme felicity.

★

. . . On the eve of their magic ceremonies, they pretend indeed to fast all the day, but they make up for this abstinence at night by a profusion of the *moukamorr*, the intoxicating poison I have described, which they eat and drink to satiety. This preparatory intoxication they consider as a duty. It is probable that they feel its effects the next day, and that they derive from it an elevation of spirits that contributes to derange their minds, and give them the necessary strength to go through their extravagant transports.

[8]

SARYCHEV, Gavriil Andreevich. Achtjährige Reise im Nordostteil Sibiriens das Eismeer und den Östlichen Ozean (1785-1793). (An Eight-Year Voyage in Northeastern Siberia, on the Arctic Ocean and the Northeast Pacific) Leipzig. 1905-15. Translated from the Russian original published in 1802 in St. Petersburg. pp. 274-5.

[There is no direct reference to the fly-agaric, but on pp. 274-5 of the reprint published in 1954 we find a curious report of the death of two reindeer

1. It is used in the Russian houses to destroy insects.

in the Chukchi country. Their death is attributed to an excessive intake of human urine. It sounds as though the urine in question was impregnated with the metabolite of the fly-agaric. – RGW]

In the last few days the Chukchi have had two dead reindeer and take the cause to be that they had given them too much human urine to drink. They give them some from time to time in order to make them strong and improve their staying-power. The fluid has the same effect on the reindeer as intoxicating drink has on people who have fallen victim to the drinking habit. The reindeer become just as drunk and have just as great a thirst. At night they are noisy and keep running around the tents in the expectation of being given the longed-for fluid. And when some is spilled out into the snow, they start quarreling, tearing away from each other the clumps of snow moistened with it. Every Chukchi saves his urine in a sealskin container which is especially made for the purpose and from which he gives his reindeer to drink. Whenever he wants to round up his animals, he only has to set this container on the ground and slowly call out 'Girach, Girach!', and they promptly come running toward him from afar.

[9]

KOPEĆ, Joseph. Dziennik . . . (Diary of a Journey Along the Whole Length of Asia . . .; in Polish) Wrocław. 1837. An amplified text was published in Berlin, in 1863, from which this translation from Polish, done by Prof. Wiktor Weintraub, was made. pp. 198-202.

[Joseph Kopeć, a Brigadier in the Polish army, was a man of letters. According to Professor Wiktor Weintraub of Harvard University, he was mentally unbalanced and prone to exaggeration but in spite of his weaknesses his good faith can be relied upon. He was not a liar and he ate the mushrooms. The year is 1797 or perhaps 1796. Kopeć is in Kamchatka. He is ill and running a fever. He arrives with his companions at a native settlement covered with snow-drifts and enters the yurt, as is customary, through the opening at the top of the roof from which the smoke of the fire emerges. He then describes what happens. – RGW]

Hardly a few moments had passed when a sudden change of air brought about a great change in my sick body. The air of this closed *yurta*, always stinking, mixed with the acrid scent of whale-fat used as lamp oil, made me so weak that I thought the hour of my death would strike. Thus abandoning the fire and tea I called for my evangelist in the hope of getting some help from this man, a bit more educated, as I understood, than other people in the art of healing. After having learned about my mishap, the evangelist comes a little later, he approaches the fire, and ordering me to sit up he tells me first to drink my tea. While I am doing this, the Kamchadals bring from the middle of the tent a large number of ermine hides and deer skins. Feeling a bit revived, I ask what is the meaning of this. To which the evangelist, complying with my curiosity, says to me:

'Before I give you the medicine, I must tell you something important. You have lived for two years in Lower Kamchatka but you have known nothing of the treasures of this land. Here,' opening some birch bark in which a few mushrooms were wrapped, 'are mushrooms that are, I can say, miraculous. They grow only on a single high mountain close to the volcano and they are the most precious creations of nature.'

'Take into account, Sir,' the evangelist goes on, 'that the hides brought by local people I receive as gifts in exchange for these mushrooms. They would even give all their possessions, had I many of them and if only I sought to take advantage of the situation. These mushrooms have a special and as though supernatural quality. Not only do they help the man who uses them but he sees his own future as well. Since you are weak you should eat one mushroom. It will give you the sleep you are lacking.'

Hearing so many strange things about the merits of that mushroom, I was in doubt for a long time whether I should make use of it. However, the wish to recover my health and above all to sleep overcame my fears, and so I ate half my medicine and at once stretched out, for a deep sleep overtook me. Dreams came one after the other. I found myself as though magnetized[1] by the most attractive gardens where only pleasure and beauty seemed to rule. Flowers of different colours and shapes and odours appeared before my eyes; a group of most beautiful women dressed in white going to and fro seemed to be occupied with the hospitality of this earthly paradise. As if pleased with my coming, they offered me different fruits, berries, and flowers. This delight lasted during my whole sleep, which was a couple of hours longer than my usual rest. After having awakened from such a sweet dream, I discovered that this delight was an illusion. I was distressed that it had disappeared, as if it had been true happiness. These impressions made pleasant for me the few hours that remained until the end of the day. Having received such bewitching support from the miraculous mushroom and even having been fortified by sleep such as I had not had for a long time, I started to have

1. When Kopeć wrote, 'magnetism' had recently caught the attention of scientists and intellectuals in Europe: this explains his usage of the word. – RGW.

confidence in its supernatural qualities (as my evangelist had taught me to do), and with the approach of night I asked my physician for a second helping. He was pleased with my courage and at once bearing the offering of his friendly benevolence, he gave me a similar whole mushroom. Having eaten this stronger dose, I fell soundly asleep in a few minutes. For several hours new visions carried me to another world, and it seemed to me that I was ordered to return to earth so that a priest could take my confession. This impression, although in sleep, was so strong that I awoke and asked for my evangelist. It was precisely at the hour of midnight and the priest, ever eager to render spiritual services, at once took his stole and heard my confession with a joy that he did not hide from me. About an hour after the confession I fell asleep anew and I did not wake up for twenty-four hours. It is difficult, almost impossible, to describe the visions I had in such a long sleep; and besides there are other reasons that make me reluctant to do so. What I noticed in these visions and what I passed through are things that I felt I had seen or experienced some time before, and also things that I would never imagine even in my thoughts. I can only mention that from the period when I was first aware of the notions of life, all that I had seen in front of me from my fifth or sixth year, all objects and people that I knew as time went on, and with whom I had some relations, all my games, occupations, actions, one following the other, day after day, year after year, in one word the picture of my whole past became present in my sight. Concerning the future, different pictures followed each other which will not occupy a special place here since they are dreams. I should add only that as if inspired by magnetism I came across some blunders of my evangelist and I warned him to improve in those matters, and I noticed that he took these warnings almost as the voice of Revelation.

It is not for me to argue about the usefulness and the influence on human health of this miraculous mushroom. But I can state that its medical usefulness, had it been known among more educated peoples, should have earned it a place among so many known remedies of nature in the matter of fighting human maladies. Can anyone deny that in spite of our vast knowledge (relative to our forces) of natural phenomena, there still exist almost countless phenomena about which we can only guess? Can one put a limit to nature at a point that delimits the possibilities of inquiries and discoveries of human research? Innumerable effects of recently discovered magnetic forces, effects that cannot be detected by physical means nor pinpointed with any degree of precision to some specification on the human body, seem to reconcile in some measure the controversy concerning this mushroom. It is then possible that in the sleep brought by the influence of this mushroom, a man is able to see at least some of his real past and if not the future at least his present relations. If someone can prove that both the effect and the influence of the mushroom are non-existent and erroneous, then I shall stop being defender of the miraculous mushroom of Kamchatka.

Whatever may be the nature and qualities of the above-mentioned mushroom, I must confess that the taking of it had a powerful effect on my mind, as well as a strong impression [on my senses], so that I grew disquieted to a certain degree, from which passing into anxiety I became finally gloomy. This credulity having even the power of faith was based first of all on the conviction of the evangelist that my truthful visions were a true warning of heaven, and, secondly, on a conviction coming from within me, when later I perceived, being awake, the confirmation of what the dreams had predicted. This led me to have confidence in the dependability of my dreams about the future. However, as time passed, during my travels, this faith started to slacken and when its influence on my mind ended, a peace of soul returned to me together with better health.

[10]

Langsdorf, Georg Heinrich von. Einige Bemerkungen, die Eigenschaften des Kamtschadalischen Fliegenschwammes betreffend. (Some Remarks Concerning the Properties of the Kamchadal Fly-Agaric) Wetterauischen Gesellschaft für die gesammte Naturkunde. Annalen, Vol. 1, No. 2, Frankfurt M. 1809. pp. 249-256. Paper submitted by the author in French to the Russian Imperial Academy of Sciences in St. Petersburg.

The plant kingdom is of immeasurable influence and usefulness for mankind, since it supplies most of our clothing, food, drink, and shelter. The medical science of primitive peoples consists entirely in their knowledge of the more or less efficacious plants, and everyday experience confirms the fact that even a number of plants native to our own regions are known to many uneducated nations almost more thoroughly than they are to us.

To demonstrate this assertion, I should like to say at this point something about the nature and effects of the fly-agaric, which we regard as extremely poisonous but which is used by various inhabitants of northeastern Asia as an intoxicant just as wine, brandy, arrack, opium, kava, and the like are used by other nations. During my stay in Kamchatka I had the opportunity to gather detailed information on the effects of this mushroom, and today I shall try to tell you the most important points as briefly as possible.

Ordinarily it would be necessary first to make a closer examination of the nature of the mushroom, in so far as possible, and thereby show whether, and to what extent, the mushrooms found in those areas differ from our own.

However, since time is too short at this moment, I shall not give any detailed description of the ordinary fly-agaric and shall confine myself to pointing out that, at least on the basis of four dried specimens which I brought with me from Kamchatka and of a drawing made there by Privy Councillor Tilesius, it does appear that some difference exists between Kamchadal mushrooms and those of our own country: the Kamchadal mushroom has a cap with a navel-like protuberance in the middle, its stalk seems to grow thicker towards the base, and, in particular, the lamellæ may be yellowish rather than white. However, since this cannot be stated with any certainty until the living mushroom has again been observed in Kamchatka, we shall regard it for the time being as a special variety:

Amanita muscaria var. *Camtschatica.*

Isolated fly-agarics grow almost everywhere in Kamchatka, in birch forests and on dry plains. They are found most abundantly in the central part of the peninsula, especially around Vishna Kamchatka and Milkova Derevna. In some years they are seen in great numbers, but in others they are extremely scarce.

The Kamchadals gather them usually during the hottest months of July and August; they maintain that those that dry by themselves in the earth, on the stalk, and that are somewhat furry and velvety to the touch on the under side of the cap have a far stronger narcotic effect than those picked fresh and strung up to dry in the air.

The size is variable, with diameters ranging from 1-1½ to 5-6 inches.

The smaller mushrooms, which are bright red and covered with many white warty protuberances, are said to be far stronger in narcotic power than the larger ones, which are pale red and have few white spots.

Since the establishment of closer contacts with the Russians, the Kamchadals have taken particularly to drinking vodka and have left the consumption of fly-agarics to their wandering neighbors, the Koryaks, for whom they gather the fly-agarics and trade them very profitably for reindeer.

The usual way to consume fly-agarics is to dry them and then to swallow them at one gulp, rolled up into a ball, without chewing them; chewing fly-agarics is considered harmful, since it is said to cause digestive disturbances.

Sometimes these mushrooms are cooked fresh and eaten in soups or sauces, since they then taste more like the usual edible mushrooms and have a weaker effect, so that when the mushrooms are prepared in this way, a larger amount can be eaten without harmful results. Occasionally, too, fly-agarics are soaked in berry juice, which one may thereafter drink at his pleasure as a genuine intoxicating wine. Juice squeezed from bilberries (*Vaccinium uliginosum*) is said to be most suitable for this purpose because it heightens the intoxicating effect, so that one may expect to achieve a more potent result with a smaller quantity.

The body's predisposition or susceptibility to the intoxicating effect of fly-agarics apparently is not the same at all times, since the same person may sometimes be very strongly affected by a single mushroom and at other times remain completely unaffected after eating twelve to twenty of them. Ordinarily, however, one large fly-agaric or two small ones are enough to make an enjoyable day.

The narcotic effect is also said to be heightened by the drinking of large quantities of cold water afterwards.

The narcotic effect begins to manifest itself about a half hour after eating, in a pulling and jerking of the muscles or a so-called tendon jump (although sometimes these effects appear only after an hour or two); this is gradually followed by a sense of things swimming before the eyes, dizziness, and sleep. During this time, people who have eaten a large quantity of mushrooms often suffer an attack of vomiting. The rolled-up mushrooms previously swallowed whole are then vomited out in a swollen, large, and gelatinous form, but even though not a single mushroom remains in the stomach, the drunkenness and stupor nevertheless continue, and all the symptoms of fly-agaric eating are, in fact, intensified. Many other persons never vomit, even after eating copiously of the mushrooms.

The nature of the ecstasy or drunkenness caused by the fly-agaric resembles the effects of wine and vodka to the extent that it renders unconscious the persons intoxicated with it and arouses in them feelings that are mostly joyful, less often gloomy. The face becomes red, bloated, and full of blood, and the intoxicated person begins to do and say many things involuntarily.

In the milder stages, as I have said, there are tendon jerks, but in the more advanced stages there are jerky movements of the limbs, and then the intoxicated persons often appear to be dancing and making the most outlandish pantomime movements with their hands. Similarly, the head and neck muscles are also in a constantly convulsive state; if a person has eaten mushrooms to excess, he goes into genuine convulsions.

According to their own statement, persons who are slightly intoxicated feel extraordinarily light on their feet and are then exceedingly skillful in bodily movement and physical exercise.

The nerves are highly stimulated, and in this state the slightest effort of will produces very powerful effects. Consequently, if one wishes to step over a small stick or a straw, he steps and jumps as though the obstacles were tree trunks. If a man is ordinarily talkative, his speech nerves are now in constant activity, and he involuntarily blurts out secrets, fully conscious of his actions and aware of his secret but unable to hold his nerves in check. In this condition a man who is fond of dancing dances and a music-lover sings incessantly. Others run or walk quite involuntarily, without any intention of moving, to places where they do not wish to go at all.

The muscles are controlled by an uncoordinated activity of the nerves themselves, uninfluenced by and unconnected with the higher will-power of the brain, and thus it has occasionally happened that persons in this stage of intoxication found themselves driven irresistibly into ditches, streams, ponds, and the like, seeing the impending danger before their eyes but unable to avoid certain death except by the assistance of friends who rushed to their aid. In this intense and stimulated state of the nervous system, these persons exert muscular efforts of which they would be completely incapable at other times; for example, they have carried heavy burdens with the greatest of ease, and eye-witnesses have confirmed to me the fact that a person in a state of fly-agaric ecstasy carried a 120-pound sack of flour a distance of 10 miles, although at any other time he would scarcely have been able to lift such a load easily.

But the strangest and most remarkable feature of the fly-agaric is its effect on the urine. The Koryaks have known since time immemorial that the urine of a person who has consumed fly-agarics has a stronger narcotic and intoxicating power than the fly-agaric itself and that this effect persists for a long time after consumption. For example, a man may be moderately drunk on fly-agarics today and by tomorrow may have completely slept off this moderate intoxication and be completely sober; but if he now drinks a cup of his own urine, he will become far more intoxicated than he was from the mushrooms the day before. It is not at all uncommon, therefore, that drunkards who have consumed this poisonous mushroom will preserve their urine as if it were a precious *liqueur* and will drink it as the occasion offers.

The intoxicating effect on the urine is found not only in the persons who have eaten the fly-agaric itself but also in every person who drinks the urine. Among the Koryaks, therefore, it is quite common for a sober man to lie in wait for a man intoxicated with mushrooms and, when the latter urinates, to catch the urine secretly in a container and in this way to obtain a stimulating drink even though he has no mushrooms.

Because of this peculiar effect, the Koryaks have the advantage of being able to prolong their ecstasy for several days with a small number of fly-agarics. Suppose, for example, that two mushrooms were needed on the first day for an ordinary intoxication; then the urine alone is enough to maintain the intoxication on the following day. On the third day the urine still has narcotic properties, and therefore one drinks some of this and at the same time swallows some fly-agaric, even if only half a mushroom; this enables him not only to maintain his intoxication but also to tap off a strong liquor on the fourth day. By continuing this method it is possible, as can easily be seen, to maintain the intoxication for a week or longer with five or six fly-agarics.

Equally remarkable and strange is the extremely subtle and elusive narcotic substance contained in the fly-agarics, which retains it effectiveness permanently and can be transmitted to other persons: the effect of the urine from the eating of one and the same mushroom can be transmitted to a second person, the urine of this second person affects a third, and similarly, unchanged by the organs of this animal secretion, the effect appears in a fourth and a fifth person.

For still another remarkable observation concerning the nature of the fly-agaric I am indebted to Steller, who, in his description of the Kamchatka region [p. 240] states the following: 'It was related to me by reliable people, among both the Russian and the Koryak nation – indeed, by a man from the lower gentry named Kutukov, who has charge of the Cassa reindeer herd – that reindeer often eat this mushroom, among others (for they have a great appetite for mushrooms), after which they fall down and thrash about for a while as if they were drunk and then fall into a deep sleep. When the Koryaks find a wild reindeer, therefore, they tie its feet until it has slept off the effects and the mushroom has lost its potency, and only then do they kill the reindeer; for if they killed such an animal while it was sleeping or still raving, everyone who ate its flesh would go into a similar frenzy, as if he had actually eaten the fly-agaric himself.'

Although I made great efforts to find out something about the harmful or possibly deadly effects of the fly-agaric, I could not obtain any satisfactory information on the subject.

The Koryaks greatly prefer fly-agarics to the Russians' vodka and maintain that after eating fly-agarics a man never suffers from headaches or other ill effects.

It is true that in extremely rare cases (of which no one could recall any specific example) persons who consumed an extraordinarily large quantity of the mushrooms are said to have died in convulsions, senseless and speechless, after six or eight days. However, it is not reported that moderate consumption ever produced any harmful after-effects.

If, contrary to expectations, immoderate consumption of fly-agarics should nevertheless be followed by pressure on the stomach or some other disturbance, two to three spoonfuls of fat, blubber, butter, or oil are reputed to be an infallible remedy that can relieve any ill effects.

There are some people in Kamchatka who drink a glass of bilberry [*Vaccinium uliginosum*] juice in which fly-agarics have been soaked whenever they have a stomach-ache, colic, or other ailment and who regard it as a universal remedy; however, I was not able to ascertain whether the consumption of fly-agaric is followed by constipation or diarrhœa or by an increase or decrease in the urine.

I was also unable to obtain any satisfactory answer when I asked whether the taste or smell of the urine had been changed – everyone was probably ashamed to admit that he had drunk his own urine or somebody else's. Nevertheless, it strikes me as not improbable that fly-agarics, like turpentine, asparagus, and other things, impart a special, possibly quite pleasant, smell and taste to the urine; by analogy, it would be worth investigating whether other narcotic substances, such as opium, *Digitalis purpurea*, cantharides, *etc.* also retain their properties in the urine.

The nature of the fly-agaric, therefore, offers the scientist, physician, and naturalist a great deal of food for thought: our *materia medica* might perhaps be enriched with one of the most efficacious remedies, and judicious physicians might find in the fly-agaric the most potent remedy to apply to the body in cases of paralysis and other diseases of the extremities.

[11]

Erman, Adolph. Reise um die Erde durch Nord-Asien und die beiden Oceane in den Jahren 1828, 1829 und 1830 ausgeführt. (A Journey around the World through Northern Asia and both Oceans in 1828, 1829 and 1830) Berlin. 1833-48. p. 223.

[The author speaks of the fly-agaric among the Kamchadal and Koryak peoples. – RGW]

My companions had eagerly been gathering fly-agaric (*Amanita muscaria* Esenb.; in Russian, *mukhomor*, *i.e.*, fly pest) both in the woods through which we had ridden in the morning and now at the foot of the Northern Baidar mountains. Because of its brilliant red color they caught sight of every one of them from afar and this always caused a sudden halt in our caravan which at first surprised me. They now confirmed what had already been told me in Tigilsk about the intoxicating properties of this mushroom and said that it was not eaten in Sedanka but only gathered for the Koryaks, who in wintertime often paid a reindeer for a single dried piece. *Mukhomor*, they said, was much rarer in northern Kamchatka and the Koryaks had only learned about its properties because the meat of reindeer which had eaten it had an effect that was as intoxicating as the mushroom itself. It was this experience that had caused them to use it most sparingly and with maximum advantage and here was why they even collected the urine of persons who had managed to come by a *mukhomor*, and mixed it with their drink as an intoxicant that was still very effective.

★

p. 259. On the same day I tasted for the first time a plant eaten by all Kamchadals, learning in the days that followed to appreciate it – in addition to other fine qualities – for the ease with which it could be carried because of its unusual lightness, and with which it could be used since it required no preparation. I am referring to a felt-like substance made from the stalks of the so-called *kiprei* (*Epilobium angustifolium*). Several of the stalks are laid on top of one another and squeezed and beaten in layers two to three inches wide so that the sweet sap inside penetrates the green bast and the pieces of woody cortex; thereafter this turns into firm dark-green pleasant-smelling strips made up into lengths of four to six feet and the width already mentioned. These are eaten raw but always smeared with butter or seal fat, and, with this addition, I too now found them palatable, easily digestible and nourishing.

The example of the Russians [from the southern tip of Kamchatka] had caused the Yelovka natives to try an agricultural experiment during this same year, for they had planted some potatoes on a plot between their houses. The plants seemed, however, in a very sad state, owing no doubt to incorrect cultivation, and on part of the plot they had already been replaced by much more carefully tended mushrooms. These belonged to Toyon's wife, who had picked them in the woods in the Spring and transplanted them when they were tiny, and now she pointed out to me with special pride the big size of their scarlet caps and their many white spots, which are considered to presage a powerful effect. She spoke with the most unrestrained enthusiasm of her love for this intoxicant, and I noticed she had a glassy look in her eyes and copper-red cheeks, which no doubt came from excessive indulgence although I did not see the same symptoms in many other mushroom-eaters I met later.

★

pp. 304-306. Thereafter in one of those open places in the woods we gathered twenty mushrooms, to the immense joy of the older of my companions who, as an enthusiastic devotee of this intoxicant, again praised its powers and its benefits. He affirmed, from his own experience, the most varied effects of this mushroom on herbivorous animals: wild reindeer that have eaten some of them are often found so stupefied that they can be tied with ropes and taken away alive; their meat then intoxicates everyone who eats it, but only if the reindeer is killed soon after being caught; and from this it appears that the communicability of the narcotic substance lasts about as long as it would have affected the animals' own nerves. He also said that this mushroom intoxication had a quite different effect from alcoholic drunkenness, since the former put the Kamchatka natives into a peaceful and gentle (*skromno*) mood, and they had seen how differently the Russians were affected by spirits.[1]

... There is no doubt, however, about a 'marvellous increase in physical strength,' which the man from Yelovka praised as still another effect of mushroom intoxication. 'In harvesting hay,' he said, 'I can do the work of three men from morning to nightfall without any trouble, if I have eaten a mushroom.' Of the various ways of using *mukhomor* he said that the best was the simplest way, *viz.*, drying it, swallowing it raw, and washing it down with water. On the other hand, the Russians of Klynchevsk, who according to him pick whole packhorse loads of this valuable plant, prepare an extract by decocting it in water, and try to take away its extremely disgusting taste by mixing the extract with various berry juices.

★

p. 312. Sept. 5. Although all we could boast of were a few exertions, but no unusual adventures, our old hunter nevertheless felt himself entitled to be rewarded with the pleasures of intoxication. Immediately after our arrival he

1.... As to the effects of *mukhomor* on the Russians who always ate too many of them (up to ten mushrooms, whereas I never saw a Kamchadal use more than two), Krasheninnikov tells of some cases in which the most intense excitement ended in the fury of the intoxicated man directed against himself. According to him, a young Cossack from Bolsheretsk, after eating many mushrooms, was restrained with difficulty from putting a knife into his abdomen, while another actually killed himself by self-castration. [*Vide* **[4]** – RGW] exchanged some of the mushrooms we had gathered for dry ones, of which he at once swallowed three small pieces (one and one-half mushrooms) and washed them down with water. When it is fresh, the *mukhomor* is so sticky and of such loose consistency that it is hard to swallow without chewing. This, however, makes the unpleasant taste of the *mukhomor* so disgusting that it is considered impossible to cope with any real quantity of them in this condition.

The strong will that is here displayed in order to obtain pleasant excitation is even more striking when one observes that this sets in only long after the narcotic is used and certainly not without troublesome transition. Thus, a good hour after he had eaten the mushrooms they had shown no effect on his mood, and he told me then that he would have to lie down quietly and sleep till the next morning in order to see the most pleasant things, some of them in dreams while asleep and others after waking next day.

[12]

MAYDELL, Baron Gerhard von. Reisen und Forschungen im Jakutskischen Gebiet Ostsibiriens in den Jahren 1861-1871. (Journeys and Investigations in the Jakutskaia Oblast′ of Eastern Siberia in 1861-1871) Published as Vol. 1-2 in Series IV of Beiträge zur Kenntniss des Russischen Reiches und der angrenzenden Länder Asiens, St. Petersburg. 1893. Vol. 1, pp. 298-300.

At the Paren′ I also obtained some dried fly-agaric (*Amanita muscaria*), which is very highly prized among the Koryaks for its intoxicating effect. It is not eaten in the fresh state, in which it is believed to be poisonous, but always hung up to be smoked until it is shriveled and quite dry so that it will keep well. It is said to occur only among birch trees and hence is confined to a few localities, among which I heard Penzhinskoye and Markovo mentioned in particular. When a Koryak consumes the fly-agaric, he chews the dried mushroom and then drinks it down with water. After a while he becomes greatly stimulated, converses with people whom he sees although they are not there at all, tells them with much satisfaction about his great wealth, and so on. He can also be asked questions by the people present, and he will sometimes answer them quite sensibly but always with reference to the things which he imagines and which, in his intoxication, seem real to him. During the intoxication he is quite capable of walking from place to place without staggering, but the mushroom seems to produce a peculiar effect on his optic nerves which makes him see everything on a greatly enlarged scale. For this reason it is a common joke among the people to induce such an intoxicated man to walk and then to place some small obstacle, such as a stick, in his way. He will stop, examine the little stick with a probing eye, and finally jump over it with a mighty bound. Another effect of the mushroom is said to be that the pupils become much enlarged and then contract to a very small size; this process is said to be repeated several times. When the drunken man has sobered up, he feels no bodily discomfort at all but only regrets that his beautiful visions have given way to harsh reality; in reply to questions he will say that he has been in pleasant company, that he was the owner of fine herds, and the like. However, the effect of the mushroom seems to differ from that of opium: the visions are not of an erotic nature; instead the mushroom produces only a feeling of great comfort, together with outward signs of happiness, satisfaction, and well-being. Thus far the use of the fly-agaric has not been found to produce any harmful results, such as impaired health or reduced mental powers; this is probably due to the fact that, in general, the Koryaks are seldom able to indulge their passion for the mushroom, since it is found only in a few places, and even there only in small quantities.[1]

The Chukchis – at least those I have encountered – were completely unacquainted with this intoxicant; however, it was said in Markovo that when the people of the town happened to be unable to get brandy, they would sometimes resort to eating fly-agarics.

1. I was told of only one fatal case that resulted from eating fly-agaric but it was explained as having been due to the fact that the man in question – a Russian – had taken fresh instead of dried mushrooms in rather large quantities.

The mushroom has a surprising effect on urine; *i.e.*, it seems as if the intoxicating effect of the narcotic contained in the mushroom passes principally into the urine. It is a well-known fact that as soon as a Koryak feels that his state of intoxication is beginning to ebb, he drinks his own urine if he has no more mushrooms, and the effect is restored. This cannot be repeated, however, for the urine has no effect a second time. I was told about a man who, while riding with a Koryak, stopped at a yurt near which a man, another Koryak, was sitting. He was thoroughly intoxicated and therefore in a very happy mood. The Koryak who had been riding naturally asked him for a piece of mushroom, which the latter deeply regretted he could not give because he had none left. However, he went and urinated, and then gave the product to his guest, who drank it all and now became intoxicated also. After a little while the host became sober while the guest's intoxication continued, and when the former complained about not being able to return to his pleasant state, the guest now gave him some of his urine, which however no longer had any effect.

[13]

Dittmar, Carl von. (Karl von Ditmar) Reisen und Aufenthalt in Kamtschatka in den Jahren 1851-1855. (Journey and Sojourn in Kamchatka in 1851-1855) Published as Vol. 8 in Series 3 of Beiträge zur Kenntniss des Russischen Reiches und der angrenzenden Länder Asiens. St. Petersburg. 1900. Part II, Section 1, pp. 98-100.

Finally, I should mention one other plant of this region of meadows and birch forests which plays a role in the life of the peoples living here. This is the fly-agaric, *Amanita muscaria*, which the local inhabitants often call *mukhomor*. These mushrooms are fiery red with many large white spots; they are not rare and can be recognized from far away in the rich green of somewhat shady and damp places. The Kamchadals themselves have little use for the mushrooms, but they like to gather them in order to sell them in a dried state to the Koryaks and Chukchis, who buy them eagerly. Although this mushroom is a great favourite in the North, it appears not to occur there at all, or at least to be a great rarity; however, the intoxicating and nerve-stimulating effect of the *mukhomor* is known far and wide among the Northern peoples, and they are very fond of it. Both Koryaks and Chukchis like to carry with them a small box made of *Betula nana* in which they carry small bits of chopped-up dried fly-agaric, so that they may have their favorite intoxicant always ready to hand. Another such box contains tobacco, which they enjoy in three different ways – smoking it, chewing it, and taking it as snuff. On the other hand, the *mukhomor* is only chewed, and the juice is then swallowed. The mushroom plays an important role at every celebration, especially among the shamans, but it is also very frequently enjoyed at other times by anyone who has become addicted to this pleasure and can no longer get along without it, just as the alcoholic cannot get along without his alcohol or the opium-addict without his opium. Anyone who indulges freely in this pleasure soon becomes a slave to it and is willing to give anything to enjoy the intoxication again.

Mukhomor-eaters describe the narcosis as most beautiful and splendid. The most wonderful images, such as they never see in their lives otherwise, pass before their eyes and lull them into a state of the most intense enjoyment. Among the numerous persons whom I myself have seen intoxicated in this way, I cannot remember a single one who was raving or wild. Outwardly the effect was always thoroughly calming – I might almost say, comforting. For the most part the people sit smiling and friendly, mumbling quietly to themselves, and all their movements are slow and cautious. Walking seems to be uncomfortable to them, although they are quite capable of it. They have a glassy, almost imbecilic look in their eyes, as if they scarcely noticed their surroundings, and their facial features are somewhat distorted. People generally claim that the effect of the mushroom poison becomes more intense and more beautiful when it has already passed through another organism. Thus an intoxicated man will often be followed by someone else who wants to collect his urine, which is supposed to possess this effect to a particularly high degree. Similarly, the meat of reindeer that have accidentally eaten of the mushroom is said to possess the intoxicating effect in a very pleasant form. The use of the *mukhomor* seems to be a very old practice, since all the early authors, such as Pallas and Krasheninnikov, tell the same and similar stories.

[Ten years earlier, in 1890, von Dittmar had published his *Historischer Bericht*, 'Historical Report', according to the diary of his travels in Kamchatka, through the same publishing channels, in St. Petersburg. On pp. 524 and 590 there are two entries pertinent to our inquiry into the fly-agaric in Siberia. – RGW]

. . . I learned that shamans are very eager to take in a certain quantity of *Amanita muscaria* in order to get themselves into a stupor resembling complete insanity. The Koryak were complaining about the fact that this drug was not available at the time and that, as a rule, it was difficult to obtain in Taigonos. The mushroom does not grow on the peninsula and has to be brought from Kamchatka, where there is much of it and it is very effective. From Kamchatka this precious merchandise has to make its way all around the Penzhinsk Gulf, from peddler to peddler, and since everybody is wild about it, not much of it gets here.

. . . As I approached her [a drum-beating widow who hopes to resuscitate her dead husband by shaman's exercises – Translator], I immediately noticed that she was drunk with fly-agaric, a fact that was corroborated by the others. As a matter of fact, it is quite common here among the Koryak, and especially among the Chukchi, to produce small, round boxes made of birch bark containing small dried pieces of fly-agaric. People sniffing tobacco take out their little boxes, so these people take out theirs containing the mushroom. They chew the pieces, keeping them in their mouths for a long time without swallowing. They assert that this practice puts them into a state of bliss during which they see the most beautiful visions. They sit peacefully, without ranting and raving, their eyes wide open and staring, as if they no longer belonged to this world . . .

[14]

KENNAN, George. Tent Life in Siberia and Adventures among the Koraks and other Tribes in Kamtschatka and Northern Asia. New York and London. 1871, pp. 202-204. The journey was made in 1865-1870.

[To one versed in the literature this account by an American a century ago of his experience among the Koryak sounds disturbingly superficial and wrong-headed. Why was he astonished by the practice of eating the fly-agaric, when every prepared traveler in the region had known of it for generations? Why does he think that the 'natives' speak Russian? 'Muk-a-moor', as he writes the word, is Russian, not Koryak. Note his prudish reticence about the urine-drinking. – RGW]

...We ... were surprised, as we came out into the open air, to see three or four Koraks shouting and reeling about in an advanced stage of intoxication – celebrating, I suppose, the happy event which had just transpired. [A wedding] I knew that there was not a drop of alcoholic liquor in all Northern Kamtchatka, nor, so far as I knew, anything from which it could be made, and it was a mystery to me how they had succeeded in becoming so suddenly, thoroughly, hopelessly, undeniably drunk. Even Ross Browne's beloved Washoe, with its 'howling wilderness' saloons, could not have turned out more creditable specimens of intoxicated humanity than those before us. The exciting agent, whatever it might be, was certainly as quick in its operation, and as effective in its results, as any 'tangle-foot' or 'bottled-lightning' known to modern civilization. Upon inquiry we learned to our astonishment that they had been eating a species of the plant vulgarly known as toadstool. There is a peculiar fungus of this class in Siberia, known to the natives as 'muk-a-moor', and as it possesses active intoxicating properties, it is used as a stimulant by nearly all the Siberian tribes. Taken in large quantities it is a violent narcotic poison; but in small doses it produces all the effects of alcoholic liquor. Its habitual use, however, completely shatters the nervous system, and its sale by Russian traders to the natives has consequently been made a penal offence by Russian law. In spite of all prohibitions, the trade is still secretly carried on, and I have seen twenty dollars worth of furs bought with a single fungus. The Koraks would gather it for themselves, but it requires the shelter of timber for its growth, and is not to be found on the barren steppes over which they wander; so that they are obliged for the most part to buy it, at enormous prices, from the Russian traders. It may sound strangely to American ears, but the invitation which a convivial Korak extends to his passing friend is not, 'Come in and have a drink', but, 'Won't you come in and take toadstool?' Not a very alluring proposal perhaps to a civilized toper, but one which has a magical effect upon a dissipated Korak. As the supply of these toadstools is by no means equal to the demand, Korak ingenuity has been greatly exercised in the endeavor to economize the precious stimulant, and make it go as far as possible. Sometimes, in the course of human events, it becomes imperatively necessary that a whole band shall get drunk together, and they have only one toadstool to do it with. For a description of the manner in which this band gets drunk collectively and individually upon one fungus, and keeps drunk for a week, the curious reader is referred to Goldsmith's *Citizen of the World*, Letter 32. It is but just to say, however, that this horrible practice is almost entirely confined to the settled Koraks of Penzhinsk Gulf – the lowest, most degraded portion of the whole tribe. It may prevail to a limited extent among the wandering natives, but I never heard of more than one such instance outside of the Penzhinsk Gulf settlements.

[15]

LANSDELL, Henry. Through Siberia. London. 1882. Vol. II, pp. 644-5.

[Note the English author's unfamiliarity with the common fly-agaric; also his prudery. – RGW]

Among the flora, however, of North-Eastern Siberia is a peculiar mushroom spotted like a leopard, and surmounted with a small hood – the fly-agaric, which here has the top scarlet, flecked with white points. In other parts of Russia it is poisonous. Among the Koriaks it is intoxicating, and a mushroom of this kind sells for three or four reindeer. So powerful is the fungus that the native who eats it remains drunk for several days, and by a process too disgusting to be described, half-a-dozen individuals may be successively intoxicated by the effects of a single mushroom, each in a less degree than his predecessor.

[16]

JADRINTSEV, Nikolai Mikhailovich. Sibirien: geographische, ethnographische und historische Studien. (Siberia: Geographical, Ethnographic, and Historical Studies) Jena. 1886. p. 337.

[The only reference that the author makes is the following: – RGW]

...Furthermore, the natives are passionately addicted to the fly-agaric (*i.e.*, to a decoction of *Amanita muscaria*), which replaces liquor and is also sold to merchants.

[17]

PATKANOV, Serafim Keropovich. Die Irtysch-Ostjaken und ihre Volkspoesie. (The Irtysh Ostyak and their Folk-Poetry) St. Petersburg. 1897. p. 121.

[The only reference to mushrooms is in a discussion of shamanism. – RGW]

The shaman must get himself into an exalted state to be able to talk to the gods. To achieve this he consumes several (either seven or fourteen or twenty-one) fly-agarics, which are capable of producing hallucinations.

[18]

SLJUNIN, Nikolai Vasil'evich. Okhotsko-Kamchatskii krai. Estestvennoistoricheskoe opisanie. (The Okhotsko-Kamchatskii Kraj. An Essay in Natural History) St. Petersburg. 1900. In two volumes. Vol. 1, pp. 654-655.

The population dislikes mushrooms and therefore does not gather them. An exception is the fly-agaric (*Amanita muscaria*), widely used by the Koryaks because it produces a peculiar inebriation. They are so fond of this mushroom that they will pay considerable sums of money (in local terms, hides of foxes and sables) to obtain it. Exploiting this weakness, some people have set up a highly profitable commerce. Thus a certain Tykaniev, a Koryak of the Olyutorskoye settlement, made a fortune (a herd of reindeer) selling these mushrooms. Although the law forbids the trade in fly-agarics and other poisonous herbs and plants, it is almost impossible to enforce this law under the local conditions of life and the distances involved. The orders of the district commander (including the explanation of the noxious effect of the fly-agaric on the human organism) are completely ignored.

The effect of the fly-agaric is quite peculiar and reminds us of the scenes of inebriation, delirium, and hallucinations which we have frequently witnessed in Chinese opium dens. An excellent antidote is a glass of strong vodka or diluted alcohol. A quarter of an hour after swallowing the vodka, the Koryak who is totally oblivious of his environment under the effect of the mushroom, completely regains consciousness. He laments the disappearance of his world of dreams, because when inebriated he loses the concept of time and all objects appear to him greatly magnified (an effect similar to the one produced by hashish), which is the source of many jokes.

The Koryaks maintain that fresh fly-agarics are highly poisonous and hence do not eat them. They are first dried for a long time in the sun and over the fire in the tent. Only then is the mushroom consumed, together with fresh water to wash it down. The Koryaks claim that continuous consumption of the mushroom has no ill effect on the person's health. We have seen addicts, however, whose emaciated aspect, yellowish colour of the skin, and uncertain gait can only be attributed to protracted consumption of the fungus. Its inebriating effect is of short duration since the poison (an alkaloid) is rapidly eliminated by the urine which, if drunk, produces an effect similar to that of the mushroom itself.

Fatal cases are very infrequent and are attributable to the intake of fresh, non-dried mushrooms.

[19]

ENDERLI, J. 'Zwei Jahre bei den Tschuktschen und Korjaken.' (Two years among the Chukchi and Koryak) Petermanns Geographische Mitteilungen. Gotha. 1903. pp. 183-184.

[This is one of the most valuable accounts that we have. – RGW]

Very little alcohol is brought up to the North, at least around Gizhiga, but the Koryak know how to make intoxicating drinks from various kind of berries. In addition, however, they have another substance with which they can produce a narcotic intoxication, *viz.*, the fly-agaric. This mushroom is seldom found in those areas. It is collected by the women in the autumn, dried, and eaten on ceremonial occasions in the winter.

At the man's order, the woman dug into an old leather sack, in which all sorts of things were heaped one on top of another, and brought out a small package wrapped in dirty leather, from which she took a few old and dry fly-agarics. She then sat down next to the two men and began chewing the mushrooms thoroughly. After chewing, she took the mushroom out of her mouth and rolled it between her hands to the shape of a little sausage. The reason for this is that the mushroom has a highly unpleasant and nauseating taste, so that even a man who intends to eat it always gives it to someone else to chew and then swallows the little sausage whole, like a pill. When the mushroom sausage was ready, one of the men immediately swallowed it greedily by shoving it deep into his throat with his indescribably filthy fingers, for the Koryaks never wash in all their lives.

The effects of the poison became evident by the time the men had swallowed the fourth mushroom.[1] Their eyes took on a wild look (not a glassy look, as may be seen in drunken men), with a positively blinding gleam, and their hands began to tremble nervously. Their movements became awkward and abrupt, as if the intoxicated men had lost control of their limbs. Both of them were still fully conscious. After a few minutes a deep lethargy overcame them, and they began quietly singing monotonous improvised songs whose content was approximately 'My name is Kuvar, and I am drunk, I am merry, I will always eat mushrooms,' and so on. The song grew more and more lively and loud, sometimes interrupted by words shouted out at lightning speed; the animal-like wild look in the eyes grew stronger, the trembling of the limbs grew more intense, and the upper body began to move more and more violently. This condition lasted about ten minutes. All at once the men – first the Reindeer-Koryak and a little later the other man – were seized with a fit of frenzy. They suddenly sprang raving from their seats and began loudly and wildly calling for drums. (Every family owns disk-shaped drums of reindeer hide, which are used for religious purposes.) The women immediately brought two drums and handed them to the intoxicated men. And now there began an indescribable dancing and singing, a deafening drumming and a wild running about in the yurt, during which the men threw everything about recklessly, until they were completely exhausted. Suddenly they collapsed like dead men and promptly fell into a deep sleep; while they slept, saliva flowed from their mouths and their pulse rate became noticeably slower.

It is this sleep that provides the greatest enjoyment; the drunken man has the most beautiful fantastic dreams. These dreams are highly sensuous, and the sleeping man sees whatever he wants to.

1. Different values have also been mentioned for this dose.

After a half hour the two men awakened at almost the same time. The effects of the poison had subsided and both men were in possession of their senses, but their walk was uncertain and convulsive. Soon, however, the effects of the poison became apparent again; the drunken men were seized with a new but weaker fit of frenzy. Then they fell asleep again; they awakened for a short period to full consciousness, which was again succeeded by another fit. The attacks continued in this way a few more times, growing less violent each time. They probably would have stopped entirely after a few hours if these intoxicated men had not used another method which renewed the intensity of the intoxication.

It appears that the poison of the fly-agaric is excreted in the urine, and this, when a man drinks it, produces the same effects as the fly-agaric itself. Since fly-agarics are relatively rare in those regions, they are highly prized by the Koryaks, and they therefore consider it too expensive to waste their urine, whose effects are entirely similar to those of the mushroom.

I noticed now that a woman brought the awakened man a small sheet-metal container, into which the man voided his urine in the presence of everybody. The container is used exclusively for this purpose, and Koryaks carry it with them even when they travel. The drunken man (or, more properly, the poisoned man) put the container down next to him; the urine was still warm and the steam was rising densely in the cold yurt when the second mushroom-eater, who was just awakening, saw the urine container near him, seized it without a word, and drank a few large gulps. Soon after this the first man, the actual 'owner of the urine,' followed the other's example. After a few moments the urine they had drunk began to do its work, and the symptoms of the intoxication grew more violent, as they had before. Sleep alternated with attacks of frenzy and moments of complete calm. The intoxication was intensified each time by drinking urine. The frenzied dances and the drinking-bouts continued in this way all through the night, and it was almost evening of the next day before the Koryaks recovered from their stupor. The remaining urine was carefully preserved for a short time, to be used again on the next occasion. Even while traveling, when the Koryak leaves his settlement in a half-drunken condition, he never squanders his urine; he continues to collect it in the container which he carries with him for the purpose.

This is the greatest enjoyment, the merriest entertainment, that the Koryak knows, and he waits for it impatiently all year long. It is true that he likes alcoholic drinks better because of their milder form (this refers only to 95% pure alcohol [sic!], which many people drink without any admixture of water), for the effects of fly-agaric poisoning, in the form of heart palpitations and nausea, often last one or two days, and immoderate consumption of the mushroom involves the danger of madness or death. Such cases, however, occur very rarely.

But the natives believe that the fly-agaric, unlike alcohol, has the power to reveal the future to the man who consumes it; if, before eating the mushroom, the man recites over it certain definite formulas stating his wish to see the future, the wish will come true in his dream.

[20]

VANDERLIP, Washington B. In Search of a Siberian Klondike. New York. 1903. pp. 214-215.

[The author places the following episode in the Koryak village of 'Kaminaw', presumably Kaminov. – RGW]

A peculiar custom sometimes to be noted among these people [Koryaks] is that of drinking a kind of liquor made from a large species of mushroom. The effect is, in some respects, similar to that produced by hashish. At first the imbiber shakes as with the ague; and presently he begins to rave as if in delirium. Some jump and dance and sing, while others cry out in agony. A small hole looks to them like a bottomless pit, and a pool of water as broad as the sea. These effects are produced only when the beverage is used to excess; a small quantity has much the same effect as a moderate amount of liquor. Curiously enough, after recovering from one of these debauches, they claim that all the antics performed were by command of the mushroom. The use of it is not unattended by danger, for unless a man is well looked after he is likely to destroy himself. The Koryaks sometimes take this drug in order to work themselves to the point of murdering an enemy. Three or four of the mushrooms is a moderate dose, but when one wants to get the full effect one takes ten or twelve.

[Vanderlip offered his book to the public as an account of the author's experiences in Siberia, but the excerpt that we have quoted ought by rights to be relegated to the secondary sources, since the statements in it can be traced without exception to others, notably Krasheninnikov [4], with only minimal changes in wording. Vanderlip could have compiled his paragraph in the Reading Room of the New York Public Library. – RGW]

[21]

JOCHELSON, Waldemar. (Iokhel'son, Vladimir Il'ich) 1. The Koryak: Memoir of the American Museum of Natural History, New York. A publication of the Jesup North Pacific Expedition. Vol. VI, Part I. Religion and Myth, New York, 1905. Part II. Material Culture and Social Organization of the Koryak, 1908. pp. 582-584.

[Vladimir Bogoraz and Vladimir Jochelson were two distinguished Russian anthropologists. On the invitation of the American Museum of Natural History, the Imperial Academy of Science designated them to collaborate with the Jesup North Pacific Expedition, they to contribute the studies of the tribes of the Maritime Provinces of Siberia and the extreme tip of Siberia opposite Alaska. They wrote more extensively about the fly-agaric practices of these tribes than anyone else has done, and with keener discernment. They were in the field at the turn of the century. We reprint in full what they had to say about the fly-agaric habit and we bring to the reader's attention a number of folk tales in which the fly-agaric figures. We have simplified their phonetic transcription of native names and words. – RGW]

The Koryak are most passionate consumers of the poisonous crimson fly-agaric, even more so than the related Kamchadal and Chukchee, probably because the fungus is most common in their territory. Some travellers, as Krasheninnikov and Dittmar, were of the opinion that the fly-agaric was bought by the Koryak from Kamchatka. Thus, Dittmar says that there is no fly-agaric on the Taigonos Peninsula,[1] and that it is brought there from Kamchatka; while Krasheninnikov[2] asserts that in general the Koryak have no fly-agaric, and that they get it from the Kamchadal. My own observations, however, have convinced me that not only is fly-agaric abundant all over the Koryak territory, but that the Koryak supply the Chukchee with it. In the middle of the month of August I saw in the valley of the Varkhalam River, not far from its mouth, an extensive field dotted with the characteristic crimson caps of the fly-agaric, with their white spots. In the villages of the Maritime Koryak, along the whole western coast of Penshina Bay, I knew individuals who were engaged in gathering and drying fly-agaric, and who carried on a very profitable trade in it. One Koryak from Alutorsk, who dealt in fly-agaric, is mentioned by Sljunin.[3]

1. Dittmar, p. 451. [Our ref. [13] p. 256 – RGW]
2. Krasheninnikov, II, p. 150. [Our ref. [4] p. 237 – RGW]
3. Sljunin, I, p 654. [Our ref. [18] p. 260 – RGW]

The Koryak do not eat the fly-agaric fresh. The poison is then more effective, and kills more speedily. The Koryak say that three fresh fungi suffice to kill a person. Accordingly, fly-agaric is dried in the sun or over the hearth after it has been gathered. It is eaten by men only; at least, I never saw a woman drugged by it.[1] The method of using it varies. As far as I could see, in the villages of Penshina Bay, the men, before eating it, first let the women chew it, and then swallow it. Bogoraz[2] says that the Chukchee tear the fungus into pieces, chew it, and then drink water. Sljunin describes in the same way[3] the Koryak method of using fly-agaric. In describing the use of fly-agaric by the Chukchee and Koryak, Dittmar[4] says that they chew it, and keep the quid in their mouths for a long time without swallowing it. Krasheninnikov[5] says that the Kamchadal roll the dried fungus up in the form of a tube, and swallow it unchewed, or soak it in a decoction of willow-herb and drink the tincture.

Like certain other vegetable poisons, as opium and hashish, the alkaloid of fly-agaric produces intoxication, hallucinations, and delirium. Light forms of intoxication are accompanied by a certain degree of animation and some spontaneity of movements. Many shamans, previous to their seances, eat fly-agaric in order to get into ecstatic states. Once I asked a Reindeer Koryak, who was reputed to be an excellent singer, to sing into the phonograph. Several times he attempted, but without success. He evidently grew timid before the invisible recorder; but after eating two fungi, he began to sing in a loud voice, gesticulating with his hands. I had to support him, lest he fall on the machine; and when the cylinder came to an end, I had to tear him away from the horn, where he remained bending over it for a long time, keeping up his songs.

Under strong intoxication, the senses become deranged; surrounding objects appear either very large or very small, hallucinations set in, spontaneous movements, and convulsions. So far as I could observe, attacks of great animation alternate with moments of deep depression. The person intoxicated by fly-agaric sits quietly rocking from side to side, even taking part in the conversation with his family. Suddenly his eyes dilate, he begins to gesticulate convulsively, converses with persons whom he imagines he sees, sings, and dances. Then an interval of rest sets in again.

1. Krasheninnikov (II, p. 150) says that the Kamchadal women do not eat fly-agaric, but Dittmar (p. 106) cites the case of a Koryak woman (a shaman) who was intoxicated by it. [Our ref. [4] p. 237; [13] p. 257 – RGW]
2. Bogoraz, The Chukchee, Vol. VII of this series, p. 205. [Our ref. [22] p. 273 – RGW]
3. Sljunin, I, p. 655. [Our ref. [18] p. 261 – RGW]
4. Dittmar, p. 506. [Our ref. [13] p. 256 – RGW]
5. Krasheninnikov, II, p. 147. [Our ref. [4] pp. 235-236 – RGW]

However, to keep up the intoxication, additional doses of fungi are necessary. Finally a deep slumber results, which is followed by headache, sensations of nausea, and an impulse to repeat the intoxication. If there is a further supply of fungi, they are eaten. At the beginning of winter, when the supply is still large, old men begin their carousals. In Kuel there are two elders of the Paren′ clans, and during my sojourn in that village I was sometimes unable to hold conversation with either of them for days at a time. They were either intoxicated by the fungi or in a bad mood from the after-effects. At the same season the Reindeer Koryak resort to the coast settlements to purchase and eat fly-agaric. To regale a guest with fly-agaric is a sign of special regard. Dr. Sljunin says a small glass of brandy or diluted alcohol serves as a splendid antidote in cases of fly-agaric poisoning.[1]

There is reason to think that the effect of fly-agaric would be stronger were not its alkaloid quickly taken out of the organism with the urine. The Koryak know this by experience, and the urine of persons intoxicated with fly-agaric is not wasted. The drunkard himself drinks it to prolong his hallucinations, or he offers it to others as a treat. According to the Koryak, the urine of one intoxicated by fly-agaric has an intoxicating effect like the fungus, though not to so great a degree. I remember how, in the village of Paren′, a company of fly-agaric eaters used a can in which California fruit had been put up, as a beaker, into which the urine was passed, to be drunk afterwards. I was told of two old men who also drank their own urine when intoxicated by brandy, and that the intoxication was thus kept up.

From three to ten dried fungi can be eaten without deadly effect. Some individuals are intoxicated after consuming three. Cases of death rarely occur. I was told of a case in which a Koryak swallowed ten mushrooms without feeling their effect. When he swallowed one more, vomiting set in, and he died. In the opinion of the Koryak, the spirits of the fly-agaric had choked him. They related that these spirits had come out with the matter vomited, in the shape of worms, and that they vanished underground.

The Koryak were made acquainted with brandy by the Russians and by American whalers. Despite the prohibition issued by the Russian Government against the importation of brandy, it often finds its way in winter into the Koryak villages and camps, being taken there on trading-trips by Russian merchants. Whalers take it to the coast settlements in summer. Like all other primitive tribes, the Koryak are passionate consumers of brandy, and dealers often obtain an arctic or red fox in exchange for one wineglassful of brandy. To my question as to which they preferred, brandy or fly-agaric,

1. Sljunin, 1, p. 654. [Our ref. [18] p. 261 – RGW]

many Koryak answered, 'Fly-agaric.' Intoxication from the latter is considered more pleasurable, and the reaction is less painful, than that following brandy. Like fly-agaric, brandy is drunk chiefly by elderly men. Old people do not give it to the young, that they themselves may not be deprived of the pleasure; and if young people or women happen to obtain brandy, they frequently give it up to the older members of the family. Two herd-owners whom I met on the Palpal were entirely unacquainted with this drink. Some Koryak in the coast villages have learned from the Russian Cossacks how to make brandy of blueberries. They subject the berries to fermentation, and by means of a pipe distil the liquid from one iron kettle into another, the latter serving as a refrigerator. The result is a rather strong liquor of such disgusting taste and odor that the mere attempt to taste it nauseated me. Krasheninnikov[1] says that the Cossacks in Kamchatka and, following their example, the Kamchadal, distilled brandy from 'sweet grass' (*Heracleum sphandilium*).

pp. 120-121. Among the objects believed by the Koryak to be endowed with particular power is fly-agaric (wa'paq, *Agaricus muscarius*). The method of gathering and the use made of this poisonous fungus will be described later on. It may suffice here to point out the mythologic concept of the Koryak regarding fly-agaric. Once, so the Koryak relate, Big-Raven had caught a whale, and could not send it to its home in the sea. He was unable to lift the grass bag containing travelling-provisions for the whale. Big-Raven applied to Existence to help him. The deity said to him, 'Go to a level place near the sea: there thou wilt find white soft stalks with spotted hats. These are the spirits wa'paq. Eat some of them, and they will help thee.' Big-Raven went. Then the Supreme Being spat upon the earth, and out of his saliva the agaric appeared. Big-Raven found the fungus, ate of it, and began to feel gay. He started to dance. The Fly-Agaric said to him, 'How is it that thou, being such a strong man, canst not lift the bag?' – 'That is right,' said Big-Raven. 'I am a strong man. I shall go and lift the travelling-bag.' He went, lifted the bag at once, and sent the whale home. Then the Agaric showed him how the whale was going out to sea, and how he would return to his comrades. Then Big-Raven said, 'Let the Agaric remain on earth, and let my children see what it will show them.'

The idea of the Koryak, is, that a person drugged with agaric fungi does what the spirits residing in them (wa'paq) tell him to do. 'Here I am, lying here and feeling so sad,' said old Euwinpet from Paren′ to me; 'but, should I eat some agaric, I should get up and commence to talk and dance. There is an old man with white hair. If he should eat some agaric, and if he were then

1. Krasheninnikov, II, p. 406.

told by it, "You have just been born," the old man would at once begin to cry like a new-born baby. Or, if the Agaric should say to a man, "You will melt away soon," then the man would see his legs, arms, and body melt away, and he would say, "Oh! why have I eaten of the agaric? Now I am gone!" Or, should the Agaric say, "Go to The-One-on-High," the man would go to The-One-on-High. The latter would put him on the palm of his hand, and twist him like a thread, so that his bones would crack, and the entire world would twirl around. "Oh, I am dead!" that man would say. "Why have I eaten the agaric?" But when he came to, he would eat it again, because sometimes it is pleasant and cheerful. Besides, the Agaric would tell every man, even if he were not a shaman, what ailed him when he was sick, or explain a dream to him, or show him the upper world or the underground world, or foretell what would happen to him."

p. 483. When the reindeer feed exclusively on lichens, they acquire a special longing for the urine of human beings. This longing attracts them to human habitations. Fig. 9 represents a vessel (the name in Koryak signifies 'the reindeer's night-chamber') made of seal-skin, which every herdsman carries suspended from his belt, and of which, he makes use whenever he desires to urinate, that he may keep the urine as a means of attraction in capturing refractory reindeer. Quite frequently the reindeer come running to camp from a far-off pasture to taste of snow saturated with urine, a delicacy to them. The reindeer have a keen sense of hearing and of smell, but their sight is rather poor. A man stopping to urinate in the open attracts reindeer from afar, which, following the sense of smell, will run to the urine, hardly discerning the man, and paying no attention to him. The position of a man standing up in the open while urinating is rather critical when he becomes the object of attention from reindeer coming down on him from all sides at full speed.

p. 565. At the present time the use of sulphur or Swedish matches is quite widespread. Even when obtaining the sacred fire, some Koryak turn the drill for a short time as a formality only, and the fire is really kindled with a match; but they cannot always obtain matches, so that the most common means of obtaining fire is the strike-a-light. Although not much used, the strike-a-light seems to have been known to the Koryak prior to their encounter with the Russians, having been introduced by the Tungus, who had received it from the Amur tribes. Even now, merchants often import from Vladivostok steel and flint of Chinese origin. Tinder is prepared from a fungus which grows on the stumps of birch-trees. [*Fomes fomentarius* – RGW] The fungus is stripped of its hard outer layer, and the inner spongy mass is boiled in water. Then it is dried; and a light, brittle, and highly inflammable punk is thus obtained.

Fire-making tools are more often needed by the nomadic Reindeer Koryak than by the sedentary Maritime people. A new fire need rarely be made in the underground houses of the Maritime Koryak. The women are very skilful in keeping up the fire of the hearth. They cover the embers with

FIG. 9. Seal-skin vessel for gathering urine impregnated with inebriating virtue derived from fly-agaric, in use among the Koryak. (After Waldemar Jochelson, *The Koryak*, p. 483; Mem., Amer. Museum of Natural History, 1908)

ashes; and when reviving the fire, they rake it up, put small chips of wood on the glowing embers, and fan them until they burst into flame. The Maritime Koryak need fire-tools only on journeys. However, when in possession of matches, they are very fond of striking them to light their lamps or pipes, even when the fire is burning on the hearth. On the other hand, if the fire goes out entirely, and neither match nor tinder is on hand, the ancient method of obtaining fire by means of the drill-bow is resorted to. This, however, happens very rarely.

[In the Koryak village where Jochelson was staying, the ceremonies attending the 'home journey of the white whale' were being celebrated. – RGW]

p. 74. The old men ate fly-agaric, and, when the intoxication had passed, they told whither the 'Fly-Agaric Men' (Wapa'qalaᵋnu) had taken them, and what they had seen. The women and the young people sang and beat the drum.

[There had been a pestilence and the community were arranging for the cremation of a child. All kinds of gifts were being sent to be placed on or near the pyre, some for the dead child and others for the child to take to relatives who had died in the pestilence. – RGW]

p. 112. Two agaric fungi were sent to one old man who had been very fond of agaric intoxication.

[In the following extract about the 'water of life' we are reminded of the fly-agaric in one of its manifestations, as a liquid, either derived directly from the mushroom, or the urine of the reindeer that has imbibed the mushroom, or human urine. Should we not consider the possibility that this conception, so widespread in Eurasian and American folklore, had its origin in the fly-agaric? Here is what Jochelson says: – RGW]

p. 351. We find in American tales some elements that occur in the myths of the Old World, but they are absent in the Koryak tales recorded here. For instance, 'the water of life,' which a hero procures to restore dead bodies to life, or to revive bones, figures frequently in Indian myths on both sides of the Rocky Mountains, and is also one of the favorite episodes of the myths of the Old World.[1] Another case in point is the cosmogonic tale about the raven, or some other bird or other animal which dives into the water to obtain some mud, out of which the earth is created. This tale is popular in many parts of North America, and is found as well among the Chukchee and Yukagir.

1. It seems that in the Koryak tales the blood of the reindeer takes the place of 'the water of life'. It must be noted here that in one Chukchee tale we find 'bladder with living water' (Bogoraz: Chukchee Materials, p. xxiv); and in one Yakut tale (Khudyakov, p. 127) 'three bottles with living water' are mentioned. As to the Chukchee, Mr. Bogoraz considers the passage as borrowed from the Russian.

[Speaking of the lack of cleanliness among the Koryak, the author says: – RGW]

p. 416. The kettles in which the food is cooked are full of reindeer and dog hair, which fall from the clothes and fill the air of the Koryak house. The Koryak kill lice, which are regarded by them as properly belonging to a healthy man, with their teeth. There is a prevailing belief among them, as well as among the Yukaghir, that when a man is deserted by lice he will soon die. They eat also the large larvæ which develop from the eggs deposited by the reindeer-flies in the hair of the reindeer. No matter how putrid food may be, the Koryak have no aversion to it, and they will even drink the urine of persons intoxicated with fly-agaric.

pp. 417-8. People addicted to the use of fly-agaric can be detected by their appearance. Even when they are in a normal condition, a twitching of the face is observable, and they have a haggard look and an uneven gait.

p. 115. In the beginning of things, at the mythological time of the Big Raven, the transformation of animals and inanimate objects into men was a natural occurrence. 'At that time, man also possessed the power of transforming himself. By putting on the skin of an animal, or by taking on the outward form of an object, he could assume its form. Big Raven and Eme'mqut turned into ravens by putting on raven coats. Kilu', the niece of Big Raven, put on a bear-skin and turned into a bear. Eme'mqut put a dog's skin on his sister, and she became a dog. Eme'mqut and his wives put on wide-brimmed spotted hats resembling the fly-agaric, and turned into those poisonous fungi.[1] The belief in the transformation of men into women after putting on a woman's clothes, and *vice versa*, is closely related to this group of ideas.'

2. The Yukaghir and the Yukaghirized Tungus. Vol. IX of the Jesup North Pacific Expedition. p. 419.

They [the Yukagir. – RGW] do not eat mushrooms regarding them as unclean food growing from dogs' urine. However, according to traditions, they used to intoxicate themselves with the poisonous fly-agaric, which is still eaten by the Koryak and Chukchee. The Yukaghir call mushrooms *can-pai, i.e.*, tree girl.

1. Tale 12.

[22]

Bogoraz, Vladimir Germanovich. (Bogoras, Waldemar) The Chukchee. Memoir of the American Museum of Natural History. Jesup North Pacific Expedition. Parts 1, 2, and 3. New York. 1904-1909. pp. 205-207.

Intoxicants. – Fly-agaric is the only means of intoxication discovered by the natives of northeastern Asia. Its use is more common in the Koryak tribe, as agaric does not grow outside of the forest border. For the same reason only the Southern Pacific Chukchee – *e.g.*, those around the Anadyr, Big River, and Opuka River – are supplied with the intoxicating mushroom. They do not compare with the Koryak, however, in their passion for agaric.

The Russianized natives of the Anadyr until recently shared in the consumption of this intoxicant, but now they have almost wholly given up its use. The reason of this change is that they consider the strong intoxication produced by this stimulant as shameful for a Christian. They also realize that the consumption of fly-agaric involves some danger. With a person unaccustomed to its use it may even cause death. The abstinence from agaric is also noticed among the northern Kamchadal, and to some degree among the Maritime Koryak of northern Kamchatka, though all of these people gather it assiduously in order to trade it to their less civilized reindeer-breeding neighbors.

Fortunately for the tribes consuming the fly-agaric, it grows only in certain places, and the supply is often limited. The mushrooms are usually dried up and strung together in threes, that number being an average dose. Some of the natives of course require much more to produce any effect. The intoxication may be followed by sickness, or the after-effect may be very slight. When eaten, the mushrooms are torn to small shreds, and these are chewed piece by piece, and swallowed with a little water. Among the Koryak the woman chews the mushroom, and offers the ready quid to her husband to swallow.

I witnessed a few times the progress of intoxication by means of agaric. The symptoms are analogous to those produced by opium or hashish. The intoxication comes on rather suddenly, in about a quarter of an hour after the consumption of the mushrooms. Usually the person remains awake; but the natives say that if a person falls asleep immediately after eating mushrooms, they will work more effectively, and in a short time he will awaken more thoroughly under their influence. The intoxication has three stages. In the first the person feels pleasantly excited. His agility increases, and he displays more physical strength than normally. Reindeer-hunters of the Middle Anadyr told me that before starting in canoes in pursuit of animals, they would chew agaric because that made them more nimble on the hunt. A native fellow-traveller of mine, after taking agaric, would lay aside his snow-shoes and walk through the deep snow hour after hour by the side of his dogs for the mere pleasure of exercise, and without any feeling of fatigue. During this period the agaric-eater sings and dances. He frequently bursts into loud peals of laughter without any apparent reason. It is a state altogether of noisy joviality. His face acquires a darker hue and twitches nervously; his eyes are now contracted, and again almost bursting from their sockets; his mouth puckers and grins or spreads into a broad smile.

Flashes of the second stage often appear early, shortly after the first traces of intoxication become visible; indeed, all three stages are frequently intermingled. This is noticeable especially among elderly inveterate agaric-eaters. During the second stage the intoxicated person hears strange voices bidding him perform more or less incongruous actions; he sees the spirits of fly-agaric and talks to them. He still recognizes surrounding objects, however, and when talked to is able to answer. All things appear to him increased in size. For instance, when entering a room and stepping over the door-sill, he will raise his feet exceedingly high. The handle of a knife seems to him so big that he wants to grasp it with both hands.

The spirits of fly-agaric have an outward appearance similar to that of the actual mushrooms, and the agaric-eater feels impelled to imitate them. For example, I saw one man suddenly snatch a small narrow bag and pull it with all his might over his head, trying to break through the bottom. He was evidently imitating the mushroom bursting forth from the ground. Another walked around with his neck drawn in, and assured every one that he had no head. He would bend his knees and move very quickly, swinging his arms violently about. This was in imitation of the spirits of fly-agaric, who are supposed to have no necks or legs, but stout cylindrical bodies which move about swiftly.

The spirits of fly-agaric are fond of playing practical jokes on men under their influence. They begin with asking for homage either for themselves or for surrounding objects, – the hills, the river, the moon, *etc.* Then they show some of the objects under a delusive aspect. When asked why this strange change has occurred, the spirits answer that it portends danger to the man's life unless he makes obeisance in a particular way. To illustrate. An intoxicated man, while talking to me reasonably enough, suddenly leaped aside, and, dropping on his knees, exclaimed, 'Hills, how do you do? Be greeted!' Then he stood up, and, looking at the full moon, asked, 'O Moon! why are you waning so fast?' He told me that the spirits answered, 'Even so will your life wane, unless you show the moon your bare buttocks.' This he did, and then, suddenly recovering his senses, began to laugh at his foolish actions.

In the third stage the man is unconscious of his surroundings, but he is still active, walking or tumbling about on the ground, sometimes raving, and breaking whatever happens to come into his hands. During this period the agaric spirits take him through various worlds and show him strange sights and peoples. Then a heavy slumber ensues, lasting for several hours, during which it is impossible, to rouse the sleeper. How persistent are the spirits' commands is shown by the following instance of a man, who, when about to retire, was ordered to lie down in the midst of his dog-team. Although he was attacked by the dogs, we could not keep him away from them. He finally succeeded in staying with the dogs all night.

On awakening, a general weakness and heavy headache ensue, accompanied by nausea, often violent vomiting. The drunken state can be renewed by a single mushroom. In this manner inveterate agaric-eaters keep up their intoxication day after day.

Drinking the urine of one who has recently eaten fly-agaric produces the same effect as eating the mushroom. The passion for intoxication becomes so strong that the people will often resort to this source when agaric is not available. Apparently without aversion they will even pass this liquor around in their ordinary tea-cups. The effect is said to be less than from the mushrooms themselves.

I have already spoken about the amount of trade in strong liquors carried on in northeastern Siberia. The Chukchee, as well as all other inhabitants of the country, are eager for a chance to drink spirits. In all my journeys through these countries I met people in only two places who knew nothing about strong liquors. In one case they were some Maritime Koryak in small villages on the northern border of the Kamchatka district. These people were far from the Kamchatka towns and from Gishiginsk trading-settlements. At the same time, they were so poor that nobody sought to bring liquors to them. The other case was that of the Kerek of the southern shore of Anadyr Bay.

p. 282. Thus, for instance, the intoxicating mushrooms of the species fly-agaric are a 'separate tribe' (*ya'nřa-va'rat*). They are very strong, and when growing up they lift upon their soft heads the heavy trunks of trees, and split them in two. A mushroom of this species grows through the heart of a stone and breaks it into minute fragments. Mushrooms appear to intoxicated men in strange forms somewhat related to their real shapes. One, for example, will be a man with one hand and one foot; another will have a shapeless body. These are not spirits, but the mushrooms themselves. The number of them seen depends on the number of mushrooms consumed. If a man has eaten one mushroom, he will see one mushroom-man; if he has eaten two or three, he will see a corresponding number of mushroom-men. They will grasp him under his arms, and lead him through the entire world, showing him some real things, and deluding him with many unreal apparitions. The paths they follow are very intricate. They delight in visiting the places where the dead live. These ideas are illustrated in a sketch (Fig. 10) drawn by a Chukchee.

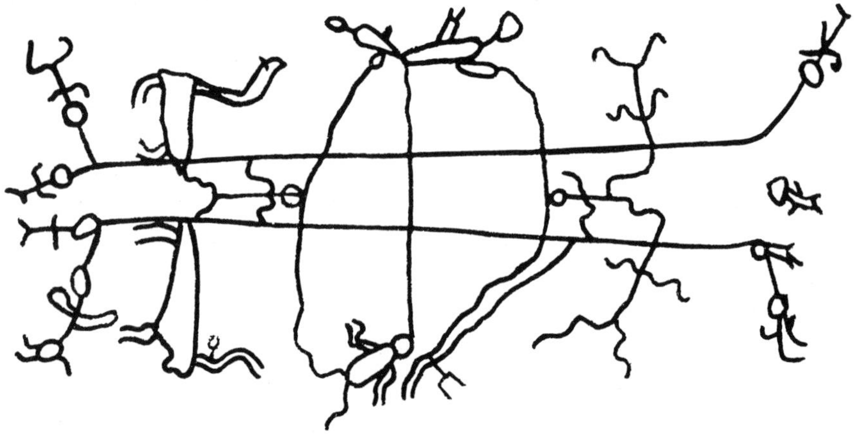

FIG. 10. Drawing made by a Chukchee of the wanderings of 'fly-agaric men'. (After Waldemar Bogoraz, *The Chukchee*, p. 282; Mem., Amer. Museum of Natural History, Vol. VII, Part 2).

pp. 322-323. Thunder is said to be produced by the passing of the thunder-bird. Others attribute it to the rattling noise made by girls playing on a spread sealskin. Rain is the urine of one of the girls. In one tale the lightning is described as a one-sided man who drags his one-sided sister along by her foot. She is intoxicated with fly-agaric. The noise caused by her back as it strikes the floor of heaven is thunder, her urine is rain. Obsidian is said to be the stone of the thunder, which falls from the sky in round balls, or even in roughly chipped arrow-heads and lances. Perhaps the idea of stone arrow-heads falling from the sky, so common in the Old World, is borrowed from the Tungus or from the Russianized natives.

Intoxicating mushrooms form a 'separate tribe' (*ya'ňra-va'rat*). We have already noted that they are very strong, and that, when coming out of the earth, they can lift a large tree-trunk on their head, or shatter a rock into pieces. They appear to intoxicated men in strange shapes.

On one sketch (Fig. 10) there are represented the tracks of a man who is led around by mushrooms. He thinks that he is a reindeer, then he is 'submerged,' and after a while he comes out laboring under the same idea. The path of his tracks connects all men and all beasts seen during the trance.

p. 414. I saw a shaman trying to recall to her senses a sick woman who had fallen into a heavy swoon. To do this he began to beat the drum with the utmost force. Then he pretended to catch something from the drum and to swallow it hurriedly. Immediately afterward he appeared to spit it out into the hollow of his hand, and then in the quickest possible way pretended to empty his palm over the head of the patient. After that, he began to mumble and gibber over the crown of the patient's head. In order to prevent the soul from leaving, he breathed into the hollow of his hand, and then applied his palm to the breast of the patient. At intervals he pretended to suck out the source of the suffering from the crown of her head. For this he made sucking motions with his mouth at some distance from her head. From time to time he made grimaces, and pretended to be choking, evidently for the purpose of showing that something bad had entered his mouth. At last he spat violently, and then began again the whole process.

From my own observation I know that a real insect is sometimes used in treatments of such a kind. This is brought near to the breast or to the head of the patient, and then vanishes, deftly abstracted by the shaman, who pretends that it has entered the body.

[Bogoraz in the following passage gives a classic description of the whalebone device used by the tribesmen of the north, in both Eurasia and America, for killing wolves. The Chukchee call it *wapaq*, which is also their name for the fly-agaric, presumably because the wolf after swallowing the *wapaq* jumps wildly around and then quiets down, exhausted. Bogoraz assumes that the primary meaning of the name is fly-agaric but does not discuss this. In the absence of further information it seems possible that the name of the device was transferred as a figure of speech to the mushroom. The mushroom with its religious associations is likely to attract to itself under tabu influence various names and *wapaq* would have been the current one in the time of Bogoraz. Here is how he describes the *wapaq*. – RGW]

p. 141. Till recent times the well-known spit of whalebone,[1] identical with that of the American Eskimo, was used to catch wolves. It consisted of a

1. Its name in Chukchee is wa'pak, which means literally 'fly-agaric' (an intoxicating mushroom). The Chukchee and the Koryak are very fond of this mushroom; and when they find it in the woods, they pick it off just as eagerly as the wolves snatch after the greased whalebone spits. The Chukchee

slender rod of whalebone, with sharp-pointed ends, folded together several times, and bound with a thin thread of sinew well saturated with oil. Afterward it was several times soaked in water and allowed to freeze. The whole object was then well covered with blubber, tallow, meat, or such like. These folded spring-spits were often joined in strings of five or six, and hung on a bush on the wolf's trail, but so high as to be out of reach of foxes; or they were laid in a hole in the ice and water was poured over them, so that it would freeze to a transparent protecting cover strong enough to resist the attacks of smaller animals. The wolf would break through the ice and swallow all the spits, which would unfold in the stomach, and, breaking through its walls, cause the speedy death of the animal. But with only one or two spits it was able to walk away for a considerable distance, even so far that it would never be found by the hunter.

[Bogoraz mentions in the foregoing footnote that mice store up the fly-agaric in their winter holes. He has more to say about this on p. 198: – RGW]

The roots of *Claytonia acutifolia* Willd, *Hedisarum obscurum*, *Polygonum viviparum*, *Polygonum polymorphum*, *Pedicularis sudetica*, *Potentilla fragiformis*, *Oxytropis*, various species of *Carex*, and several others, are used by the Chukchee. They are the only vegetable food that is really relished. During the summer women often go digging roots. They use a digging-pick, which in former times consisted of a handle with bone point or simply of a sharp-pointed piece of antler, while at present it has an iron point tied to a wooden handle. Nests of mice are also robbed. It is considered dangerous, however, to take all the roots from the nests, because the owner might retaliate by means of magic. Moreover, the Chukchee believe that some of the roots and herbs found in the storehouses of mice are poisonous, and are gathered by the mice partly for the purpose of poisoning the robbers, partly as an intoxicant, like fly-agaric (*Agaricus muscarius*), which is used by man.

[On p. 148 Bogoraz quotes in Chukchee a riddle: – RGW] 'I have a headache. I am bleeding from my nose. Stop my nose bleeding! . . . What is it?' Answer: 'Fly-agaric.' He explains this riddle with the following note: 'The eating of the fly-agaric causes, after the intoxication has passed, a violent headache, which may be assuaged by a new dose of the same drug.'

believe, moreover, that mice, when gathering roots for the winter, bring in some unknown intoxicating herbs which they use in their ceremonials. These herbs also serve to protect their stores from intruders, because they are said to act as poison on most other animals, including man. These herbs are called by a name derived from that of the intoxicating mushroom, – elhi-wapak ('white agaric'), – and a similar name is given to the whalebone spit on account of its power of killing the animal that swallowed it.

[Bogoraz devotes considerable space to recording the life of the reindeer herdsmen and their tribulations. He says that one of their hardships comes with the mushroom season in the fall, when presumably the herds become unruly. Another is the insects, two species of which lay their eggs under the skin of the animals and another in the nostrils. Here is what he says about the nostril infestation: – RGW]

Another fly, of smaller size and darker color (*Œdemagena tarandi* Slunin), lays its eggs in the reindeer's nostrils. The larvæ go up to the throat, and penetrate the cartilage. The next year, when the maggots are full-grown, they cause a constant cough, which continues until the last one drops to the ground. The Yukaghir and the Tungus, following the example of Yakut cattle-breeders, try to protect their reindeer from obnoxious insects by a smudge of smouldering dung, or of a fire covered with green leaves. But with the wild and large herds of the Chukchee such fires are of little value, and not without danger. Thus, some five years ago, Tungus herdsmen who were tending the herd of a rich Chukchee on the Alaseya River tried to surround it with fires, and finally burned the whole pasture, and injured half of the animals.

[23]

ITKONEN, T. I. Heidnische Religion und späterer Aberglaube bei den Finnischen Lappen. (Heathen Religion and Late Superstitions of the Finnish Lapps) Mémoires de la Société Finno-Ougrienne LXXXVII. Helsinki. 1946. p. 149.

[The author is a well known and reliable Finnish scholar. – RGW]

. . . When speaking of sorcerers, reference must be made to the custom of Siberian shamans of eating fly-agarics to get into an ecstatic stupor; the Ob-Ugrian sorcerers, for instance, consumed each time three or seven mushrooms. It is interesting to note that according to a tradition among the reindeer Lapps of Inari, Lapp sorcerers used to eat fly-agarics with seven dots.

[24]

LEHTISALO, T. Entwurf einer Mythologie der Jurak-Samojeden. (Outline of Yurak Samoyed Mythology) Mémoires de la Société Finno-ougrienne. Vol. LIII. Helsinki. 1924.

The forest Yurak magicians also knew the custom of eating fly-agarics. These were eaten when they were dry and fully grown; the small capless mushrooms were too potent, and it is said that a female magician died from eating them. Only someone who is familiar with the origin of the fly-agaric can eat it 'with fortunate results,' but if in his intoxication he does not see the mushroom spirits properly, they may kill him, or he may go astray in the dark. The number of mushrooms eaten is usually two and a half, *i.e.*, only half of the last mushroom is eaten. The magician sees man-like creatures appearing before him, as in a dream; they number as many as the mushrooms he has eaten, and the half-mushroom is represented by a half-man. They run away quickly, along the path which the sun, after it has set in the evening, travels in order that it may rise again in the morning, and the magician follows them. He is able to stay close on their heels only because the half-spirit runs slowly and keeps looking back as if it were waiting for its other half. It is dark there, and the magician cannot see anything. Along the way the spirits of the fly-agaric tell him what he wants to know, *e.g.*, the possibilities for curing a sick person. When they come out into the light again, on the spot where God created the fly-agarics, there is a pole with seven holes and cords. After the magician ties up the spirits, the intoxication leaves him and he awakens. Now he sits down, takes in his hand the symbol of the Pillar of the World, the four-sided staff with seven slanting crosses cut into each side at its upper end, and he sings of what he has seen and heard.

[25]

DUNIN-GORKAVICH, A. A. Tobol'skij Sever. (The Northern Region along the Tobol) St. Petersburg. 1904. p. 95.

[The author is discussing the Ob-Ugrians. – RGW]

The shaman first eats some *panga* (dried mukhomor [fly-agaric]) and becomes drunk on it. After this he works his magic, that is to say, he utters peculiar cries and plays on the tambourine.

[26]

KARJALAINEN, K. F. Die Religion der Jugra-Völker. (Religion of Ugrian Folk) Helsinki. 1927. pp. 278-280.

The use of the fly-agaric, *paŋχ*, as an intoxicant is a wide-spread custom among the Ugrians and is most firmly established in the southern regions. The custom was mentioned even in the Berezov area, but with the additional comment that it was rare; in the Obdorsk area, on the other hand, it is unknown. The significance of the *paŋχ* is also confirmed to a certain extent by the fact that in the songs of the North Voguls the spirits are described as consuming it as a delicacy. When the Man who Observes the World is called, he may be in a state of 'intoxication caused by the one-footed notched-edged sevenfold *paŋχ*.' However, this is not said to his shame; on the contrary, it is supposed to indicate his power and wealth. This intoxicant has been used by magicians[1] for their own special purposes; they use it along the Tremyugan, and even more along the Vasyugan; it is an ordinary stimulant along the Irtysh, and it is also used for the same purpose in some places in the Vogul region. The only edible part of the *paŋχ* is the upper part of the cap, stripped of the stalk and covering; it may be eaten raw, direct from the forest, but in most cases – and along the Tremyugan almost exclusively – it is eaten after it has been dried in the sun or in an oven; winter supplies, of course, are always dried. Even when it is eaten in the ordinary manner, various precautionary rules must be followed; these rules arose from the relatively high toxicity of the mushroom, but in later times they took on a religious character. If the magician eats of the mushroom, this always has *a cultic significance*, which is only natural, since by eating he creates helpers for himself, or, as the people of Tsingala put it, '*paŋχ* enters into him' through the eating. Along the Irtysh fly-agarics are usually eaten in the evening. The number of mushrooms or bites eaten is three or seven, and according to Patkanov, even fourteen, twenty-one, or more; these numbers should be taken *cum grano salis*, since the effect of such a large number of whole mushrooms can seldom be tolerated even by a person accustomed to eating them. Sometimes the mushrooms are spread with butter or fat before eating, but usually they are eaten simply with bread, and water is drunk to make them easier to swallow. According to the Vasyugan people, the power of the mushroom derives from the fact that it was created from the spittle of the God of Heaven, and the mushroom is so potent that the Devil was unconscious for

1. Shamans. – RGW.

seven days and nights after eating it. For this reason, men also must not eat much of it. If anyone consumes too much of it, his teeth become clamped together, foam comes out of his mouth, his eyes bulge, and he can be saved only by the forcible administration of milk or salt, for '*paŋχ* wants nothing to do with' these substances. The most potent is the 'king fly-agaric,' which is small in stature, with a high stalk and only one single white spot in the center of the cap; ordinary, smaller specimens grow in a circle all around it. These ideas of the Vasyugan people are legendary, but we see from their ceremonies that they maintain a careful reserve with regard to the *paŋχ*. When the Vasyugan magician eats the mushroom, he always leaves the second half of the last mushroom and hides it, for then the *paŋχ* cannot use its full power to harm him. Eating the mushroom results in intoxication, and in the case of the *paŋχ* this includes a compulsion to sing which only a few can resist. The effect lasts 'from morning to sundown,' according to one informant. The Tremyugan magician eats *paŋχ* at any time of the day and does not hide the other half; instead, he cuts out a piece of the middle of each mushroom, 'the crown of the *paŋχ's* head,' and throws it into the fire or onto a clean spot in the yard. The man who is intoxicated from eating fly-agarics sees the *paŋχ* 'dancing' before his eyes, invisible to others; that is to say, they move in the direction of the sun and sing a song, which the intoxicated Ostyak repeats after them word for word, so that the *paŋχ* act as 'singing-leaders' for the prophesying magician. At the same time the *paŋχ* tell the magician what he wants to know.[1]

Drum, zither, and *paŋχ* are the 'great' material means by which the Ugrian magician attempts to communicate with the spirits and obtain the information he needs. There are also many other means, both material and mental, used for uncovering mysteries. One such means which is found everywhere and is certainly very ancient is the *dream-vision*, possibly the natural forerunner of artificially stimulated ecstasy; another is soothsaying based on dreams, and this is what soothsaying with the aid of the *paŋχ* essentially amounts to.

pp. 306-8. The ceremonies of the Irtysh Ostyaks are very different from the foregoing in certain respects. When *t'ərtəŋ-χoi* is brought into the house where he is needed (so it was said along the Demyanka), he takes resinous tree bark and fills the hut with smoke, waving everywhere in the hut the

1. Concerning a stimulant of his own, the Swedish officer in 1714 (?) reports: 'We gave them *šaar* (tobacco); they smoked it and inhaled the smoke and thereupon fell to the ground quite unconscious, as if they were dead; afterwards they said that Shaitan had tormented them.' However, to judge by the name used for tobacco, this description most probably refers to the Ostyak Samoyeds [= Selkup].

cloth which has been hung up as a sacrifice to Säŋkə. At his direction home-made beer is brewed and in the evening the bath is heated for him. After the bath the magician eats three or seven fly-agaric caps on an empty stomach – for he has fasted all day long – swallowing them either fresh or, in most cases, after they have been dried in the sun or sometimes in an oven, and then he lies down. When he has slept for a while, he springs up and begins to shout and walk to and fro, his whole body trembling with excitement. As he shouts, he reports what the spirit has revealed to him through his emissaries,[1] which spirit should be offered a sacrifice, what is to be sacrificed, which man has spoiled the luck of the hunt, how the luck can be regained, and so on. After the 'emissaries' have told everything, they depart and the magician sinks into a deep sleep, from which he does not awaken until morning. In the morning the appropriate spirit is entertained with home-made beer, porridge, and bread at the rear window, *i.e.*, in the icon corner, and is given the promise that the desired sacrifice will be made as soon as the animal designated has been procured.

The procedure is essentially the same in Tsingala; *t'ərtəŋ-χajat* eats three dried *paŋχ* with bread, half a mushroom at a time, and then goes to sleep; after he awakens, he shouts out what has been revealed to him. Before this proceeding, 'three seven kopeks'[2] (= three copper two-kopek pieces) are placed on the table, 'in the way of the *paŋχ*,' as a sacrifice for Säŋkə, and while the magician begins to tell his story, both he and the money are smoked with *pikhta* [spruce bark]. This questioning does not take place until evening, when night is falling. In the same village Schultz personally witnessed a questioning of the following nature: The magician was an old woman who, before beginning her magic, placed candles in front of the icons and a loaf of bread on the table; she made seven indentations in the loaf, naming Astanai and other Ostyak epic folk heroes. An old man who served as the sacrificer or priest in the village smoked the old woman, the bread, and all those present, using spruce bark smoke. The old woman ate three dried fly-agarics, a bite at a time, taking a swallow of water after each bite. After three or four minutes she began to hiccup; the hiccupping was following by shouting, and this in turn by a kind of singing. This went on for about half an hour. The hiccupping and singing gradually died away, and the whole proceeding ended when the old woman drank water and bowed before the icons. The others followed her example. The subject of the old woman's song was not she herself but 'the spirit of the fly-agaric.' – In the Zavodniya yurt there

1. 'Little spirits' appear in person, while 'great spirits' only send their messengers, about whom nothing more specific is known.
2. 'drei sieben Kopeken.' – RGW.

exist today women who undertake such investigations; they too eat *paŋχ* and then tell their revelations in song after a short sleep, during which they visit Säŋkə.

pp. 315-316. The procedure followed by the Vasyugan fly-agaric soothsayer is quite simple, simpler than it is, for example, in the Irtysh region. Such a man was asked to give information about the mental disturbance suffered by a woman. Towards evening, he ate two and a half *paŋχ* and slept for a little while; after awakening, he sat down in the corner of his birch-bark yurt and began to sing, keeping his eyes closed and shaking his body to and fro. The intoxication did not seem to be very strong, since after he stopped singing, he was able to speak clearly with the spectators and take snuff into his nose. He continued singing in this way until morning, narrating the events of his journey, telling how far the *paŋχ* had brought him, through many districts and different countries, how it had led him into a church, and so on. – In spite of all his efforts, however, this time he did not reach the goal of his journey, *viz.*, the place where the information was available. The reason for this unfortunate conclusion was that ill-behaved people had shown the hidden half-*paŋχ* to me, and this had angered those *paŋχ* that had been eaten.

As can be seen from the foregoing, the manipulations of the Ugrian shamans may be quite different in the various regions, and special features in customs and ideas are found particularly in the southern Ostyak regions. Customs that may doubtless be regarded as late additions from the Irtysh region are *bathing in the bath-room, the deposition of sacrifice gifts, smoke-curing, and the entertaining of the Saŋkə*, although the latter, *e.g.*, in Tsingala, is not asked directly for information at all. The entire questioning of the *paŋχ*, for example, is also something that came into use in later times; the Voguls and Ostyaks know this to some extent from the Surguts, it has spread to some extent along the Tremyugan, and it has become predominant among the Irtysh Ostyaks.

It is not impossible that this custom is a coalescence of the dream-vision and the earlier intoxication; this is suggested by the fact that after consuming the *paŋχ* the magician always sleeps for a while; to my mind, this is needed not for evoking and intensifying the effects of the fly-agaric, but for evoking dream-visions in order to give information and make prophecies. The representation of the magician's assistants that has become established along the Vasyugan, a representation which is late in its present form, was also apt to influence the prevailing ideas concerning the magician's method, contribute a new nuance to them, and give rise to additional features, even though a great deal that is native may essentially be found in the ceremonies, as a comparison with North Ostyak customs shows.

A method that was already used by magicians in early times but was not their only method was magic *in the dark*, which is still practiced today in many regions, both among the Voguls[1] and among the Ostyaks. I am inclined to believe that the Tsingala custom of questioning the *paŋχ* in the evening, when night is coming on, is also a remnant of this. In such cases, the magician is often tied up with ropes, a custom for which Novitskij and Müller cite examples from the West and which is still prevalent along the Vasyugan.

[27]

BERGMAN, Sten. Vulkane, Bären und Nomaden. (Volcanoes, Bears, and Nomads) Stuttgart. 1926. pp. 159-160.

On the day after the celebration a Koryak visited the Lamut camp. He came with his dog and apparently had no other purpose than to sit and chat for a while. When the loud barking of the dogs announced his arrival, we went out of the yurt to welcome him. It was a Koryak named Akei, who owned a large herd of reindeer and a yurt in the vicinity. He was small and sinewy, his face was leather-colored, and his features were not unlike those of an American Indian. His movements were slow and awkward. He walked to the yurt with a gently rolling gait.

'Priyatel´, have you any fly-agarics for me?' was the first thing he asked after exchanging greetings. Priyatel´ is the form of address used by the Koryaks and Lamuts to everyone and means simply 'friend.' He apparently took me for a trader, for when these men travel to the mountains, they usually carry fly-agarics with them; the Koryaks and Lamuts[2] are passionately fond of using these as an intoxicant.

1. Munkácsi mentions that in the case of sacrifices the magic is practiced during the day in front of the spirits' food-storehouse or at some other place that is sacred to a spirit, while in other cases it is usually practiced at night in a dark yurt, where the fire on the open fireplace is extinguished while the magic is going on.

2. Bergman alone among our sources includes the Lamut among those who use the fly-agaric. The Lamut are closely related linguistically and culturally to the Tungus. – RGW.

[28]

DONNER, Kai. 1. Ethnological Notes about the Yenisei-Ostyak (in the Turukhansk Region) [Yenisei-Ostyak = Ket] Mémoires de la Société Finno-ougrienne. Vol. LXVI. Helsinki. 1933. pp. 81-2.

H*aŋgo* is a mushroom, the flybane [= fly-agaric], eaten by the shamans. Seven such mushrooms are eaten, whereupon human beings become 'mad'. Those who are not, or are not going to be, shamans die from eating these mushrooms. The shamans, specially those of the Ostyak-Samoyed (= Selkup) in the Narym district, were known for consuming flybanes as a means of intoxication before starting the shamanizing. At present (1912-1914) it is not so much practised.

[In the preceding extract, note the telltale marks of a folk belief: those who are shamans or who are going to be shamans eat the mushrooms with impunity, but others die from them. Donner mentions the same belief as being present in the Selkup culture, a Samoyed group that are neighbors to the Ket. We quote the passage in the third of our extracts from Donner; *vide infra*. It seems to have lingered on among the Samoyed and Ket of the upper Yenisei. – RGW]

2. Bei den Samojeden in Sibirien. (Among the Samoyed in Siberia) Stuttgart. (Our copy is dated 1926. First published in Swedish in 1918). p. 110.

The performance always takes place in the evening, when darkness is falling, and may last all night long. When the performance is about to start, the shaman calls his assistants, who bring out the drum and slowly warm it at the fire in order to stretch the skin as tight as possible. Along the Ket the shaman makes no preparations of any kind, but in other places he often eats several fly-agarics in order to go into a trance more easily. These mushrooms contain a very strong poison, and I can say from personal experience that it is highly intoxicating. The natives often use it to get drunk on when they have no alcohol.

3. La Sibérie: La Vie en Siberie: les Temps anciens. Translated from the Finnish by Léon Froman. Paris, Gallimard, Nouvelle Revue Française. 1946. p. 225.

Even one who otherwise would find it easy to enter into rapport with the world of spirits would not enter into ecstasy in a state of trance indispensable for this purpose without having acquired the habit. Nor must it be so easy to absorb a sufficient dose of fresh fly-agarics, or preferably dried specimens, to enter into the necessary state of drunkenness or stupor. Among the Samoyed-Ostyak of the Yenisei [Selkup], the mushrooms should number from two to seven, but one says that he who is not a shaman and who eats them falls sick and dies, which is probably true for the poison of the mushrooms is violent and one could not absorb a strong dose without being used to it. The absorption of fly-agarics – as of all direct stupefying agents – must have been especially current among the Samoyed of the region of Narym. Nevertheless there are many shamans who feel that they have no need of inebriating substances or other similar products to provoke a change of state permitting them to pass into the other world, where they will meet later with all the spirits that influence the condition and life of mankind.

[29]

Koryak and Kamchadal Tales, taken down by Waldemar Jochelson in the original languages.

[Among Jochelson's mss. found after his death there were a number of tales in the Kamchadal language. Professor Dean S. Worth took them in hand and has published them in a beautiful edition in 1961, produced by Mouton in The Hague. Professor Worth has edited them in the Kamchadal language with literal interlinear translation in English, followed by a smooth translation. Jochelson also took down many Koryak tales. His mss. in the original language do not survive, but he turned many of them into English and published them in his work on the Koryak. The fly-agaric figures in eight of these tales, four of them among those in the Kamchadal language and four translated into English from the Koryak, and with the permission of Mouton & Co. we reproduce them here.

The only alteration that we have made in the text of these tales is a simplification of the spelling of native words, or in some cases the suppression of the native name where it is irrelevant to the theme of our book. – RGW]

KAMCHADAL TALES

A

CZELKUTQ AND THE AMANITA GIRLS

There lived Czelkutq. He wooed Kutq's daughter Sinanewt and worked for her. He brought in much wood. Czelkutq married Sinanewt. They began to live. They amused themselves well. Sinanewt gave birth; a son was born. Czelkutq set off into the woods, where he met the beautiful Amanita girls. Czelkutq stayed with the girls and forgot his wife. Sinanewt thought of her husband and waited for him. She thought: 'Where is he, long ago he was killed!'

With them there lived an old aunt of hers, Kutq's sister, who said: 'Well, Sinanewt, stop waiting for your husband; long ago he stayed with the Amanita; send your son to his father.'

The little boy set off to his father. He began to sing: 'My father is Czelkutq, my mother is Sinanewt, father has forgotten us.'

Czelkutq heard his son singing and said to the girls: 'Go and burn him with burning brands, and tell him that I am no father to him.'

The girls took the burning brands and burned the boy all over, they burned his little hands all over. 'It is hot! Mother, they burn me!' he cried, and back he went to his mother. She asked: 'Well, what did your father say?'

'He said: "I am no father to you;" he ordered the Amanita girls to burn me with hot brands; he burned my hands all over; it is hot and hurts me. I will not go to my father again, or they will burn me with hot firebrands.'

The next day his grandmother sent him to his father again, saying: 'Go once more, sing again, say "Father, tomorrow all of us will leave, you will stay in the forest here with the Amanita; afterwards you will surely starve"'.

The little boy set off to his father and began to sing: 'Father, tomorrow all of us will leave together. You will remain in the forest with the Amanita; afterwards you will surely starve.'

Czelkutq, hearing his son singing, became angry and said: 'Go, girls, and beat him thoroughly with a leather strap and burn him with fire; tell him to stop coming here.'

Thus the girls took firebrands and a leather strap and began to beat him and burn him; thus they drove him away. The boy cried, and started back to his mother. He was burned all over when he arrived, but his grandmother blew at him and made him well. The old woman said: 'Well, Sinanewt, let us get ready to go, we shall go into the woods.'

They began to get ready; they tied up all the animals and took them all, leaving nothing. They started into the woods. When they arrived, they picked out a high mountain, climbed up to the top, and poured water all over the mountain, making a sheet of ice.

Czelkutq began to go into the woods. He did not kill any animals at all; all the traces had gotten lost. He and the girls began to starve. What could they eat? Then Czelkutq remembered his wife and son and went home. He came to his house, but did not find his wife and son. He began to weep: 'Where did all my people get lost? I am starving, Sinanewt, I am hungry. Where did you and our son go away?'

He followed his wife by their traces, and reached that high mountain. 'How to get up? The ice is very slippery.' From below he called up: 'Sinanewt, pull me up!'

Sinanewt threw down a leather cord and called out: 'Well, Czelkutq, catch the cord!'

He caught the cord. She began to pull him up to the top of the mountain, but when he was ready to step onto the top, she cut the cord with a knife. Czelkutq flew downwards, he fell, he died, he revived, and again he called out: 'Sinanewt, pull me up, I am starving!'

'Why don't you live with the Amanita? Why don't you live with the Amanita? Why do you come to us? You tortured your son, and now you are being paid back; it's you yourself who began that sort of life.'

'Sinanewt, stop being angry, pull me up, I am hungry!'

She threw down the cord again, saying: 'Well, catch, I shall pull you up now.'

Czelkutq caught the cord and she pulled him up. When he got near the top, she again cut the cord with a knife; he flew down, he fell, he died, he lay there, he came to life again, and cried out: 'Sinanewt, stop being angry!'

'If I pull you up, will you go on living like that afterwards?'

'No, I won't, Sinanewt, I shall stop living like that.'

She threw down the cord, he was pulled up, dried out, and became happy; he ate, he became satiated. Again they began to live as before, amusing themselves. The Amanita dried up and died.

B

EMEMQUT AND HIS WIFE YELTALNEN

Kutq lived with his wife Miti. They had a girl-child, Yeltalnen. There was also an old woman, Kutq's mother, who lived with them. Many suitors came, but the old woman ate them all up, letting nobody pass by. The old woman was a cannibal. Ememqut heard of the very pretty girl Yeltalnen. He made ready to go there, and caught a wild reindeer. When he went, he took the reindeer with him. When he came near, he drove the reindeer on ahead of him. The old woman ate up the reindeer, but Ememqut passed by without her noticing him. He came to Kutq and asked: 'Kutq, where is your girl?'

Kutq said: 'We have no girl.'

Yeltalnen was in another house. Ememqut began to live with Kutq. He certainly wanted to get to Yeltalnen, but she did not accept him. Ememqut thought this over, turned himself into an old woman, and made a violent snowstorm. He came to Yeltalnen again and began to plead, 'Let me in, Yeltalnen, I am suffering from the cold.'

She let him in. Yeltalnen did not recognize Ememqut, but thought he was really an old woman and said: 'Old woman, sit down there by the door.'

Ememqut lulled Yeltalnen to sleep. She fell fast asleep and felt nothing. Ememqut did what was needed and then left. Thus Yeltalnen became pregnant. Yeltalnen realized she was pregnant and made baby clothes. Miti came in and said: 'Well, Yeltalnen, what are you doing? What are you thinking about?'

Yeltalnen answered: 'Yes, Mother, I am pregnant. I certainly slept with nobody. I only let a little old woman in here once during a violent snowstorm.'

Miti said: 'That must have been Ememqut.'

Yeltalnen gave birth to a very pretty child. She said to her mother and father: 'Well, tell my suitor that Yeltalnen says, "All right, I accept."'

Kutq and Miti answered: 'Marry Ememqut if you wish.'

Ememqut married her. He began to live well. Ememqut said: 'Well, let us start home.' They began to get ready. Ememqut went out into the yard and whistled; three pair of reindeer arrived, and they started home. Yeltalnen's friends told her: 'You are happy now, but later your snot will dangle on your whip.'

They came home. Many ravens had soiled Ememqut's house with excrement. Ememqut fixed up the house. He began to arrange a feast and invited everybody. The cossacks came. Cickimcican came, acting as if he had eaten Amanita. He said: 'Yeltalnen, urinate into a scoop of horn, and I shall drink your urine, since we used to sleep under the same covers.'

Yeltalnen said: 'Cickimcican, you are lying.'

Ememqut became angry with his wife and abandoned the feast; all the guests went back home without having feasted. Ememqut lay down all the time, he stopped getting up altogether; he was angry because of Cickimcican's remarks. He stopped looking at his wife and was angry all the time. Yeltalnen said: 'Ememqut, you are angry with me all the time; I shall go home to my father.'

Ememqut said: 'Well, go ahead.'

Yeltalnen began to cry and went out into the yard. She whistled, and two pair of reindeer arrived. She went back into the house and said: 'Well, good-by, Ememqut, I am going.'

Ememqut tried to grab his wife by the skirt, but he couldn't hold her. Then Yeltalnen drove off with the reindeer and disappeared. She began to weep; a snot was dangling on her whip. She said: 'My girl-friends were really telling the truth.'

She came to her father and mother, and began again to live in the other house.

It grew warm. In the yard the sun was warm. Sinanewt said: 'Ememqut, it is warm in the yard; I shall carry you out there.' Ememqut didn't want to go out, but Sinanewt carried him out anyway, together with his bedding. Ememqut's whole side had been fouled by lying down. He began to sit there in the yard, and said: 'Sinanewt, get my arrows. I shall count them and see whether any got lost.'

Sinanewt brought him the arrows and Ememqut began to count them. A piece of grass was dangling on the arrows; he could not untie it, so he cut it off with a knife and threw it behind him. Then somebody began to cry behind him, saying: 'I take pity on you, Ememqut; you have cut me loose with a knife.'

Ememqut looked back and saw a little old spider-woman. The spider-woman got up and went around Ememqut three times. Then he improved and got well. He remembered his wife, and started off to see her. He arrived there, but was not let in. He began to live there. He worked for Kutq three years, but his wife was not given back to him. Ememqut dug a passage under the earth to his wife. He went through the passage secretly and slept there with his wife. Three years went by, and Ememqut's wife was given back to him. They started home. They came home, and began to live well again. Again he arranged a feast, and invited all the people. Many guests came. The bad man Cickimcican came again. Ememqut grabbed him and threw him somewhere far away. He began to feed the guests. When they stopped eating, they began to wrestle. Nobody could compete with Ememqut; they all fell. They stopped wrestling and began to toss each other on a skin. Again nobody could compete with Ememqut; they all fell down. They stopped playing this game. They all began to urinate. Ememqut urinated very far: nobody could compete with him, and he beat them all. Ememqut began to live and to rejoice.

C

KUTQ, MITI, AND THE LITTLE LOUSE

Kutq lived with his wife Miti. Their children were Ememqut and Sinanewt. Ememqut used to go hunting in the woods. He started out to spend the autumn in the woods. At home, Kutq used to carry in wood. He grew tired, and once he said: 'Miti, you give birth to a small louse.'

'Eh, stop talking nonsense, Kutq!'

Kutq went to get wood, and Miti gave birth to a small louse. She looked and saw that it was very bright, like the sun. Then Miti began to sneeze. She swaddled the louse and hid it. Kutq came home and said: 'Ah, I am tired; take the load of wood, Miti.'

Miti said: 'Put it up yourself, Kutq: I already gave birth to a louse.'

'Yes, you gave birth; well, I shall have a look at the child.'

'It is very ugly.'

'Well, Miti, I shall see.'

Miti unwrapped the louse. Kutq looked, and fell on his back. He said: 'Hide the child, Miti, it is too bright.'

Then Miti hid the louse. Kutq said: 'I shall make another dwelling for the louse.'

'Well, make it.'

Kutq began to work and made the dwelling. The louse began to live there secretly.

Ememqut came home. Nobody told him anything, and he did not know about the louse. Miti used to bring food to the louse. Ememqut watched where his mother went with the food. Once Ememqut ate and then went out to the yard and hid among the piled-up logs. Miti took food to the louse and opened the door of the dwelling. A light shone out; then Miti covered up the entrance and went back into her house. Ememqut found the dwelling, opened the door, and saw his sister, who shone like the sun. Then Ememqut fell down and died. Kutq and Miti said: 'Where did Ememqut go off to?'

Miti got up early and looked for Ememqut. She found him, but he had already died. Miti began to weep, and Kutq also wept; they grieved for Ememqut. Kutq said: 'Miti, let us go and look for pleasure.'

They got ready to go. Ememqut was carried into the house and covered over. Outside they blocked up the whole door with wood. They set out. When they looked back at the house, they again began to weep. They met the people called Raven's Berries, who invited them to stay there to live with them. Kutq did not want to do this. They went on, and met the storks, who also asked them to live there with them. Kutq said: 'The heart does not want to live here.'

They went farther on, crossed mountain ridges, and met the Amanita. There they found pleasure. They stayed there: they were well met. Kutq began to live there with his family. They were rejoicing. However, when they remembered Ememqut, then they wept. The sun and the moon said: 'Where did Ememqut get lost? He is not to be seen. Maybe he has died somewhere.'

The sun said: 'Moon, let us go look for Ememqut.'

The moon said: 'Well, let us go.'

They began to get ready, and set out. They went around the whole universe, but Ememqut was nowhere to be found. The sun said: 'Well then, let us go to Kutq's house.'

They set off to Kutq's house, but when they got there they found the house all blocked up with wood. They threw down the wood and went into the house. They saw that Ememqut was dead. The sun said: 'Moon, let us revive Ememqut. You go around him beating the drum.'

The moon said: 'I am not strong: I appear, wax, and wane, without warming anybody,' and added: 'You warm everyone better, sun.'

The sun said: 'You try first.'

Then the moon began to go around Ememqut, beating his drum and kicking him, but he could not revive him. Again he went around him, beating the drum and kicking him, but again he did not revive him. A third time he went around him beating the drum and kicking him, but Ememqut only moved his little finger. The moon became tired. Then the sun began to go around Ememqut beating the drum and kicking him; Ememqut opened his eyes. The sun went around a second time, beating the drum and kicking him; Ememqut sat up. The sun went around a third time beating the drum and kicking Ememqut; Ememqut then got up and said: 'Yes, I have been asleep for a long time.'

The sun and the moon said: 'If we hadn't come you would have slept forever.'

The three of them began to make a summer-hut. They made it on three portable posts. All kinds of animals came and sat down: migratory geese, swans, cuckoos, and grebes, all came and sat down to sing songs. The only animals they did not seat were the bears. They set out to Kutq; they sat down to amuse themselves. The bears stretched up against the house; they too wanted to sit down. The cuckoo began to laugh at the bears. The sun said: 'Cuckoo, stop laughing, or we will throw you out.'

Wherever they passed by the people heard them and came out to see. Old men and women were carried out in their bedding; for the first time they heard of such pleasure. They came to Kutq. Kutq and Miti saw that everyone was having a good time, and began to weep, saying: 'If only Ememqut were alive he too would be sitting here enjoying himself!'

Kutq wept again; they all began to climb down from the summer-hut. There they saw Ememqut. Kutq and Miti began to be very happy. Then the sun married the louse and the moon married Sinanewt. Ememqut married the Amanita. All the people began to rejoice and started back to Kutq's house. They all became Amanita. They all began to live there and to rejoice.

D

THE CODFISH, THE RAMS, AND THE AMANITA

Codfish lived with her son Ilaqamtalxan. The Amanita girls lived there too, and gathered berries. The rams courted the girls, but the girls did not wish to marry. Codfish slept all the time. When it rained the Amanita set out to gather berries. They got soaking wet. When they came back they went into the codfish's ear and built a fire to dry themselves. The codfish woke up and said: 'Ow, it's hot, there's a fire in my ear.'

His mother said: 'Come here; I shall see what's the matter.'

She looked, but nothing was burning. 'You are lying,' she said.

Again the codfish fell asleep. The girls went out and ran farther away. Codfish woke up and saw the girls, saying: 'So, it was you who built a fire in my ear.'

'Well, we were only drying ourselves out.'

Codfish said: 'Come here and eat.'

The girls said: 'We don't want to eat.'

The girls went home. Again the rams came, and then the girls married them. They invited the codfish to the wedding. It ate up everything and became satiated; then it went home and fell asleep. The rams began to live well and to rejoice.

KORYAK TALES

A

LITTLE-BIRD-MAN AND RAVEN-MAN

Raven-Man said once to Little-Bird-Man, 'Let us go to Creator's to serve for his daughters.' Little-Bird-Man consented, and they started off to go to

Creator. 'What have you come for?' he asked them. 'We have come to serve here,' they answered. 'Well, serve,' he said. Then he said to Miti, 'Let Little-Bird-Man serve at our house, and Raven-Man at sister's.' – 'No,' replied Miti, 'let Raven-Man serve here, and Little-Bird-Man there.' Raven-Man and, Little-Bird-Man began to serve. A violent snowstorm broke out, which lasted several days. Finally Creator said to the suitors, 'Look here, you, who always keep outside, stop the storm.' Raven-Man said, 'Help me get ready for the journey.' They cooked all sorts of food for him. He took his bag, went outside, stole into the dogkennel, and ate all his travelling-provisions. When he had finished eating, he returned to the house, and said, 'I have been unable to stop the snowstorm.' Creator said to Little-Bird-Man, 'Now it is your turn to go and try to put a stop to the storm. The women shall cook supplies for your journey too.' Little-Bird-Man replied, 'I don't need anything. I will go just as I am.' He flew away to his sisters. They asked him, 'What did you come for?' He answered, 'I am serving at Creator's for his niece, and he has sent me to stop the snowstorm.' Then his older sister knocked him over the head and stunned him. Little-Bird-Man broke in two, and the real Little-Bird-Man came out from within. His sisters brought him a kettle of lard and some shovels, and went with him to the land of the sunrise. There they covered up all the openings with snow, caulked the cracks with fat, and it stopped blowing. It cleared up. Little-Bird-Man went home with his sisters, caught some reindeer, and drove to Creator's. On his way he ate some fly-agaric which his sisters had gathered, and became intoxicated. He arrived at Creator's, and noticed that his entire house was covered with snow. He shovelled off the snow, and shouted to his bride, 'Kĭlu', come out! untie my fur cap.' The people came out of the house to meet him, and saw that it had cleared up.

Soon after that, Raven-Man and Little-Bird-Man married, and on that occasion ate some fly-agaric. Raven-Man said, 'Give me more. I am strong, I can eat more.' He ate much agaric, became intoxicated, and fell down on the ground. At the same time, Creator said, 'Let us leave our underground house, and move away from here. The reindeer have eaten all the moss around here.'

They called Raven-Man, but were unable to wake him. They struck his head against a stone, and it split, so that his brain fell out. Creator left him in that condition, saying to a post in the house, 'When he recovers his senses, and calls his wife, you answer in her place.' Thereupon Creator wandered off.

When Raven-Man came to, he cried, 'Yiñe'a-ñe'ut!' The Post replied, 'Here I am.' – 'Have I become intoxicated with fly-agaric?' – 'Yes, with fly-agaric,' the Post replied. Then he noticed his brain, and asked, 'Have you made a pudding for me?' – 'Yes, I have,' the Post replied again. Raven-Man took his brain and ate it. Then he came to his senses. He felt of his head, and discovered that his skull was split, and that there was no brain in it. 'Whither shall I fly now?' he thought. He flew up to a mound and sat down. 'My sister Mound,' he said, 'I have come to you. Give me something to eat.' She replied, 'I have nothing. All the birds sit here upon me, and they have eaten all the berries.' – 'You are always stingy!' said Raven-Man. 'I will fly to a place from which the snow has thawed off.' He arrived at another place, and said, 'Sister, give me some berries to eat.' – 'I have nothing,' that place replied. 'Every bird sits here, and they have eaten everything.' – 'You, too, are stingy,' said Raven-Man. 'I will go to the beach.' He flew down there, and said, 'Sister, give me something to eat.' – 'Eat as much as you please,' said the Beach. 'I have plenty of seaweed.'

And Raven-Man remained on the seashore. That's all.

Told by Kučä'ñin, a Reindeer Koryak woman,
in camp on Chaibuga River, April. 1901.

B

EMEMQUT AND SUN-MAN'S DAUGHTER

It was at the time when Creator lived. There was no village and no camp near him. One evening his son Ememqut was returning home. It was getting dark. Suddenly he noticed sparks coming out of a marmot's hole. He went into the hole, and saw Marmot-Woman sitting there. He married her, and took her home. On the following day he again went hunting, met Sphagnum-Woman, took her for his wife, and also conducted her home.

Ememqut's cousin Illa' envied his success in having found pretty wives for himself, and conceived a plan to kill him in order to take away his wives. Illa' said to his sister Kĭlu', 'Go and call Ememqut. Tell him that I have found a tall larch-tree with gum. Let him go with me to take out the gum; and while there, I will throw the tree upon him and kill him.' She went and called Ememqut, and he and Illa' started off to the woods. They began to pick out the gum. Suddenly Illa' threw the tree down upon Ememqut and killed him.

Illa' ran home, singing and repeating to himself, 'Now Marmot-Woman is mine, and Sphagnum-Woman is also mine.' He came running home, and said to Kĭlu', 'Go into Creator's house and tell Ememqut's wives, your future sisters-in-law, to come to me.'

Kĭlu' came into Creator's underground house, and saw Ememqut lying in bed with his wives, and all of them chewing larch-gum. She returned to her brother, and said, 'Ememqut is at home alive, and lying with his wives.' – 'Well,' said Illa', 'now I will kill him in another way.'

On the next day Illa' sent his sister to Ememqut to tell him that he had found a bear's den. Illa' added, 'He shall go with me to kill the bear.' Kĭlu' delivered the message to Ememqut. Ememqut came, and went to the woods with Illa'. As soon as they reached the den, the bear jumped out, rushed upon Ememqut, and tore him into small pieces.

Illa' ran home again, singing and repeating, 'Now Marmot-Woman is mine, and Sphagnum-Woman is also mine.' He came running home, and said to his sister, 'Go and call your sisters-in-law.' She went into Creator's house, and saw Ememqut sitting at the hearth, and his wives cooking bear-meat. Kĭlu' came home, and said to her brother, 'Why, Ememqut is alive at home, and his wives are cooking bear-meat.'

'Well,' said Illa', 'now I will put an end to him.' He dug a hole in his underground house, and made an opening which led to the lower world, and put a reindeer-skin on top of the hole. 'Go and call Ememqut to play cards with me.' Thus said Illa' to his sister. Ememqut replied, 'I am coming.' When Kĭlu' was gone, Ememqut said to his wives, 'He is likely to kill me this time, for he has made a hole for me which leads to the lower world. I shall go now. If I do not come back for a long time, go out and look at my lance which is standing there. If it should be shedding tears, then I am no longer among the living. Then tie some whalebone around your bodies, which will wound him when he lies down to sleep with you.'

Ememqut went away. When he entered Illa''s house, Kĭlu' said to him, 'There is a skin spread for you: sit down on it.' As soon as Ememqut stepped on the skin, he fell down into the lower world.

Soon his wives went out, and, seeing that tears were running from his lance, they said, 'Our husband is dead now.' Then they tied some whalebone around their bodies. After a while, Kĭlu' came and said to them, 'Come, Illa' is calling you.' They went. Illa' said to his sister, 'Make a bed for us: we will lie down to sleep.' Kĭlu' made the bed, and Illa' lay down with Ememqut's wives. They tried to lie close to Illa', and pricked and wounded him all over. After a while, when they went outside, both stepped accidentally upon the skin, and fell down into the lower world.

Having fallen into the lower world, Ememqut found himself in a vast open country. He walked about, and came upon a dilapidated empty underground house. This was the abode of Sun-Man's daughter. Her name was Mould-Woman. Sun-Man covered her with a coating of mould, and let her down into the lower world, that the people on earth might not be tempted by her dazzling beauty. Ememqut stopped near the house, and began to cry. Suddenly he heard Mould-Woman's voice behind him, saying, 'You are such a nice-looking young man, why do you cry?' Ememqut answered, 'I thought that I was all alone here. Now, since I have seen you, I feel better. Let us live together. I will take you for my wife.' Ememqut married her, and they settled down to live together.

When Ememqut's wives fell down into the lower world, they also found themselves in a vast open country. They wandered about, and soon fell in with Mould-Woman. They said to her, 'We are Ememqut's wives.' She replied, 'So am I.' – 'Well, then don't tell your husband that we are here. You are bad-looking; and when he finds out that we are here, he will desert you and come to look for us.' Mould-Woman returned home. After she had met the two women, she used to go out to visit them; and Ememqut noticed her frequent absence. He asked her, 'Is there some one near our house?' – 'No, there is no one there,' she replied.

Once when she went out, Ememqut followed her stealthily. She sang as she went, 'My husband is a valiant man: he kills all the whales; he kills all the reindeer!' and Ememqut walked behind her, and laughed. She heard his laughter, turned around, but there was no one to be seen, for Ememqut had suddenly turned into a reindeer-hair. Then she said to her buttocks, 'Buttocks, why do you laugh?' She went on singing. Ememqut again laughed behind her. She looked back again, but Ememqut had turned into a little bush.

Thus she reached the place where Ememqut's former wives were. Ememqut suddenly jumped out in front of her. She was so much frightened that she fell down dead. Then the coating that covered her cracked, broke in two, and the real handsome and brilliant daughter of Sun-Man appeared from it. Ememqut took all his three wives and settled down.

Once Ememqut said to his wives, 'The Fly-Agaric-Men (Wapa'qala$^{\varepsilon}$nu) are getting ready to wander off from here into our country: let us move with them.' His wives prepared for the journey, and made themselves pretty round hats with broad brims and red and white spots on them, in order to make themselves look like agaric fungi. Then they started, and the Fly-Agaric people led them out into their country, not far from Creator's underground house.

Illa' and Kĭlu' went to gather agaric fungi. Suddenly Ememqut and his wives jumped out from among the fungi. Then they took Illa' and Kĭlu' home. Ememqut put them upon the Apa'pel,[1] on which they stuck fast. Ememqut

1. Apa'pel (from A'pa, 'grandfather' or 'father' [Kamenskoye]) is the name given to sacred rocks or hills.

said to his wives, 'Boil some meat in the large kettle, and scald Illa' and Kĭlu' with the hot soup. In the morning pour out over their heads the contents of the chamber-vessels. Put hot stone-pine-wood ashes from the hearth also on their heads.'

They did as they had been told. Finally Ememqut's aunt Hanna said to him, 'You have punished them enough; now let them off.' Ememqut let them off, and they lived in peace again.

Ememqut took his wife to Sun-Man's house, then he came back with Sun-Man's son, who married Yiñe'a-ñe'ut. Thus they lived. That's all.

Told by Kuča'ñin, a Reindeer Koryak woman,
in camp on Chaibuga River, April, 1901.

C

EMEMQUT AND WHITE-WHALE-WOMAN

It was at the time when Big-Raven lived. A small spider was his sister, and her name was Amı'l̤l̤u. Pičvu'čin wished to marry her. At that time Big-Raven became very ill, and was unable to leave his bed. 'Pičvu'čin,' he said, 'you are my brother-in-law-to-be. Do something for me, go in search of my illness.' Pičvu'čin beat his drum, found the illness, and said to Big-Raven, 'Take your team to-morrow and go to the seashore.' In the morning Big-Raven started with his team of dogs. After a while he was able to sit erect upon the sledge; then he tried to stand up; and soon he was able to run along, and direct his dogs. At the mouth of the river he saw a water-hole, and in that hole he found a White-Whale woman, Miti by name, whom he took for his wife. He carried her home. In due time she gave birth to Ememqut, who soon grew to be a man, and also took a White-Whale woman for his wife. Then Ememqut went for a walk, and found there Withered-Grass-Woman whom he also took for his wife. After that he brought home Fire-Woman, and then Kıněesa'tı-ña'wut.

These four women lived together without quarrelling, until finally Ememqut found Dawn-Woman. She began to quarrel with all the others. The White-Whale woman said, 'I am his first wife. I am the oldest woman. I will go away.' Big-Raven's people sat up for several nights watching, to prevent her leaving the house. At last Big-Raven's lids dropped, and he said, 'I want to sleep.'

Then she ran away. She reached a lake, and there her heart was swallowed by a seal. She transformed herself into a man, and married a woman of the Fly-Agaric people. Ememqut went in search of her. While on his way, he found a brook from which he wanted to take a drink of water. He smelled smoke coming up from beneath. He looked down, and saw a house on the bottom. His aunt Amɪ′ḷḷu, and her servant Kɪhɪ′ḷḷu, were sitting side by side in the house. While he was drinking from the brook, his tears fell into the water, and dropped right through into his aunt's house, moistening the people below.

'Oh!' they said, 'it is raining.' They looked upward, and saw the man drinking. 'Oh!' they said, 'there is a guest.' Then Kɪhɪ′ḷḷu said, 'Shut your eyes, and come down.' He closed his eyes, and immediately found a ladder by which he could descend. 'Give him food,' said Amɪ′ḷḷu. The servant picked up a tiny minnow from the floor, in the corner, all split and dried. She brought also the shell of a nut of the stone pine and a minnow's bladder not larger than a finger-nail. Out of the latter she poured some oil into the nutshell, and put it before Ememqut with the dried fish. 'Shut your eyes, and fall to.' He thought, 'This is not enough for a meal;' but he obeyed, and with the first movement dipped his hand into the fish-oil, arm and all, up to the elbow. He opened his eyes, and a big dried king-salmon lay before him, by the side of the oil-bowl. He ate of the fish, seasoning it with oil. Then his aunt said 'Thy wife is on the lake, and her heart has been swallowed by a seal. She has turned into a man and wants to marry a woman of the Fly-Agaric people.' He went to the lake and killed the seal. Then he took out his wife's heart, and entered the house of the Fly-Agaric people. An old woman lived in the house. He put the heart on the table, and hid himself in the house. His wife, who had assumed the form of a man, lived in that house; and in a short time she came in from the woods, and said, 'I am hungry.' – 'There is a seal's heart on the table,' said the old woman. 'Have it for your meal.' She ate the heart, and immediately she remembered her husband. He came out of his hiding-place. They went home, and lived there. That's all.

Told in the village of Palla′n.

D

RAVEN AND WOLF

Raven said to his wife, 'I want to go coasting. Give me a sled!' She gave him a salveline. He refused to take it, and said, 'It is too soft: it will break into pieces.' Then she gave him a seal. He rejected it also, saying, 'It is too round: it will roll away.' Then she gave him an old dog-skin. On this he coasted down hill. A Wolf passed by, and said, 'Let me, too, coast down hill.' – 'How can you? You have no sled: you will fall into the water.' – 'Oh, no! My legs are long: I will brace them against the stones.' Wolf coasted down the hill, fell into the water, and cried, 'Help me out of this! I will give you a herd of water-bugs!' – 'I do not want it!' – 'Help me out, and I will give you a herd of mice!' – 'I do not want it!' – 'Help me out, and I will give you my sister, the one with resplendent (metal) ear-rings!' Then Raven helped him out. Wolf said, 'Fare thee well! I am an inlander. I will go inland, far into the country. Where are you going?' – 'I belong to the coast. I will stay here, close to the seashore.' Wolf went his way. Raven transformed himself into a reindeer-carcass, and lay down across Wolf's path. Wolf ate of it. Then Raven revived within his belly, and cried, 'Qu!' Wolf started to run. Raven tore out his heart, and dashed it against the ground. Wolf died. Raven dragged the body to his house, and said to Miti, 'I have killed a wolf! Dance before the carcass!' Miti began to dance, and to sing, 'Ha′ke, ha′ke, ka ha′ke! Huk, huk! My husband killed one with a long tail!' Wolf's brothers followed the trail; but Raven dropped on the trail a couple of whalebone mushrooms.[1] They swallowed them, and were killed. Raven's people dragged them into the sleeping-room of Raven's daughters, Yiñ′ía-ñe′whut and Čann·a′y-ña′wut, pretending that these were the girls' bridegrooms. The oldest of Wolf's brothers, whose name was Long-Distance-between-Ears (literally 'large-[between-the]-ears-interval'), followed Raven's trail. Again Raven dropped a couple of whalebone mushrooms. Wolf, however, did not swallow them, but took them to Raven's house. 'What are these?' he asked Raven. 'These are my children's toys.' – 'And where are my brothers? Their trail seems to lead here.' – 'No, they did not come here.' Wolf and his hosts went to sleep. In the night-time Wolf stole into the girls' sleeping-room, wakened his dead brothers, and they led the girls away.

Next morning Ememqut said, 'Now I will at least steal the Wolves' sister.' He asked The-Master-on-High to let down for him the ancestral old woman. Then he killed the old woman, skinned her, put on the skin, and sat down on the snow, weeping, and his teeth chattering with the cold. The Wolf people passed by. 'What are you weeping for?' – 'My children lost me in the snow-storm, and now I am freezing to death.' They took her along and put her into the sleeping-room of Wolf's sister. 'Ho! make her warm!' But in the morning the girl was with child. That's all.

Told in the village of Opu′ka.

1. A well-known contrivance, made of a slender spit of whalebone bent around, tied with sinew, and then covered with hard, frozen tallow. When swallowed by a wolf, the tallow melts, the sinew string gets loosened, and the sharp ends of the spit break through the walls of the stomach.

[30]

VOGUL Hymns and Heroic Songs. Antal Reguly and Bernát Munkácsi. Vogul Népköltési Gyűjtemény. (An Anthology of Vogul Folklore) Vol. I. Regék és Énekek a Világ Teremtéséről. (Sagas and Songs about the Creation of the World) Published in fascicles from 1892 to 1902. Vol. II. Istenek Hősi Énekei, Regéi és Idéző Igéi. (Heroic Songs, Sagas, and Invocative Spells of Gods) 1892. Budapest.

[There survives a substantial corpus of texts in Vogul and Ostyak taken down by Antal Reguly and Bernát Munkácsi from native singers. Reguly, a man of remarkable character, was working among the Vogul in 1843-1846, and he died prematurely in that decade without having published the Vogul texts that he had recorded. In 1888-1889 Munkácsi, with Reguly's notes, visited the Vogul country and, seeking out the same singers or their successors, had them repeat the same songs. With the lapse of time there were naturally variations. In the two volumes before us Munkácsi published Reguly's texts, his own texts, and translations of his own texts into Magyar. He also brought out volumes of textual exegesis. Hungarian scholars merit our gratitude for having preserved these texts for posterity and annotated them, publishing a large part of them.

[In these native texts there is a category that scholars call in German *Fliegenpilzlieder*, 'fly-agaric-songs', songs composed under the influence of the fly-agaric. I have unfortunately not had an opportunity to explore these songs, which lie hidden behind linguistic barriers for the English-speaking world. Apart from their general interest for the world of scholarship, they may well contain important treasures for the ethno-mycologist; and if the thesis of this book turns out to be right, it will become imperative that the West gain access to this store-house of fly-agaric poems. After all, of the Siberian peoples who knew the fly-agaric, the cult among the Vogul in recent centuries was the strongest: it pervaded their religious life, their vocabulary, their songs. The sanctions for the abuse of the fly-agaric still held the people in thrall into this century: anyone not a shaman who ate the fly-agaric did so at peril of death.

[In May 1967 I spent three days in Budapest with Dr. János Gulya, of the Linguistic Institute of the Hungarian Academy of Sciences, the outstanding Vogul scholar. With the help of Tamás Radványi, university instructor in English, we concentrated our attention on passages from one of the three

Heroic Songs that Munkácsi had discussed in his paper on mushroom inebriation. (*Vide* [32]) This song is of particular interest for the light it sheds on the question of the bellicosity of the man under fly-agaric inebriation. Our Hero, the Two-Belted One, has eaten three sun-dried fly-agarics. News comes that the Mocking-bird Army from the north is invading the country and our Hero is desperately needed to lead the fight against the invaders. But he is in a fly-agaric stupor and sends the messengers to his younger brothers. Later they return and implore him *to throw off his stupor* and come and fight. This he does: he sallies forth and slays the enemy right and left. The Song exists in two recensions, Reguly's and Munkácsi's: the meaning of both is the same. Here is the English translation in prose of Munkácsi's recension of the Song (Vol. 1, pp. 113, 115, 117) as we worked it out in Budapest. – RGW]

The men keep going on. Whether for a long time or a short time, they keep going on. To their fortress Jäkh-tumen they returned, they got home. Their mother, a woman of the Kami [river], sets up a kettle [for brewing beer] so big that it could not be used up by the whole town. For three nights and three days people keep drinking. The ecstasy of the ecstatic would not come to the eldest man, the inebriation of the inebriated one would not come. The Sovereign of the Lake to his daughter, to his wife, goes home, comes up to them, says, 'The inebriation of the inebriated man has not come to me; listen, woman, go out, fetch me in my three sun-dried fly-agarics!' She answers, 'Perhaps in your folly you wish to drink the blood of your paternal line, perhaps in your folly you wish to drink the blood of your maternal line!' He says, 'Why have you vexed me, the Two-Belted One, so long as I was calm? Do I ask you whether I wish to drink the blood of my paternal line? I do not ask you. Woman, now fetch me in my three sun-dried fly-agarics!' She throws them before him. He, putting them into his mouth with the ten bear-teeth, chews them, and the ecstasy of the ecstatic man comes upon him.

★

The big larch-wood door is kicked open. 'Oh, Uncle, don't carouse with the drunkenness of the drunken man! From northern regions the mocking-bird army with the red rump has flown here and they have all occupied your seven silver-headed posts that you yourself set up in the age of your increasing manhood [childhood]'. 'I have not strength enough, because of my inebriation [heat] of the inebriated [heated] man. Carry the news to the two younger sons of my father!'.

★

The big larch door is again opened, 'Oh, Uncle, don't carouse with the drunkenness of the drunken man! From northern regions the mocking-bird army with red rump has flown here and they have all occupied, surrounded, your seven silver-headed posts! 'The man says, 'Bring me my armour . . .'

[The Hero then goes forth and wreaks havoc on the enemy. . . . The reader will have noticed that the Hero has asked for his three *sun-dried* fly-agarics. When he is urged to leave off his 'carousing with the drunkeness of the drunken man', in the Vogul original the three operative words are all derivatives from *pa:ŋχ*, 'fly-agaric'. Dr. Gulya informs me that the Vogul no longer think of the fly-agaric when they use these words: the root gives them their everyday word for inebriation and its source is not present in their minds. One can become 'bemushroomed' on alcohol. – RGW]

B. The Linguistic Aspect

Preliminary Note

In this section we do not translate our authors. We paraphrase them, eliminating passages where they quote their predecessors, simplifying (sometimes radically) their phonetic representation of the original words, and standardising their various systems. We have reduced the number of dialectal differences or eliminated them. So far as we know, we have included every philologist who has dealt with the *poŋ* cluster. Linguists who interest themselves in the problems that the cluster raises will wish to consult the original sources.

The special characters that we use are:

ə : 'shwa', pronounced like the vowel in 'but'.
ɔ : an open *o*, pronounced like the vowel in 'awl'.
ŋ : a nasal; thus we write 'siŋer' for 'singer' but 'fiŋger' = 'finger'.
ł : a 'dark' *l* as in American English, rather than the French *l*.
χ : the Greek chi is pronounced as the *ch* in Loch Lomond.
ñ : as in Spanish; *cf.* *ny* in 'canyon'.
ʔ : glottal stop; as the word 'bottle' is pronounced by Scots, — *botʔl*.
ä : the sound of the vowel in 'let'.

A vowel long in quantity is indicated by the colon that follows it.

One of our linguistic contributors, Artturi Kannisto, contributes important, even sensational, ethno-mycological data. His information is dependable, and he dug out the fact that the Voguls of the Sosva and the upper Lozva use mushrooms other than the fly-agaric for shamanistic ends. In the valley of the Pelymka it is not clear whether the sexual distinction between male and female corresponds to different species of mushrooms or to some conventional distinction in the specimens of *A. muscaria*. Kannisto gathered his information in the first decade of this century, though it saw the light of day only in 1958. Even now it may not be too late for Russian mycologists to learn the precise species that the Vogul shamans were utilizing in Kannisto's day. It is to be hoped that they will not let these clues go unexplored.

Independently of Kannisto, I have received confirmation from Ivan A. Lopatin, the authority on Siberian cultures, that mushroom species other than the fly-agaric are used for their psychic effects. His personal communications to me were dated January 28, 1963, and July 19, 1966.

In these pages 'the Chukotka' will occasionally be found as a convenient geographical term. It is used in the Soviet Union to embrace all lands of the Chukchi, Koryak, Kamchadal, and Yukagir, in the Far Northeast of Siberia.

[31]

Boas, Franz. Handbook of American Indian Languages. Part 2. Washington, Government Printing Office. 1922.

On p. 693 Boas arrives at the stem *poŋ* as meaning 'mushroom in the Chukchi, Koryak, and Kamchadal languages. There is a duplication of the syllable in those languages, *poŋpoŋ* and *poŋpo*, etc., and this is discussed on p. 688, where however the meaning of the word receives a specific sense: the fly-agaric. Boas gathered his data at the turn of the century.

[32]

Munkácsi, Bernát. 'Pilz' und 'Rausch'. ('Mushroom' and 'Intoxication') Keleti szemle. (Oriental Review) Vol. viii. Budapest. 1907. pp. 343-344.

[This paper by Munkácsi was the first of a number that discuss the *paŋχ* cluster of words in some of the Altaic languages. We call attention to a sentence in it that we print in italic. This sentence seems to give justification for the Scandinavian belief that the fly-agaric can incite furious behaviour – berserk-raging – in the eater, and as we were dealing with a traditional Heroic Song issuing from the very entrails of the Vogul culture, rather than with the questionable observations of foreign travelers, it was vital to determine the facts. This led me to Budapest in the spring of 1967, where I worked out a translation from Vogul into English with the help of Dr. János Gulya, which we give in [30]. As the reader will perceive, the sense of the poem is utterly different from Munkácsi's one-sentence synopsis of it: in a stupor from three sun-dried agarics, our Hero is unable to respond to the call to arms. But time passes and the urgency grows, and when the messengers press their appeal to throw off his stupor he finally calls for his arms. The distinguished Magyar scholar had certainly never heard of the debate in Scandinavia and wrote out his summary without regard to it: even today it seems to be unknown to Altaic specialists in Hungary. The poem exists in two recensions, Reguly's and Munkácsi's, but the sense of both is identical. – rgw]

In Vogul the fly-agaric (*Amanita muscaria*) is called in the northwest dialect *pa:ŋχ*, and in the Middle Lozva West dialect *pə:ŋk*. This word is probably identical with the North Ostyak *poŋχ*, 'mushroom, fly-agaric', the Irtysh

Ostyak *pa:ŋχ, paŋχ*, 'fly-agaric'; the Yugan Ostyak *paŋga*, 'fly-agaric'; the Mordvinian *paŋga, paŋgo*, 'mushroom'; the Cheremis *poŋgo, pongə*, 'mushroom'. In North Vogul the word *paŋχ* also means 'intoxication', 'drunkenness'; whence *pa:ŋχeŋ χum*, 'drunken man'. In the Lozva *pa:ŋχli, pa:ŋχeta:li*, and in the Pelymka *pə:ŋkˀli* mean 'he is drunk, intoxicated' (*jolpa:ŋχles*, 'he has gotten drunk'; *pa:ŋχlim χɔtpä*, Lozva, *pə:ŋkˀlemkˀɔr*, 'a drunken man').

The connection between these similarly pronounced words can be explained by the fact that among the Voguls and Ostyaks in earlier times the fly-agaric played the same role as brandy plays today. As Patkanov reports ('A type of Ostyak Epic Hero,' pp. 5, 39: Irtysh Ostyaks I: 121), it is said that singer of heroic songs or a shaman would, in order to bring himself to a state of exaltation, consume 7, 14, or even 21 fly-agarics, which had been dried for this purpose and either soaked in water or spread with butter or fat before consumption. After consuming this narcotic, a person becomes almost crazy, undergoes severe hallucinations, and sings in a loud voice all night long until, completely exhausted, he finally falls helpless on the ground and lies unconscious for a long time. Gods also find pleasure in this narcotic. Thus a hymn addressed to the 'Man-Who-Observes-the-World' (*mir susne-χum*) contains the following lines:

> While you, in the corner of your seven-sided holy house that came from your father *Numi-Tɔrem* [Upper Heavens], your seven partitioned golden house, on your seven golden-footed tables, on your seven golden-lidded chests, run about in ecstasy caused by seven one-footed glasses, as you are running about in ecstasy caused by seven one-footed notch-edged fly-agarics (akw laˀylep lar'siŋ χu:rpä sa:t pa:ŋχ se:ŋwen χajtnen χalt), may your holy little ear, which is as big as a lake, as big as the Ob, nevertheless hearken here. May your golden eyes, which reflect the Ob, sparkle in this direction! (Vogul Népköltési Gyűjtemény, II:314)

This thought is expressed in another hymn to the same deity as follows:

> While you run about in the ecstasy of your intoxication caused by seven fly-agarics with spotted heads (kumliŋ puŋkep sa:t pa:ŋχ kus'män, se:ŋχwen χajtne χalt), may your mind, demanding blood sacrifice, may your mind, demanding food sacrifice, like the swelling water of the Ob, like the swelling water of the Lake, direct itself here. (*Ibid.*, II:362)

In the song about the creation of the heavens and the earth, recorded by Reguly, we are told of the Kami woman's oldest son, who later turned into a bear, that when after a three-day beer feast he had not yet achieved the intoxication he had been longing for, *he asked his wife to bring him three fly-*

agarics that had been dried in the sun (χɔ:tel tɔ:sem χu:rem pa:ŋχ), *and that after he consumed these, he flew into such a rage that he mercilessly slaughtered a great crowd of people.* (*Ibid.*, 1:114)

In the name *paŋχ* of this narcotic we recognize the Old Persian word *baŋha-*, whose meaning, according to Bartholomæ (Altiranisches Wörterbuch, 925), is the following: '1. Name of a plant (and its juice) which was also used for producing abortions; 2. Name of a narcotic made from that plant and also a designation of the state of narcosis produced thereby.' Other instances of this word are: Sanskrit: *bhaṅga-*, *bhaṅgā-*, meaning 'hemp; a narcotic prepared from hemp seeds'; modern Persian: *bang*, 'henbane' [Hyoscyamus niger] (*bangī*, 'senseless'); Armenian: *bang*, 'hyoscyamus'; Afghan: *bang*, 'hemp' (Horn, Modern Persian Etymology, 53; Uhlenbeck, Etymological Dictionary of the Old Indian language, 194). According to Wilhelm Geiger (East Iranian Culture in Ancient Times, p. 152), hemp is used in Persia for making 'the notorius hashish, the use of which deranges the human organism in the most frightful manner'. From all this we conclude that the Vogul word *paŋχ*, Ostyak *poŋχ*, 'mushroom, fly-agaric', as well as the Mordvinian word *paŋga*, and the Cheremis *poŋgo*, 'mushroom', according to their original etymon, have properly the meaning 'intoxicating', 'narcotic', and that the knowledge of this culture product among the Finnish peoples comes from the Aryans, just as this may be assumed to be true with regard to 'beer' (Vogul, Zyrian, Votyak *sur*, Vogul *sor*, Ostyak *sɔr*, Hungarian *ser*, *sör*, Sanskrit *súrā*, Avestan *hura*, 'a spirituous drink, chiefly brandy' or 'beer'.)

[33]

KANNISTO, Artturi. E. A. Virtanen and Matti Liimola, editors. 'Materialien zur Mythologie der Wogulen gesammelt von Artturi Kannisto.' (Materials on the Mythology of the Vogul Gathered by Artturi Kannisto) Mémoires de la Société Finno-ougrienne. Vol. 113, pp. 419-420.

In order to become intoxicated, the Vogul shaman employs or used to employ, above all things, the fly-agaric (*pa:ŋχ*, *pə:ŋk*,[1] *etc.*); after having eaten them, he becomes intoxicated. He must be in this state when he begins to exercise his functions [as shaman]. From the area around the river Sosva we have the report that fly-agarics grow out of a single 'foot,' six or seven of them together; when they are dry, they are yellowish brown.

1. *pa:ŋχ* and *pə:ŋk* are the same term in two dialects.

Around the Upper Lozva, the term *pa:ɒχ* refers to a small mushroom (not the beautiful many-colored fly-agaric) that grows at the base of a tree stump and the like in clusters, as if out from a single root; the *ña:it-χum*, 'shamans', dry and eat them. In the area around the Lower Konda the *pa:ɒχ* are collected after St. Peter's Day and dried in the hut. The shaman *ñoait-χar* or *käiləɒ-χar* eats seven of them when he begins his séance. Before they are eaten they are soaked in water; if there is butter on hand, they are eaten with butter; otherwise without. Once the fly-agaric has begun to affect the shaman, one says: *pa:ɒχɒə joχtwəs*, 'the fly-agaric has come into him'. After the shaman has eaten the fly-agaric he walks around the room, sings, and continues the séance. Around the river Pelymka, the fly-agaric is called *saittoal pə:ɒχ*, 'the fly-agaric that causes loss of sense'. There are male and female fly-agarics (respectively *kum-pə:ɒk* and *ne:-pə:ɒk*); the former are eaten by male shamans and the latter by female shamans. One must eat either three or seven of them. Nowadays fly-agarics are no longer used around the Pelymka. When the shaman 'makes the room dark' and eats fly-agaric before his séance, he must don *wa:rim u:lamt*, 'sacrificial clothes' (Upper Lozva).

p. 286. Around the area of the Lower Konda he [*käiləɒ* 'the shaman'] says which sacrifice is to be made after he has eaten fly-agaric, and he sleeps after having eaten them.

pp. 429-430. Around the area of the Lower Konda there are shamans who are married, unmarried, or who have been married in the past, as well as male and female shamans. The *käiləɒ* (shaman) also performs the following: he eats seven fly-agarics on the preceding evening, walks about the room, leaves the room a number of times, looks at the sky, yells something, enters, lies down, and remains in that position until morning. Then he tells what he knows and gives advice.

[28a]

DONNER, Kai. 'Ethnological Notes about the Yenisey-Ostyak (in the Turukhansk Region)' Mémoires de la Société Finno-ougrienne LXVI. Helsinki. 1933, pp. 81-82.

Donner here says that *haɒgo* means 'fly-agaric' in the dialect of Ket spoken in the Turukhansk region. We have already given the quotation under [28].

[24a]

LEHTISALO, T. 1. Juhlakirja Yrjö Wichmannin kuusikymmenvuotispäiväksi. (Publication in honor of Yrjö Wichmann's 60th birthday) Mémoires de la Société Finno-ougrienne. Vol. LVIII. Helsinki. 1928, p. 122.

On this page there is a note on a word reported by Castrén in the Tavgi tongue, one of the Samoyed languages, called in the Soviet Union today the language of the Nganasan people. That word is: *faŋkáˀam*, 'to be drunk'. He sees parallels with this word in various Finno-ugrian languages:

Mordvinian:	paŋgo, 'mushroom, lichen'
Cheremis:	poŋgə, 'mushroom'
Vogul:	paŋχ, piŋka, 'fly-agaric'
Ostyak:	pa:ŋχ, pɔ:ŋk, paŋχtəm, puŋkłem, depending on the dialect; the meaning is 'fly-agaric'

From the wide dissemination of this word the author concludes that intoxication from eating the fly-agaric goes back probably to pre-Uralic times.

The similarity of this word to the Latin *fungus*, Greek σπόγγος may be merely accidental. Munkácsi's assumption that the Finno-ugrians borrowed the word from Indo-Iranian is not likely, the author holds, giving as the example in Sanskrit *bhaṅga-s*, 'hemp and the narcotic substance made from it'.

2. 'Sampa, sammas'. Virittäjä. Helsinki. 1929. pp. 130-132. Translated from the Finnish.

Our language has many words phonetically reminiscent of the term *sampo* in the Kalevala. They may all belong to the same word family. Of these only one has been compared with an equivalent in the more distant related languages, *viz.*, the word *sammakko, sammakka* ('frog', 'toad') [and other related words – RGW], which has been linked to Lapp *cuobo*, gen. *cubbu*, 'frog', 'toad'. (*Vide* E. N. Setälä, 'Zur Etymologie von *Sampo*, FUF II, pp. 146 ff.) In this paper I will present a form from Samoyed that corresponds to another word in this Finnish family. Since the word I am writing about must be kept distinct phonetically from *sammakko, etc.*, we must reckon with two word families of the type *sampa*.

Let us look at the following Samoyedic forms. Yurak Samoyed: *sa:mpa:*, 'to carry the ghost-soul of the deceased to the hereafter to the accompaniment of the shaman's drum', ... [and other dialectal forms] meaning 'to sing an incantation to the accompaniment of the shaman's drum', 'he who sings an incantation to the accompaniment of the shaman's drum', 'shaman', etc. – On the Finno-Ugric side, we are immediately reminded of the Ostyak word (Konda dialect) *paɒχtəm*, 'to sing and shamanize after having eaten fly-agaric; to cure with incantations'. This word is of course related to Ostyak *paɒχ*, 'fly-agaric'. (*Vide* NSFOu Dol. 58, p. 122.) Bearing this in mind, the next step is to look for a word meaning 'fungus' or 'mushroom' that might be phonetically linked with the Samoyed words mentioned above. Thus we come upon the Finnish words *maansampa*, 'puffball', and *puunsampa*, 'a white pinhead-sized efflorescence on birches, alders, and willows', in which we propose that the element *sampa* originally meant 'fungus' or 'mushroom'. The Samoyed words go back to the initial palatalized sibilant **s*-, which distinguishes them, on the evidence of Lapp, from the word *sammakko*, which goes back to **č*-.

Closely related to the meaning 'fungus', 'mushroom', 'efflorescence', are the disease names *sampa*, *sammas*, *sampaat*, 'childhood disease affecting the mucus membrane of the mouth; swollen glands in the jaw' (Mikkeli area); 'glands in the jaw' (Häme, Savo); 'a disease of horses' (Viitasaari). From this we proceed further to the word *sammasvesi*, 'water from a certain spring used to cure *sammas*' (Sääksmäki); 'water from a hole in a rock used to wash out the mouth of a child suffering from *sammas*' (Luopioinen). According to Gottlund, the word *sammaslahde* is used to designate medicinal springs where a specified amount is paid when water is taken from them.

It might be mentioned also in passing that it would be tempting to link *tatti*, 'fungus', 'mushroom', with the following Samoyed word: Yurak Samoyed ... [The author supplies various words from Samoyed dialects meaning 'shaman'. But he concludes that this is rendered somewhat uncertain as *tatti* might be related to a different Yurak Samoyed word, though in turn there are counter-arguments that the author gives. – RGW]

We may furthermore bear in mind that I have linked the Tavgi Samoyed *fankáˀam*, 'to be inebriated', with the words in the Finno-Ugric family of languages meaning 'mushroom' and 'fly-agaric'; *cf.*, *e.g.*, Mordvin *paŋgo*, 'mushroom', Ostyak *pa:ɒχ*, 'fly-agaric' (*vide* NSFOu LVIII, p. 122). [The writer now quotes Donner and others whom we have translated. – RGW] We may regard it as certain that in the dim antiquity of the Proto-Uralic period the Finnic shaman ate intoxicating fly-agaric during his sorcery sessions. We now understand how it was that later, when the shaman had stopped eating the fly-agaric and was beating his drum, he relapsed in the course of his sorcery session into a state resembling madness and finally fell unconscious to the ground ('falling into a trance').

[34]

UOTILA, T. E. 'Etymologioita.' ('Etymologies') Virittäjä. Helsinki. 1930, pp. 176-7.

Some Asian and notably Finno-Ugric peoples use the fly-agaric for inebriating purposes. Here are the members of this word family:

Mordvinian (Mokša)	-paŋga, 'mushroom'
(Erza)	-paŋgo, 'mushroom, lichen'
Cheremis	-poŋgo, paŋgə, 'fungus'
Ostyak (North)	-poŋχ, 'mushroom, fly-agaric'
(Irtysh)	-paŋχ, 'fly-agaric'
Other Dialects	-paŋχ, "
	-pɔŋk, "
	-paŋχtəm, -puŋkłəm 'to shamanize while singing after having eaten the fly-agaric; to cure by shamanizing'
Vogul (North)	-paŋχ, piŋka, 'fly-agaric'
Tavgi	-faŋkáˀam, 'to be drunk'
Selkup	-pöŋer, 'drum'
	-pəŋgar, 'a special Selkup musical instrument'

These words belong to the Uralic family *_p8nk8_ (the '8' = back vowel). To it belong the following Zyrian forms:

(Sysola dialect)	-pagalny, 'to lose one's consciousness'
	-pagavny, 'to poison oneself, to kill oneself'
(Luza)	-pagyr, 'bitter taste in beer', 'sour, sharp, penetrating'

After giving some more derivatives, the author says that they reflect the effect of the fly-agaric on those who have eaten it. From the Uralic root *_p8nk8_ Uotila derives the *_pag_ form meaning the fly-agaric; and _pagyr_, bitter, from 'having the taste of fly-agaric', and _pagal_, 'the power to inebriate'.

[35]

BOUDA, Karl. *Das Tschuktschische.* (The Chukchi) Published as Part 4 in Beiträge zur Kaukasischen und Sibirischen Sprachwissenschaft, by Deutsche Morgenländische Gesellschaft. Leipzig. 1941.

In this philological study Bouda on p. 35 under entry 20 discusses Chukchi *poŋpoŋ, pompoŋ,* meaning mushroom. By comparing it with Koryak *pˀonaw,* mushroom, and *a-pˀonake,* 'without mushrooms', he arrives at the Chukchi root *poŋ*. He compares these words with

Cheremis:	*poŋgo,* 'mushroom'
Mordvinian:	*paŋgo,* "
Ostyak:	*poŋχ, paŋχ,* 'fly-agaric'
Vogul:	*pɔŋk, pa:ŋχ,* "

In a footnote he calls attention to the use of the fly-agaric by the Ob-Ugrian shamans so that, having been entranced by its poison, they may communicate with gods and spirits.

[36]

STEINITZ, Wolfgang. Geschichte des Finnisch-ugrischen Vokalismus. (History of Finno-Ugrian Vowel Structure) Stockholm. 1944. p. 37.

The author includes in his list of words the Vogul *paŋk,* fly-agaric, which in Mordvinian is *paŋgo.*[1] In the Ostyak of Surgut he supposes that there has been an alternation of the vowel from **a* to **u* to **i,* leading to these three forms: *pɔŋkełłem; puŋkłəm; pyŋkła.* Only **y* in Vogul, with *pə:ŋk-* in one dialect. Changing vowel in Cheremis: *poŋgə.* It is unclear whether Zyrian *pagal-, pagyr* comes from this source.

1. The Mordvinian and Cheremis word means 'mushroom'.

[37]

Hajdú, P. Von der Klassifikation der samojedischen Schamanen. (On the classification of Samoyed Shamans) Glaubenswelt und Folklore der sibirischen Völker. (Religion and Folklore of the Siberian Peoples) Edited by V. Diószegi. Akadémiai Kiadó, Budapest. 1963. pp. 161-190.

p. 170. T. Lehtisalo is of the opinion that the family of the Nenets (Samoyed) word *sa:mpa:* may be related to the Finnish word *sampa*, fungus, sea-foam.[1] This Finnish word occurs in the names of diseases, as, for example, *sampa*, *sammas*, *sampaat*, a disease of the mucous membrane of the mouth in children; swelling of the submaxillary glands; a disease of horses. The word is also found in compounds such as *maansampa*, puffball; *puunsampa*, white mushroom the size of a pinhead on birches, willows, and alders. According to Lehtisalo, the meaning 'to work magic' originally referred to magic that was performed in the state of ecstasy brought on by fly-agarics. For a semantic confirmation of this, he cites the Khanti verb *paŋχtem*, to work magic while singing after consuming fly-agarics, to heal by magic (<: *panχ*, fly-agaric) On phonetic grounds, Lehtisalo distinguishes the Finnish *sampa*-Nenets *sa:mpa:* from the Finnish *sampo* and its derivatives. This is certainly an imaginative explanation, but there are many reasons which compel us to disagree with it. In the first place, this comparison seems doubtful to us because the agreement between the words is found only in the two most widely separated members of the Uralic language family. Another argument against it is that the meaning 'mushroom' has not been found in Samoyed and the meaning 'to work magic in the state of ecstasy brought on by fly-agaric' is unknown in these languages. One could suppose, of course, that the verb 'to work magic' was expressed by a derivative of a word meaning 'mushroom'; however, in Samoyed the form *sa:mpa:* is without a suffix, in so far as we can consider it to be the form corresponding to the Finnish *sampa*.

Because of these many problems, I have attempted to find a different explanation for the Samoyed *sa:mpa:*.

[The author then refers the reader to an article that he wrote, 'Etimológiai megjegyzések' ('Etymologische Bemerkungen') ['Etymological Notes'], in *Nyelvtudományi Közlemények*, Vol. LVI, pp. 53-56.]

1. Lehtisalo, T. '*Sampa, sammas*'. *Virittäjä*, Vol. XXXIII, pp. 130-132. [Our ref. [24a] pp. 310-312 – RGW]

[38]

BALÁZS, J. 'Über die Ekstase des ungarischen Schamenen.' (On the Ecstasy of the Hungarian Shamans) Glaubenswelt und Folklore der Sibirischen Völker. (Religion and Folklore of the Siberian Peoples) Edited by V. Diószegi. pp. 57-83. Akadémiai Kiadó, Budapest. 1963.

[Sections 1 and 2 (pp. 57-59) discuss the Hungarian word *rejt*, 'hide,' the Old Hungarian word variously spelled *rüt, rőt, röjt, réüt, rit, riüt*, 'entrance, enrapture,' and the problems whether these two words are related in meaning and whether one or both may be traced back to the shamanism of the Ugric period. – RGW]

pp. 59-67.

3. We may come closer to a solution of this problem if we investigate the methods by which the shamans worked themselves into a state of ecstasy; in this process we shall make use of the latest linguistic, ethnographic, and archæological results, as well as a critical analysis of earlier data. The semantic explanation of the Hungarian verb *rejt*, 'ecstasize,' is based, to this day, chiefly on Hunfalvy's arguments and on Munkácsi's commentaries on the Russian academician P. S. Pallas's description of his travels.

Concerning the ecstasy of the shamans among the Mansi and the 'northern' shamans in general, Munkácsi writes: 'The visible reality in the magical performances of the northern shaman is that he transports himself into a state of unconsciousness resembling a trance, *i.e.*, into a state of ecstasy. Even when he begins the magical performance, he is in a befogged condition owing to the fly-agaric or brandy he has previously taken. His drunkenness and his agitated nervous state are intensified by the foot-stamping, the loud, excited singing, the wild howling, and the noise of clashing kettles and pots, as well as by the illusion of being in communication, or even in combat, with the spirit.' Here we see that the shaman eats the poisonous fly-agaric or becomes drunk with brandy and in this way achieves a state of trance. In connection with the shamanism of the Ugric peoples, Karjalainen mentions that the shamans of the Irtysh-Khanty often used the fly-agaric as a narcotic. Many travelers describe how the Siberian shamans eat fly-agarics and thus fall into a state of ecstasy.

A systematic compilation of such data has been made by A. Ohlmarks. He states that the fly-agaric (*Amanita muscaria* L. Pers.) is the stupefacient most commonly used by shamans among the Koryak, Kamchadal, Nentsy (Yurak Samoyed), and Selkup (Ostyak Samoyed), and among the Khanty along the Yenisei and the Irtysh. Ohlmarks devotes a separate chapter to alcohol, which is used as a stupefacient by the shamans of many Siberian tribes. He states that the shamans of the Irtysh-Khanty, the Tungus, the Lapps, the Buryats, and other peoples often drink alcoholic beverages until they lose consciousness. Nevertheless alcohol cannot compare with the fly-agaric as an effective narcotic. The designations for the fly-agaric form a very widespread family of words in the Ob-Ugric languages and in a number of Finno-Ugric languages. The corresponding derivations are extraordinarily revealing from a semantic point of view because they indicate the method (quite likely a very old and very widespread method) by which the Finno-Ugrian shamans achieved a state of ecstasy.

Here is a list of the etymologically related words for the fly-agaric in the various Finno-Ugric languages:

Mansi: *paŋχ*, *piŋka*, 'fly-agaric'; Northern dialect: *pa:ŋχ*; Middle Lozva dialect: *peŋk*, 'fly-agaric, *agaricus muscarius*'; Khanti Northern dialect: *poŋχ*, 'mushroom, fly-agaric'; Irtysh dialect: *pa:ŋχ*, *paŋχ*, 'fly-agaric'; Yugan dialect: *paŋga*, *pa:ŋχ*, 'fly-agaric'; Irtysh dialect: *poŋk* (*puŋkəm*), 'the fly-agaric'; Lower-Demyanka dialect: *paŋχ*; Verkh, Kalymsk, Vartokovsk dialect: *paŋk*; Tremyugan dialect: *pəŋk*; Nizyam-Berezovo dialect: *poŋχ*; Kazum dialect: *ponk*, 'fly-agaric'; Mordvinian: *paŋga*, *paŋgo*, 'mushroom'; Cheremis: *paŋgə*, *poŋgo*, *poŋγə*, 'mushroom'.

The importance of the poisonous fly-agaric in the ecstasy of the Mansi shamans is proved by the fact that in the Northern Mansi dialect the word *pa:ŋχ* means 'drunkenness'.

The following words are derivatives by which the concepts of 'drunkenness, ecstasy' are expressed in Mansi: *pa:ŋχli*, *pa:ŋχeta:li*; Lozva Pelym dialect: *pəŋkli*, 'he is drunk, intoxicated.' Likewise, *paŋχtem* in the Northern Khanti dialect and *puŋkłəm* in the Irtysh dialect mean 'to work magic while singing after eating fly-agarics, to heal by enchantment'; in the Upper Demyanka dialect *panγətta* means 'to shout and make noise after eating fly-agarics'; in the Tremyugan dialect *pəŋkəlta:γə* means 'to become intoxicated with fly-agaric, to sing through the effects of fly-agaric.' The singing of the shaman who is transported into ecstasy by eating fly-agarics is expressed in Khanti by the following words: Tremyugan dialect: *pəŋk a:rəχ*, *pəŋkəłsauγə*, 'fly-agaric song, a song which the shaman sings after eating fly-agarics'; Vasyugan dialect: *panləmnəŋ*, 'song which is sung after eating fly-agarics'.

According to T. E. Uotila, the following also belong to this word family: in Komi (Wichm.), Sysola dialect: *pagal-*, to lose consciousness (*i.e.*, through drinking alcohol); (Wied.): *pagal-* (*pagav-*), to lose consciousness, to be overwhelmed (?) [or: to be deluded (?)]; (Sachow [or: Zakhov]): *pagav-*, to poison oneself, to kill oneself; Udorka-Vashka dialect: to be unsteady, to stagger, Sysola dialect. According to Uotila the primary noun root of these words is *pag-, which we can relate to the above words denoting the fly-agaric. He believes that this word must also have existed in Komi and cites as evidence the following Komi adjectives: (Wichm.) Luza dialect: *pagyr*, sour, sharp (like the taste of beer); (Wied.): *pagyr*, sour, sharp, penetrating; (Wichm.): *pagyra*, penetrating (beer), *pagyra*, sharp-tasting. Uotila assumes that these adjectives are derived from the noun **pag*, 'fly-agaric,' by adding the suffix *-r* and that their original meaning is 'tasting like fly-agaric.' He believes that a special significance must have been attached to the fly-agaric in the Komi language also. Steinitz, on the other hand, doubts that the reference to these Komi words is justified in this connection. They may have acquired their meaning of 'falling into ecstasy, into a trance' through the same semantic development as the above-mentioned Mansi and Khanti verbs with similar meanings.

According to Lehtisalo the following words also belong to this category: Nganasan (Castr.) *faŋká ʔam*, to be drunk, and even Selkup (Castr.), Kamas, Kheya, Kha, Upper-Ob dialect *pöŋer*, drum, Tym dialect *pəŋgər*, a special Samoyed musical instrument, the Russian *domra*; Narym, Lower Vasyugan, Middle-Ket dialect *pəŋgər*, Upper-Ket dialect *py:ŋga:r*, Khaya dialect *pəggər*, the same.

Lehtisalo compares the Nenets verb *sa:mpa:*, to sing the shade of a dead person down into the underworld to the accompaniment of the magic drum, *etc.*, with the Finnish noun *sampa: maansampa*, puff ball, *puunsampa*, a whitish mushroom the size of a pinhead, found on birches, alders, and willows. If the root of this Samoyed verb is a noun meaning 'mushroom', then we have here a semantic relationship similar to that between the following nouns and verbs: Khanti *pa:ŋχ paŋχtəm*, Nenets, Obdorsk dialect *ja:pʔe:*, to be drunk, Nenets (Reg.) *javebś*, fly-agaric, Lyamin dialect *wi:ppi:*, Nyalina dialect *wʔi:ppi:*, Pur dialect *wi:pi:*, fly-agaric, Enets, Khantaika dialect *jebiʔe:-rro*, Bayikha dialect *jebiʔedo*, to be drunk.

On this basis, Lehtisalo assumes that the Finnish shamans also used the fly-agaric as a stupefying agent in earlier times and that people probably 'became intoxicated by eating fly-agarics as early as proto-Uralic times.'

In every case the above-mentioned Finno-Ugric words meaning 'ecstasy, intoxication, drunkenness' are similarly traceable to a noun which means 'fungus, fly-agaric' and was used figuratively to denote the intoxicated, ecstatic, or drunken state itself (as in Mansi), whereas the verbal derivatives of this same noun in many Finno-Ugric languages have the meaning 'to fall into a trance.' From the semantic viewpoint this requires no further explanation, since it is obvious that the root-word of derivatives which mean 'to come into a condition' is the same word which is used to denote the substance producing the condition. There is a causal relationship between such root-words and their derivatives. The verbs formed from these nouns mean 'to provide somebody with something' or 'to be provided with something.' Thus we find not only from descriptions given by travelers but also from definite linguistic facts (derivatives of Finno-Ugric words which mean 'fungus' or 'fly-agaric') that the Finno-Ugrian shamans used the fly-agaric as a stupefying agent in ancient times.

4. It might be asked, however, whether all verbs meaning 'to fall into ecstasy' in the Finno-Ugric languages are derived from a noun meaning 'fungus' or 'fly-agaric,' and whether these Finno-Ugric words are not loan-words.

Munkácsi believes that these Finno-Ugric words are of Old Iranian origin. He bases his opinion on the fact that the semantic development of the Old Iranian word *baŋha* is remarkably similar to that of the above-mentioned Mansi word for the fly-agaric. According to Bartholomæ this Old Iranian word has several different meanings: '1. Name of a plant (and its juice) which was also used for producing abortions; 2. Name of a narcotic made from that plant and also a designation of the state of narcosis produced thereby.' Other references in connection with this word are: Old Indian *bhaṅga-*, *bhaṅgā-*, 'hemp; a narcotic prepared from hemp seeds'; modern Persian *bang*, 'henbane,' (*bangī*, 'senseless'); Armenian loan-word *bang*, '*Hyoscyamus niger*'; Afghan *bang*, 'hemp.' In Persia, according to W. Geiger, hashish is made from hemp.

Whether Munkácsi is right in making this correlation is a matter for experts in Iranian to decide. As far as the semantic development is concerned, it is rather difficult to imagine that a word which originally meant 'hemp' would be used by the Finno-Ugric peoples to describe the fly-agaric, since these two plants do not resemble each other in the slightest, except for their narcotic effect. (Lehtisalo considers the borrowing of these words unlikely.) Nevertheless, it must be borne in mind that hemp seeds might have played, and perhaps still play, a role in the ecstasy of the shamans. We shall return to this point later.

[39]

Pedersen, Holger. 'Przyczynki do gramatyki porównawczej języków słowiańskich'. (Contributions to the comparative grammar of Slavic languages) Materiały i Prace Komisyi Językowej Akademii Umiejętnosci w Krakowie. Vol. 1, No. 1, pp. 167-176. Kraków. 1901.

There follows an abstract of the original paper in Polish, an abstract prepared by Roman Jakobson as a working paper.

I

Old Bulgarian gąba 'spongia', Polish gąbka. The Old Bulgarian gąba is usually compared with the Lithuanian gum̃bas, 'illness of the uterus, excrescence on a tree', (Mieżinis), 'an excrescence on an organic body or on a stone, colic, cramps of the stomach' (Kurschat), Latvian gumba 'tumor', gumpis, 'colic'. This etymology is improbable.

II

It presents difficulties when one compares the accent of the Lithuanian gum̃bas with the Serbian gȕba, Slovenian góba: the Serbian short falling accent ̏ corresponds to the Lithuanian ˊ, and the Slovenian ˊ corresponds to the Lithuanian ʹ.

III

Lorentz' law, surmising the change of um̃ to the Slavic ą is erroneous.

IV

To the Slavic gąba there corresponds etymologically the Old High German swamb. Kluge compares swamb with the Greek somphós, 'porous', but it is not plausible. Swamb could represent the Indo-European *sg``hombho-s. The Indo-European g``ho gave in Germanic wa, as in the Gothic warms: Latin formus. In Baltic and Slavic languages the consonant s in the initial cluster sg had to disappear; compare Old Bulgarian gasiti and Lithuanian gesýti with Greek sbénnȳmi.

V

Also the German swamb could be easily explained from the original *sgᵘhombho- as well as from *sgᵘhōmbho-. But if the Lithuanian gum̃bas does belong here, one must accept *sgᵘhombho- as the original form and the Lithuanian gum̃bas represents the reduction *sgᵘhm̥bho-. The same reduction occurs in the German sumpf, but English swamp.

VI

Also *sgwombho- could be admitted as the original form. Then the Slavic gąba either lost the semivowel w during the Baltic-Slavic stage, or it represents an Indo-European variant without w.

VII

In this case it is possible to combine gąba: swamb with Greek sfóngos, Latin fungus. The Greek and Latin expressions spring from the original forms beginning with *sph-. Here also belong Armenian sunk, sung, 'sponge, pumice', but Armenian s can represent only the Indo-European *sp and not *sph. However it is possible to suppose *sphwongo-. From *sphwongo there could arise *sgwompho- and also *sgwombho-; the alternation ph∾bh is not unusual after the nasal consonants.

C. Secondary Sources

[40]

Hartwich, Carl. Die menschlichen Genussmittel: ihre Herkunft, Verbreitung, Geschichte, Anwendung, Bestandteile und Wirkung. (Human Stimulants: Their Origin, Distribution, History, Use, Components, and Effects) Leipzig. 1911. pp. 255-260.

[Hartwich was a distinguished toxicologist and pharmacologist of his day. He had read widely, he was intelligent, and he expressed himself well and forcibly. What he wrote about the fly-agaric in *Die menschlichen Genussmittel* reflected the best opinion of his day and in pharmacology it exerted wide influence; but on the fly-agaric in major features it was wrong or misleading. As late as 1911 he gave expression to the ancient European superstition that the fly-agaric is dangerously toxic: 'four mushrooms can kill a man'. On the contrary, it is difficult to find case histories of healthy adults who have died from it, which is in striking contrast to the deadly amanitas! He perpetuates the notion that the Tungus and Yakut tribesmen – both speaking Altaic tongues – use the fly-agaric. He quotes 'a Zürich man's letter dated 1799' reporting that the Russian troops occupying Zürich in that year had gathered fly-agarics on the Zürichberg and eaten them, allowing his reader to suppose that it was for their intoxicating effect. In historic times no Russian – whether Great Russian or Little Russian or White Russian – has taken the fly-agaric for its inebriating effect. This statement will not be successfully gainsaid. The Russians have produced numbers of excellent specialists in Slavic folklore and folk practices, and today when the Russian world stands revealed to us, thanks to numbers of persons in the West who have learned the language, it would be impossible for a pharmacologist of standing to say that the Russians eat the *mukhomor*, as they call the fly-agaric, for the inebriation that follows. True, there have been reports that some of the Russian civil servants and Cossacks stationed among the Palæosiberian tribes have taken the habit from the natives, but it is not in this sense that ignorant people lay the charge of fly-agaric eating. A West, where only yesterday there were elements in the educated classes that regarded the Russians as *Untermensch* and not worthy of study, was prepared to believe anything that, according to Western prejudices, was degrading about their neighbors in Eastern Europe. It is not that consuming the fly-agaric would have been disgraceful, any more than consuming alcohol or tobacco. But there were Westerners who believed the

Russians ate the fly-agaric, as Hartwich seems to have done. It was not so. Diverse cultures should not be confused: each stands on its own bottom, and to associate the Slavic culture with that of the Koryak is like asking an American tourist whether Iroquois is his native tongue.

[Hartwich of course gives some measure of credence to the belief, common in Scandinavia, that berserk-raging was produced by the fly-agaric. On the use of the fly-agaric in Siberia Hartwich quotes Enderli, who is excellent, and Kennan, who is childish. He ignores Strahlenberg, Maydell, Langsdorf, Georgi, Erman, and Dittmar, who wrote in German; Krasheninnikov who was translated into German; and Bogoraz and Jochelson, whose works had already been published in English in Leiden. Had he done his homework properly, he would have found some answers to the questions that baffled him, such as a possible explanation for the urine drinking of the reindeer folk. He also draws on Ernst von Bibra, a pioneer pharmacologist whose *Die narkotischen Genussmittel und der Mensch* appeared in Nüremberg in 1855, though he was only a secondary source relying on writers whom Hartwich should have read. He quotes Kennan to the effect that the continued use of the fly-agaric has harmful results, adding that Kennan no doubt was right. Kennan had been in the Koryak country, but the man on the spot can be a bad observer and Kennan's text gives no ground for reliance on his word. The fly-agaric may be harmful if its use is continued, but it is the function of a man of science to reduce the area of guessing and not himself reach conclusions that are only guesses. The excessive use of alcohol is harmful, but not the continued moderate use. May this not be true of the fly-agaric? – RGW]

The fly-agaric is one of the most remarkable stimulants. Except for one otherwise unknown poisonous mushroom which was consumed as a delicacy[1] by the ancient Mexicans, it is the only stimulant in the broad subkingdom of cryptogamic plants, and no other stimulant ranges so far north. Only some unusual forms of alcohol production from milk are found in the region of Siberia bordering on the area where fly-agaric is used, and in the extraordinary method of its use it is unparalleled by any other stimulant on earth.

The fly-agaric (*Amanita muscaria* L. Pers.) grows in forests and its distribution is circumpolar in the northern hemisphere; it is found in Europe, Northern Asia, and North America, but it also occurs in South Africa.[2]

1. Today we know that there are more than a dozen species of these mushrooms, and we have given each of them names. I would take exception to Hartwich's use of the word 'poisonous' unless we are ready to apply it also to alcohol and tobacco. 'Delicacy' is hardly the word to describe the attitude of the Indians toward their holy eucharist. – RGW.

2. Engler-Prantl, *Pflanzenfamilien* (Plant Families) I, I, 275.

It is highly poisonous, and the splendid red colour of its cap, with its sharply contrasting white warts, sometimes entices the inexperienced to eat it. A great many deaths traceable to this cause have been reported. Four mushrooms can kill a man; sheep, on the other hand, appear to be immune to it. The symptoms observed after such poisoning are nausea, vomiting, thirst, colic pains, mucous and bloody stools, flow of saliva, and fainting spells; sometimes there are also intoxication-like states and lethargy, dilation of the pupils with disturbed vision and even temporary blindness, delirium, hallucinations, raving, cyanosis (blue colouration of the skin), difficulty in breathing, loss of consciousness, and cramps. Death has occurred after ten hours, or sometimes after eight. A victim may recover after 5-24 hours.[1] It derives its name from the fact that an extract of this mushroom was used for killing flies, particularly in earlier times, before fly-paper saturated with arsenic, quassia, and the like became widely known. Its use in medicine has always been insignificant. According to Kosteletzky,[2] it has been recommended for use against nervous attacks, swollen glands, ulcers, and – in Kosteletzky's time – for use as a powder or tincture against consumption.

At present it is used as a stimulant only in Siberia, and its range is very extensive. It begins with the Ostyaks, whose territory extends from the Ob to the Yenisei. From here it continues uninterrupted to the easternmost part of Asia. After the Ostyaks come the Samoyeds and Tunguzes, then the Yakuts, the Yukagirs, the Chukchis, the Koryaks, and the Kamchadals. The area of fly-agaric use probably does not extend far south of 60° north latitude at any point.

According to some reports, the use of the mushroom extended farther west in earlier days and was perhaps forced back by the spread of alcohol. A Zürich man's letter dated 1799,[3] the year in which a Russian army under Korsakoff was in Zürich, mentions with amazement the fact that the Russians gathered fly-agarics on the Zürichberg and ate them. These Russians, of course, must have become acquainted with the mushrooms in their own homeland.

At Polotsk on the Dvina (presumably in 1812) French soldiers ate fly-agarics, and four of them died within a short time.[4] No doubt they had seen the local inhabitants eating the mushrooms. While the Russians in these last two

1. Lewin, *Lehrbuch der Toxikologie* (Textbook of Toxicology), second edition, 1897, p. 410.

2. Kosteletzky, *Med. Pharm. Flora*, Vol. 1, 1831, p. 13.

3. In recent years we have made strenuous efforts to locate this letter but without success. The fly-agaric can be prepared properly for the table. Whether the Russians in some parts of that vast country may have known how, we cannot say. – RGW.

4. Vadrot, 1824. After Orfila, *Allgemeine Toxikologie* (General Toxicology), translated into German by Hermbstädt, 1818, Vol. 4, p. 40.

cases were probably not Mongols, as in Siberia, but Slavs, there are conjectures that go even further, and according to some, it may be assumed that even Germanic peoples used the fly-agaric in earlier times. For example, it is sometimes reported[1] that the berserkers – those Germanic warriors who attacked the enemy in a wild frenzy, naked and without adequate defensive weapons, and who in some ways resembled Malays who run amuck – worked themselves into such a state of senseless fury by drinking fly-agaric mixed with an alcoholic drink. These are merely conjectures; nevertheless, the possibility must be considered that the use of fly-agaric was more widespread in earlier times than it is today.

The mushroom is used in a good many different ways in Siberia, but it is certainly not true that it is fermented into a beverage; assertions to this effect are obviously due to a confusion with the production of alcohol through fermentation. According to von Bibra, it is eaten fresh in soups or sauces, but it is reputed to be less effective when used in this way than in the dried state; this is readily understandable in view of the high water content of the fresh mushroom. It is called by the names *muchumor*, *mukamor*, and *mucho-more*.[2] It seems to be a fairly common custom to eat the mushroom in the juice of berries of *Vaccinium uliginosum* L. or in an extract made from the leaves of the narrow-leafed willow herb (*Epilobium angustifolium* L.). (I should point out that these berries are believed by some to have an intoxicating effect, as is suggested by the German names *Rauschbeere* and *Trunkelbeere*.) Even though there are a few reports of cases in which an effect of this kind has been observed, there are many others of cases in which the berries were eaten without any such effect. Ascherson believes that the reports of intoxication are due to a confusion of the berries with those of *Empetrum nigrum* L. (crowberries), which are also called *Rauschbeere* in German. But the fact seems to be that these also can be eaten without ill effects. I remember that in describing his crossing of Greenland, F. Nansen writes that he and his companions ate tremendous amounts of this fruit. (See also Schübeler, *Pflanzenwelt Norwegens* [Flora of Norway], pp. 276 and 324 . . .)

The most common practice, however, is simply eating the mushroom in a dried state. It has been observed that the active component passes unchanged into the urine, and this remarkable fact has been utilized to obtain a long series of successive effects with a relatively small quantity of mushrooms.

[Hartwich then quotes J. Enderli in full; *vide* our text, [19] – RGW]

The most striking thing in this detailed account is that the active principle of the mushroom evidently is excreted unchanged in the urine, so that the

1. *e.g.*, in E. Krause, *Tuisko-Land* (Tuisko Land), 1891, p. 379. Lewin, *l. c.*
2. Why did not Hartwich look up *mukhomor* in a Russian lexicon? – RGW.

latter retains the effectiveness of the mushroom. There is no doubt whatever that the urine is drunk for purposes of intoxication, and this has also been reported by other observers. According to Kennan,[1] however, it is only the settled and more primitive Koryaks who know this custom. Von Bibra reports that the efficacy is not lost even after repeated consumption, so that four or five people in succession may be intoxicated by the same quantity of mushrooms, first directly and then indirectly.

Apart from the fact that this custom strikes us as highly unappetizing, it seems strange at first glance that the Koryaks could ever have hit upon the idea of drinking urine. Of course, the intoxicating effect of the urine might have been noticed by accident when a man intoxicated with fly-agaric mistakenly drank the urine which he had collected in a container in the close quarters of the yurt. It must also be remembered, however, that urine was not always and everywhere regarded as something repugnant. Among the Chinese it is still used as a medicine, both in its original state and in the form of the residue left after evaporation. Not only among the Chinese, however, but even in our own part of the world, it occasionally has a rôle in folk use, although the rôle is that of a sympathetic agent rather than a recognized medicament. (*Vide*, for example, Kristian Frantz Paulini, *Neu-Vermehrte Heylsame Dreck-Apotheke*, 'On how almost every kind of . . . illness, and harm done by sorcery . . . is successfully cured by means of excrement and urine – Franckfort am Mayn, 1734.' A new edition of this book was printed about the middle of the last century. I also have before me a new edition of another *Dreck-Apotheke*, which turns out to be an extract of Paulini's book.) If we go back several centuries, we find that science regarded urine not as merely a substance excreted by the body but as something with a special mystery about it. It was ideas of this kind that led Brand in Hamburg in 1670 to seek the philosopher's stone in urine and to discover that it contained phosphorus.

It is also noteworthy that the mushroom-eater discovers the future in his intoxication if he follows certain prescribed formulas. Thus we see that there is a religious factor involved. In this same connection, von Bibra reports that the shamans, the Siberian sorcerer-priests, use the mushroom to transport themselves into a state of ecstasy.[2] (This is not explicitly stated on page 136 of von Bibra's book, but it is unmistakably clear from the context.) His des-

1. George Kennan, *Tent Life in Siberia*, p. 156 of D. Halk's German translation, *Zeltleben in Sibirien*, Leipzig, Reclamsche Bibliothek.

2. O. Stoll, *Suggestion und Hypnotismus in der Völkerpsychologie* [Suggestion and Hypnotism in Folk Psychology], second edition, 1904. He discusses the Siberian shamans and their tricks in detail but does not say that they use the fly-agaric in order to achieve a state of ecstasy; instead, he attributes this state to loud noises, dance movements, and auto-suggestion developed by practice.

cription of the effects differs in some ways from Enderli's: the symptoms described by von Bibra are, in general, not so severe, and the people in his account act in a more lively manner – those with musical talent sing without interruption, while others converse, laugh, and tell their secrets to all the world. They lose their sense of space and leap into the air in order to get over a straw or other small object. Frequently, however, muscular strength is also remarkably increased: one man intoxicated with fly-agaric carried a 120-pound weight for about 10 miles.

According to Kennan (p. 156 of the German translation) – who says, no doubt quite rightly, that continued use of the fly agaric has very harmful results – the practice has been prohibited by the Russians in Siberia; as we have seen, this prohibition has had as little effect as the prohibition of other stimulants. An indication of the Koryaks' craving for the mushroom may be found in the high prices they pay for it. Since the mushroom does not grow on the Koryaks' own steppes, it is brought to them by Russian dealers, and Kennan reports that he saw furs worth 25 rubles paid for a single mushroom.

[41]

Eliade, Mircea. 1. Yoga: Immortality and Freedom. Pantheon Books. New York. 1958. A translation from: Le Yoga: Immortalité et Liberté. Paris. 1954. pp. 338-9 (French text: p. 335).

[In the following excerpts from this writer's books we have printed in italic certain passages on which we will comment. – RGW]

In the sphere of shamanism, strictly speaking, intoxication by drugs (hemp, mushrooms, tobacco, *etc.*) seems not to have formed part of the original practice. For, on the one hand, shamanic myths and folklore record a decadence among the shamans of the present day, who have become unable to obtain ecstasy in the fashion of the 'great shamans of long ago'; on the other, it has been observed that where shamanism is in decomposition and the trance is simulated, there is also over-indulgence in intoxicants and drugs. In the sphere of shamanism itself, however, we must distinguish between *this (probably recent) phenomenon of intoxication for the purpose of 'forcing' trance*, and the ritual consumption of 'burning' substances for the purpose of increasing 'inner heat' . . .

2. Shamanism: Archaic Techniques of Ecstasy. Pantheon Books. New York. 1964. A translation from: Le Chamanisme et les Techniques archaïques de l'Extase. Paris. 1951.

p. 477. (French text: p. 415) . . . But closer study of the problem gives the impression that the use of narcotics is, rather, indicative of the decadence of a technique of ecstasy or of its extension to 'lower' peoples or social groups. In any case, *we have observed that the use of narcotics (tobacco, etc.) is relatively recent* in the shamanism of the far Northeast.

pp. 400-1. (French text: 360-1). The importance of the intoxication sought from hemp is further confirmed by the extremely wide dissemination of the Iranian term through Central Asia. In a number of Ugrian languages the Iranian word for hemp, *bangha*, has come to designate both the preeminently shamanic mushroom *Agaricus muscarius* (which is used as a means of intoxication before or during the séance) and intoxication;[1] compare, for example, the Vogul *pânkh*, 'mushroom' (*Agaricus muscarius*), Mordvinian *panga, pango*, and Cheremis *pongo*, 'mushroom.' In northern Vogul, *pânkh* also means 'intoxication, drunkenness.' The hymns to the divinities refer to ecstasy induced by intoxication by mushrooms.[2] These facts prove that the magico-religious value of intoxication for achieving ecstasy is of Iranian origin. Added to the other Iranian influences on Central Asia, to which we shall return, *bangha* illustrates the high degree of religious prestige attained by Iran. It is possible that, among the Ugrians, the technique of shamanic intoxication is of Iranian origin. But what does this prove concerning the original shamanic experience? Narcotics are only a vulgar substitute for 'pure' trance. *We have already had occasion to note this fact among several Siberian peoples; the use of intoxicants (alcohol, tobacco, etc.) is a recent innovation* and points to a decadence in shamanic technique. Narcotic intoxication is called on to provide an *imitation* of a state that the shaman is no longer capable of attaining otherwise. Decadence or (must we add?) vulgarization of a mystical technique – in ancient and modern India, and indeed all through the East, we constantly find this strange mixture of 'difficult ways' and 'easy ways' of realizing mystical ecstasy or some other decisive experience.

p. 223. (French text: 202-4) Intoxication by mushrooms also produces contact with the spirits, but in a passive and crude way. But, *as we have already said, this shamanic technique appears to be late and derivative.* Intoxication is a mechanical and corrupt method of reproducing 'ecstasy', being 'carried out of oneself'; it tries to imitate a model that is earlier and that belongs to another plane of reference.

pp. 220-1. (French text: 201-2) Summoned to a house, the shaman [among the Ostyak of the Irtysh – RGW] performs fumigations and dedicates a piece of cloth to

1. Bernhardt Munkácsi, «'Pilz' und 'Rausch'»; *Keleti szemle*, Budapest, VIII, 1907, pp. 343-344. I owe this reference to the kindness of Stig Wikander. [*Vide* Exhibit **[32]** – RGW]
2. *Ibid.*, p. 344.

Sänke, the celestial Supreme Being.[1] After fasting all day, at nightfall he takes a bath, eats three or seven mushrooms, and goes to sleep. Some hours later he suddenly wakes and, trembling all over, communicates what the spirits, through their 'messenger', have revealed to him: the spirit to which sacrifice must be made, the man who made the hunt fail, and so on. The shaman then relapses into deep sleep and on the following day the specified sacrifices are offered.[2]

Ecstasy through intoxication by mushrooms is known throughout Siberia. In other parts of the world it has its counterpart in ecstasy induced by narcotics or tobacco, and *we shall return to the problem of the mystical powers of toxins.* Meanwhile, we may note anomalies in the rite just described. A piece of cloth is offered to the Supreme Being, but communication is with the spirits and it is to them that sacrifices are offered; shamanic ecstasy proper is obtained by intoxication with mushrooms – a method, by the way, which allows shamanesses, too, to fall into similar trances, with the difference that they address the celestial god Sänke directly. These contradictions show that there is a certain hybridism in the ideology underlying these techniques of ecstasy. *As Karjalainen already observed,*[3] *this type of Ugrian shamanism appears to be comparatively recent and derivative.*

Today Professor Eliade enjoys renown and his word carries weight in certain circles interested in the origin and history of religions, and since he seems to hold the view that the use of divine inebriants is probably a 'recent phenomenon' among shamans, or at least among Siberian shamans, a few words of comment on his treatment of this vital aspect of his subject seems called for. As we have seen on pp. 165 ff, there is valid linguistic evidence that the use of inebriating mushrooms in Siberia goes back to the Uralic period, at a time when the Ob-Ugrian and the Samoyed languages had not yet evolved out of their mother Uralic tongue, more than 6000 years ago. Professor Eliade does not tell us what he means by the term 'recent phenomenon', but presumably he would not carry shamanic inebriation back more than, say, five centuries. The linguistic evidence seems to contradict his conclusions, and there is no evidence supporting them.

1. We have presented the quotations from Professor Eliade's books in their inverse order, the last one first, with deliberate intent, to reveal a trait of his thinking.

1. The original meaning of *sänke* was 'luminous, shining, light'. (K. F. Karjalainen: *Die Religion der Jugra-Völker*, II, p. 260)

2. *Ibid.*, III, 306. A similar custom is attested among the Tsingala (Ostyak). Sacrifices are offered to Sänke, the shaman eats three mushrooms and falls into a trance. Shamanesses employ similar methods; achieving ecstasy by mushroom intoxication, they visit Sänke and then sing songs in which they reveal what they have learned from the Supreme Being himself (*ibid.*, p. 307). Cf. also Jochelson, *The Koryak*, II, 582-583.

3. *ibid.*, III, 315 ff. [*Vide* Exhibit [26] – RGW]

The extract that we have quoted from his book on Yoga gives no source or authority for his statements, not even himself. He speaks *ex cathedra*. The reader is not told where one may consult the 'shamanic myths and folklore', nor is it clear how 'myths and folklore' would document a decadence in shamanism, nor do we know whom he quotes when he cites the '"great shamans of long ago"'. We do not know why divine inebriation is a '(probably recent) phenomenon'. That there is decadence in shamanism among the minor nationalities of the world is clear and self-explanatory, as their feeble cultures founder in the mælstrom of the modern world. It is also true that in their disarray these peoples have taken to hard liquor. But this proves nothing as to the antiquity of divine inebriation, and in particular divine inebriation from the indigenous mushrooms of Siberia.

On p. 477 of Professor Eliade's book on shamanism he repeats the same thought: the use of 'narcotics (tobacco, etc.) is relatively recent in the shamanism of the far Northeast'. He says 'we have observed' this, by which he probably does not mean that he has observed this in the far Northeast. He probably means that he has discussed the matter and arrived at a conclusion earlier in the volume. He gives no page reference. But, contrary to what he says, in this volume there is no discussion of inebriation in far Northeastern shamanism. There is discussion of Ob-Ugrian shamanism, in the far West of Siberia.

On p. 401 Professor Eliade says: 'We have already had occasion to note this fact among several Siberian peoples; the use of intoxicants (alcohol, tobacco, etc.) is a recent innovation'. We turn back, and on p. 223 he says, '... as we have already said this shamanic technique appears to be late and derivative', referring to intoxication by mushrooms. Again we look back and on p. 220 we find the subject mentioned once more: 'As Karjalainen already observed, this type of Ugrian shamanism appears to be comparatively recent and derivative'. In the same passage he includes this surprising phrase, '... we shall return to the mystical powers of toxins'! And so the reader has found himself shunted back, and back, and back, and again back, only now to be shuttled forward. His statements that the shamanic use of mushrooms is a recent innovation stand unsupported. His reference to Karjalainen gives him no comfort, as the reader will see from [26] where we quote in full the relevant passage. Karjalainen is discussing the fine points of shamanic technique. It seems that the Ostyak shamans of the Irtysh regions have recently influenced the other comunities in certain particulars. Karjalainen has nothing to say about the span of time during which the shamans have used the fly-agaric to achieve ecstasy.

2. In the excerpt that we have quoted from page 400 of his work on shamanism, Professor Eliade makes a number of flat assertions about the etymon of the word used among certain Finno-Ugric peoples, especially the Ob-Ugrians, for 'mushroom' and 'fly-agaric'. He says that this word is derived from *bangha* meaning 'hemp' (= hashish, marijuana) in Iranian. He quotes as his authority for this derivation the Hungarian philologist Munkácsi, but Munkácsi turns out to have been more cautious than Professor Eliade. Munkácsi said in 1907 that he sees in *bangha* and in *pango* (I cite only one of the Finno-Ugric variants) the same word, and he advances the hypothesis that the Uralic peoples took their words for inebriation from the Aryans. (We quote Munkácsi in full in [32].) Munkácsi leaves the question open when the borrowing took place and his wording even permits the reader to ask whether the borrowing may not have taken place before the Aryans emigrated to what has since been called Iran. Other philologists specializing in the Finno-Ugric languages have been inclined to disagree with Munkácsi, but Professor Eliade does not quote them. (*Vide* Lehtisalo [24a] and Balázs [38].) Professor Eliade says, 'These facts prove that the magico-religious value of intoxication for achieving ecstasy is of Iranian origin.' But he has given us no facts, he has asserted as facts what is his own philological speculation, and from these speculative remarks he has gone on to draw religious conclusions that are non-sequiturs.

Apparently the Iranian sources offer no support to Professor Eliade's case. His *Shamanism* appeared in French in 1951. In the same year the Oxford University Press brought out Professor Walter B. Henning's Ratanbai Katrak Lectures delivered in 1949 under the title *Zoroaster: Politician or Witch Doctor?* At the end of his second lecture this eminent Iranian scholar found that the inebriating derivatives of hemp were unknown in Iran before the 11th century at the earliest, that the Persian word *bang* in the sense of 'Indian hemp' is a borrowing from India, that in Iran this Indian word collided with a homonym *bang*, a word used in Iran since Avestan times for a number of Hyoscyamus species (the English 'henbane'), most or all of which are lethal. Persian *bang* in the sense of 'hemp' appears for the first time in medical writings of the 13th century. It is surprising that the English edition of *Shamanism* appeared in 1964 without taking into account the discoveries made since the first edition.

Even without the evidence that Professor Henning adduces, an Ob-Ugrian borrowing of the name for the fly-agaric from Iran in the recent past, as Professor Eliade alleges, is on its face unlikely. Are we to suppose, contrary to all probability, that the tribesmen discovered only late in the day the peculiar inebriating virtue of the familiar fly-agaric, and that having discovered it they crossed rivers and mountains to reach the distant Iranian plateau where it does not grow, and there borrowed the name of another plant that outwardly does not resemble it in the slightest, and brought the name home, and that thereafter everyone called the spectacular mushroom by a new, foreign name?

Professor Eliade suggests that the Ob-Ugrian shamanic practices may have been borrowed from Iran along with the word. But the use of the mushroom is certainly indigenous. A rôle for inebriants in shamanism, including hallucinogenic mushrooms, has sprung up spontaneously in many unrelated parts of the New and the Old Worlds. There is every reason to think that the inebriating mushroom in its religious rôle is millennia old, long outdating the emigration of the Aryans to the Iranian plateau.

Professor Eliade then gives expression to his feelings about the shamanic use of inebriants:

> Narcotics are only a vulgar substitute for 'pure' trance. We have already had occasion to note this fact among several Siberian peoples; the use of intoxicants (alcohol, tobacco, *etc.*) is a recent innovation and points to a decadence in shamanic technique. Narcotic intoxication is called on to provide an *imitation* of a state that the shaman is no longer capable of attaining otherwise.

His preference on moral grounds for other techniques to attain ecstasy has affected his critical faculty when he discusses what is purely an historical question: how old is mushroom inebriation?

3. Professor Eliade lumps all inebriants together. Here is his habitual way of referring to them, as quoted from his books:

> ... intoxication by drugs (hemp, mushrooms, tobacco, *etc.*)
> – (from *Yoga: Immortality and Freedom*)
> ... the use of narcotics (tobacco, *etc.*)
> – (from *Shamanism*, p. 477)
> ... the use of intoxicants (alcohol, tobacco, *etc.*)
> – (from *Shamanism*, p. 401)

For the modern toxicologist this lumping together of diverse psychotropic drugs, obliterating the enormous differences among them, is crude. Let the reader note the catch-all expression, 'etc.'. Professor Eliade does not even take the first step, which is to distinguish between fermented drinks and distilled alcohol. Apparently his translator ignores the difference between the French *alcool*, which is limited to the distillate, and the English 'alcohol', which also embraces fermented beverages such as beer and wine. The discovery of the distillation technique was a sensational development in the social history of the world, never adequately documented and commented on. The technique of distilling potable alcohol seems to have been devised only once, by the school of Salerno, in about A. D. 1100.[1] ('Alcohol' is an Arabic word but in Arabic it meant 'mascara'.) After leading an obscure existence for some centuries in alchemical laboratories and monastic establishments, it leapt into prominence and importance in the 16th century. Schübeler [42] says that brandy (*i.e.*, 'burnt wine') first became known in Norway in 1531. In England it seems not to have been widely known as late as A. D. 1530 but by 1550 it was cheap. It was everywhere called *aqua vitae*, the water of life. In Russia the sources say that the art of distilling penetrated to Moscow through the Black Sea route, perhaps brought from Italy by Italian merchants, in the early 16th century. Sigismund von Herberstein on a diplomatic mission to Muscovy describing a banquet in the Kremlin[2] in 1526 says:

> . . . At length, the servers going out for food . . . first brought in *aqua vitae*, which they always drink at the commencement of dinner.

Yermak's invasion of Siberia in 1580 marked the beginning of the conquest of that land, and it is probable that hard liquor became known to the tribesmen through the Russians. Has hard liquor ever played a rôle in the religious life of any people? The outlying races of the world may have tried it out in recent centuries sporadically and spasmodically, but Professor Eliade is right in saying that its use in shamanism marks the dying phase in the indigenous religious life of the Siberians.

Tobacco was a gift of the New World to the Old. In the American Indian cultures it was (and still is) a holy plant used in the religious life of the Indians and on other solemn occasions. Among Europeans and their descendants elsewhere it became a habit and an addiction but played no rôle in religion. But after tobacco reached Siberia, probably also in the latter part of the 16th century or at the latest in the 17th century, it is astonishing how quickly the tribesmen adapted it to shamanism, thus recapturing for it the religious meaning that it has always had for the American Indians. This religious connotation was seized upon and quickly integrated into the ways of com-

1. *Vide* R. J. Forbes: *Short History of the Art of Distillation*. Brill, Leiden. 1948. pp. 32, 88-89.

2. In the Latin text, *Rerum Moscoviticarum Commentarii*, Bâle, 1551, p. 134, *aqua vitae* appears; and in the Italian translation, *Commentarii della Moscovia* . . . Venice, 1550, folio 77 verso, *acqua de vita*. Apparently from Herberstein's words the pre-prandial vodka was already in 1526 established as a custom. We do not know when it came to be called *vodka*, from *voda*, an affectionate diminutive of 'water'. (The English and Russian words are cognate.) Already we find *vodka* in Kamieński [1] in 1658.

munities that were in a stage of their cultural evolution similar to the American Indians' when these were first seen by Europeans. Let it be noted that tobacco became adapted to Siberian Shamanism without the influence of the American Indian cultures. Here lies a wholesome object lesson for those who would lightly draw inferences of trans-Pacific contacts merely on the strength of parallel usages.

When Professor Eliade lumps the fly-agaric with hard liquor and tobacco, he is committing an anachronism. He forgets that the fly-agaric is indigenous to the forests of Siberia and that, for the eater, the effects are utterly different from hard liquor. He says that 'narcotics are only a vulgar substitute for "pure" trance.' Would he have said this in 1951 about the Soma that inspired the hymns of the ṚgVeda? We know incomparably more about the world of psychotropic drugs than was known even as late as 1951, when Professor Eliade's work on shamanism was first published. The abuses of these drugs by unbalanced or childish people that are reported in the press do not speak for their use. The West is on the threshold of penetrating their secrets. There is an unexplored world before us, and we should not prejudge it.

All over the world, wherever anthropologists have been making their way, they have been finding the native peoples utilizing as shamanic inebriants natural plant products. With astonishing resourcefulness untutored folk, or rather their herbalists, in ages past discovered these 'drugs', as we call them, and how best to prepare them for medico-religious ends. The plants themselves and the methods of treating them are often secrets of the shaman, not to be had for the asking. Sometimes the inquirer is initiated into the mystery, only to suspect that he has not been told the full story. In this book we present the testimony of many witnesses on the use of the fly-agaric in Siberian shamanism. But Kannisto [33] tells us expressly that mushrooms in addition to the fly-agaric serve the same purpose in the Vogul area, a remarkable fact confirmed for the Chukotka by Ivan Lopatin. In the Vogul case the deficiency in our knowledge is attributable rather to the lack of zeal in the searchers than to concealment by the natives. To this day no one in the Western world can tell us what those other mushrooms are. The inebriants used in food-gathering communities seem to be myriad, their use going back to pre-history. Each presents a problem to our biochemists and pharmacologists, whose abilities are taxed to isolate the active agents, to describe their molecular structure, to synthesize them, and to explore their potentialities.

The thinking of the West is obsessed with alcohol as the sole inebriant: we down-grade and ignore the natural intoxicants that primitive man discovered long ago by himself: we are loath, we seem to be afraid, to discover their possibilities. The distillate of alcohol – what the French call *l'alcool* – is a late comer on the stage of history, but in the perspective of the past of *homo sapiens* even fermented drinks are relatively recent. It is hard to see how the peoples of Siberia could have mastered the art of fermentation before they acquired the technique of making pottery or wineskins. Even today in the tropics there are primitive peoples (such as the Kumá of New Guinea) who did not know fermented drinks until the white man brought them in. It is significant that the Uralic word for 'inebriation' is taken from the name for the fly-agaric: in the sequence of history the food-gatherers must have discovered the natural inebriants long before they learned the art, tricky for those who performed it first, of fermentation.

[42]

Brekhman, I. I., and Sem, Y. A. 'Ethno-pharmacological Investigation of some Drugs of Siberian and Far-Eastern Minor Nationalities.' Paper submitted to the Symposium on the Ethnopharmacologic Search for Psycho-active Drugs held at the University of California, San Francisco Medical Center, on January 28-30, 1967.

[This paper by two Soviet scientists based in Vladivostok is of interest for the information that it contains and also because it is, I think, the first utterance out of the U.S.S.R. on the use of the fly-agaric among the Siberian tribesmen. The attitude of the Soviet Union toward this practice finds expression in the concluding paragraphs of the paper. It will be observed that they do not write from personal experience: they rely almost entirely on what Krasheninnikov [4] wrote more than two centuries ago – a source superseded by many others today. – RGW]

Various narcotics and stimulants had been used as intoxicating liquors or in popular medicine by the minor nationalities of Siberia and the Far East. Appertaining to them is the use of fly-agaric, tobacco, and its substitutes, alcoholic drinks, root of ginseng, young antlers of the maral, etc.

Fly-agaric (*Amanita muscaria*) had been used mostly by the palæoasiatic peoples of Kamchatka and Chukotka (Itelmen, Koryak, Chukchi, Yukagir). The use of fly-agaric on a vast scale was unknown to all the Tungus world. To it belong Evenki, Eveni, Udegay, Neguidaltsi, Nanai, Ulchi, Orochi, Solon, Manchu, Oroki; and of the palæoasiatic peoples, the Nivkhi [Gilyak] of the Amur and Sakhalin.

I. THE FLY-AGARIC IN SIBERIA

Fly-agaric had been used with aims of: 1) merry-making, 2) overcoming certain difficulties, 3) instilling one's nerve in the time of inter-tribal clashes and wars, and 4) during a performance of rituals. Stepan Krasheninnikov, one of the first explorers of Kamchatka and its population, thus described the use of fly-agaric by the Itelmen (Kamchadal) and Koryak: 'For good cheer they at times use fly-agaric, too, that is familiar among us; herewith we exterminate flies'. At some other place he remarks that 'Kamchadal and sedentary Koryak eat fly-agaric when they design to kill some one'.[3]

There were known two ways of the use of fly-agaric, in its natural state and in the way of infusion. Fly-agaric was picked in spring, in summer, not often in autumn. For the use of fly-agaric in its natural state the gathered fungi were kept in a dry cool place and were slightly desiccated. 'As need be', wrote S. Krasheninnikov, 'dry mushrooms when rolled are swallowed whole, the which way is in great usage'. For making an infusion the caps of the fungi were soaked in water. After 5-6 days the infusion could be used.[3] More drastic was an infusion of fly-agaric in willow-weed wash. They prepared the latter from willow-weed (*Epilobium angustifolium* L) having boiled it down to a sweet and thick wash. In days of old fly-agaric had been used in Siberia and Kamchatka for a homebrew or added to underproof vodka, which led to intense excitement frequently ending in murder or suicide and now and again in death as a result of poisoning.[1,5]

A twofold use of fly-agaric and its infusion was known to the minor nationalities of North-East of Asia – the so-called 'moderate' and 'immoderate'. The Itelmen [Kamchadal] themselves considered the use of the fungus up to four mushrooms at a time as moderate, which contributed to an increase of organism resistance to fatigue, took off weariness, acted tonically. After such an application, wrote S. P. Krasheninnikov, 'they feel within themselves extraordinary ease, mirth, valour and nerve'.

After an immoderate use, from 5 to 10 fungi at a time, would come a second stage of fly-agaric effect which was accompanied by intoxication and hallucinations. S. P. Krasheninnikov,[3] who observed in person the effect of fly-agaric on Itelmen and Kamchadal, wrote: 'The first and usual sign whereupon ye can apprize of a person being wrought up with the mushroom is a twitch of limbs which would come in an hour or less, thereafter the drunk rave as in a fever; and they dream apparitions, ugly or cheery, as their temper be: from this cause some go a-hopping and some a-dancing, all being in great horror. To some, a slit taketh a view of a big door and a spoon of water of a sea . . . But this needs must be thought of those that use it beyond measure'. The use of more than 10 fungi at a time led to a fatal end.

The second stage of intoxication was, probably, accompanied by a temporary partial paralysis and exuberant hallucinations, by involuntary actions. 'All their actions are so harmful that of persons left heedless scarcely any would escape with life.' During S. P. Krasheninnikov's stay in Kamchatka the Cossacks and Russian servants of government also acquired from the Itelmen the habit of fly-agaric use with the aim of intoxication. It served them as a substitute for alcohol. Its effect was known among the inhabitants of Kamchatka as 'extravagancies', 'visions of infernal regions and fire-spitting abyss', and an inclination to confess one's sins; attempts at suicide, etc.

Everybody addicted to fly-agaric accounted for his actions as if they had been in obedience to the order of Omnipotent Mushroom. In the opinion of the Chukchi 'inebriant mushrooms are a special tribe (*anra-varat*). They are strong and when growing they break through thick roots of trees and dissect them into halves with their sturdy heads. They shoot through stones and crush them to bits. Fly-agaric appears to drunken men in a shape strange and man-like. Thus a certain mushroom appeared in the shape of a single-armed and single-legged man, and another was like a stump. They are not ghosts, they are mushrooms themselves. Their number seen by a man corresponds to the number of fungi eaten by a man. If a man ate one mushroom, he would see one man-mushroom; if he ate two or three, then he would see their respective number. The mushrooms take a man by the hand and lead him to the other world, they show him all that is there, make him do all kinds of unbelievable things. The ways of mushrooms are tortuous. They visit the land where the dead live.'[2]

The effect of the fly-agaric continued until the products were evacuated from the system of a man. Even the urine of this man would have an inebriating effect. 'With sedentary Koryaks the fungus is in so high esteem', wrote S. P. Krasheninnikov, 'that a drunken man is not allowed to pass water on the floor, but they put a vessel before him, and his urine they drink and run mad like those that had eaten the mushroom'.

By the customary law of the minor nationalities of Chukotka and Kamchatka the use of fly-agaric was permitted only to men. Its use by women was prohibited.

Interesting are observations of fly-agaric action on deer, they after eating fungi and the following violent excitement fall into a deep sleep; the meat of such deer, the Koryak vouched, acted on a man inebriatingly.[4]

The northern group of nationalities, especially the Itelmen of Kamchatka, had had another stimulant – wine from 'sweet herb'. The secret of its production passed from them to Russian Cossacks and sedentary Koryaks in the 16th and 17th centuries. For making wine Itelmens used the 'sweet herb' *Heracleum dulce* Fisch. sem. *Umbelliferæ*). For making grass sugar and wine were used young spring stems, which were usually gathered by women. To avoid the influence of the poisonous juice of the plant they put on gloves. The stored bunches of stems were put in grass bags and kept there until sugar was educed on the stems. Grass sugar was used for making various kinds of Itelmen dainties, beverage, etc.

Sometimes the 'sweet herb' was eaten, like betel, in its fresh natural state. The effect of its chewing was similar to that of alcoholic intoxication. The 'sweet herb' was mostly used for making wine. First was made wine dough. For this aim the 'sweet herb' was put in warm water and leavened with honeysuckle or bog bilberry. The mixture prepared thus was kept in warmth, which contributed to fermentation. As soon as fermentation ceased the wine dough was considered to be ready. Usually, this required about a day. For making wine the 'sweet herb' now in a large quantity was soaked in warm water and leavened with dough, and in a day they began distillation. At first, as S. P. Krasheninnikov notes, comes wine of proof similar to that of vodka, then comes a softer wine. The strength of vodka is such that it is possible to mordant even iron. Itelmens attest that this wine 'presses on the heart very much' and 'decreases sex appeal'.

G. V. Steller, another explorer of Kamchatka, thus described the effect of 'herbal wine': '... whosoever partaketh of it but a few goblets he is harassed by queer phantasies all night and the next day he feels melancholy just as if he had committed some crime'.[3] So the effect of the wine from the 'sweet herb' was similar to that of fly-agaric in some measure.

During a shamanic rite the Nanai, Udegay, Ulchi, and Orochi used to employ some plants that made a specific influence on the psychic state of a person, which, probably, promoted to an arrival of mass hypnosis so needed during a rite. Usually it was the leaves of Ledum, which passed under the name of 'senkura' with the Nanai, 'sengkuro' with the Ulchi, 'sengkia' with the Udegay, 'synkiu 'with the Orochi, etc. They used to employ *Ledum palustre* L and *Ledum hypoleucum* Kom. The desiccated leaves stored beforehand were put on the hearth or on a frying pan. A strong active smell in a small dwelling would stupefy those present at the rite. Probably, in some measure it calmed the sick as it was employed at ritual exercises over the sick.

Undoubtedly, the psychoactive drugs employed by the minor nationalities of Siberia and the Far East are not studied yet completely and require further investigations by ethnographers and pharmacologists.

The minor nationalities of Siberia and the Far East now do not use any psychoactive drugs, which have been relinquished because of radical changes in economy, culture, mode of life, and ideology of the population of this part of the Soviet Union. After the October Socialist Revolution and the establishment of the Soviet Power these not numerous nationalities have embarked on a new way of historical development. Formerly backward nationalities of one of borderlands of Russia, with the help of the Russian people and Soviet Power, passing through intermediate stages, they have soon reached the socialist phase of social development.

LITERATURE USED

1. Balov, J. Poisonous fungi. Medicine Gazette No 40, 1428, 1912.

2. Bogoraz, V. G. The Chukchi. Part 2. Religion. Authorized translation from English. Glavsevmorput Publishing Co., Leningrad, 1939.

3. Krasheninnikov, S. P. Description of the land of Kamchatka, Vol. 2. St Petersburg, 1775.

4. Orlov, N. I. Edible and poisonous mushrooms. Medguiz Publishing Co., 1953.

5. Tikhomirov, V. A. Edible and poisonous mushrooms. Moscow, 1879.

2.

THE FLY-AGARIC IN SCANDINAVIAN WRITINGS

NO one who discusses the fly-agaric in Europe can ignore the debate that has been carried on for almost two centuries in Scandinavia on this issue. First Samuel Ödman in 1784 and then Fredrik Christian Schübeler in 1886 propounded the thesis that those Viking warriors known as 'berserks' ate the fly-agaric before they 'went berserk'; in short, that 'berserk-raging' was deliberately caused by the ingestion of our spotted amanita. Apparently the argument went to Ödman and Schübeler by default, because today we find their thesis incorporated in the Scandinavian encyclopædias and the school history books. In fairness to them I here present in full what they said in presenting their views, as well as the utterances of a number of their followers, eminent professors all of them. I wish I could agree with them.

I add also a part of the account of a tragic episode that happened in Hungary at the end of the second World War, in 1945, written by a Swedish woman who was there at the time. There is in it a curious statement about a fungal decoction said to have been used by the Russian troops that occupied briefly the town of Fehérvár.

[43]

ÖDMAN, Samuel. 'Försök at utur Naturens Historia förklara de nordiska gamla Kämpars Berserka-gång.' (An attempt to Explain the Berserk-raging of Ancient Nordic Warriors through Natural History) Nya Handlingar, published by the Kungliga Vetenskaps Akademien, Vol. 5. Stockholm. 1784. pp. 240-247.

In the oldest chronicles of our country we not infrequently come across incidents which, for want of a knowledge of natural history, are either explained quite erroneously or risk being relegated, through the injustice of the scholar, to the myths of the dark ages. I am sure that I am not mistaken if I include in this category the accounts, preserved in the old Norse Sagas, concerning the *Berserks* of ancient times, and the celebrated frenzies which, under the name of *Berserka-gång* (Going Berserk), are described in such curious wise.

It is not for me, nor is this the place, to discuss these prodigies of History at length. A brief outline will suffice to pave the way for the present enquiry.

These warriors, according to the military thinking of their times necessary adjuncts for the aims of their warlords, but also often feared by the Prince in whose service they were employed, are depicted more as wild beasts than human beings. As soon as the Berserk-fury came over them, they were seen to rage like ravening wolves, recoiling from neither fire nor iron, braving the direst perils, rushing at the most redoubtable enemy, biting their own shields, *etc.*, or, if no enemy were at hand, venting their fury on inanimate objects, uprooting trees, overturning rocks, and in their exalted state scarcely distinguishing friend from foe. King HALFDAN'S Berserks are portrayed in HROLF'S Saga in the following interesting manner: 'On these warriors', the account reads, 'there at times fell such a fury that they could not control themselves but slaughtered man and beast, and all that stood in their way and minded not what they did. While this fury lasted they stopped at nothing; but when it left them they were so powerless that they did not have half of their normal strength, and were as weak as if they had just recovered from some sickness, and this fury lasted about one day.' An illuminating example of this is also given in Hervarar Saga, at the battle that took place at Sámsey.

The respect shown towards these heroes was always mingled with a sort of secret hatred, even in heathen times, for which their arrogance no doubt gave good cause. With the peaceful principles that the first Christian preachers tried to instil in the Barbarians of the north, the Berserks soon lost all their prestige, as an alleged connection with the devil caused their skill and their trade to be viewed with even greater horror. To be sure, fighting did not cease altogether, but the difference in attitude made it no longer permissible to use such sinister aids for the purpose. So the science died out with its practitioners; for a long time no further clue to the mystery was sought other than ascribing it to the assistance and co-operation of unclean spirits, so that not only did Prof. VERELIUS call it a *diabolical art*,[1] but even in this century two dissertations published at Uppsala propounded the same theory,[2] probably because the no doubt exaggerated accounts of these fighters which have been handed down to us, have been taken all too literally.

I am not of the opinion that these transports should be viewed solely as the effect of some special quality of temperament whereby such extraordinary motions can be induced by a mind in a state of violent ferment for, although we are not entirely without examples which could support such an assumption, the persons who suffer from such afflictions, between paroxysms, are not able to maintain the defiant arrogance which made the Berserks so hated even in times of peace. On the other hand, since the Plant-kingdom affords several means of creating such disorder in the mind and inducing the most frenzied attacks of folly, I am inclined to believe that the Berserks knew of some such intoxicating substance which they used when occasion demanded and which they kept as a secret among themselves so that their prestige with the public would not be diminished on account of the simplicity of the means employed.

That *Opium* could produce exactly the same effects is a matter of common knowledge. What we are told on this subject about the inhabitants of the island of Celebes and their opium-induced frenzies when they go to battle matches in every detail the accounts of the Norse Berserks of old. KEMPFER'S reports of the frenzies which were common practice in his day, in Java, and went by the name of *Hamuk*,[3] so strongly support the possibility that opium could turn people into Berserks, that the testimony of ALPINUS can very well be omitted.[4] But since as yet no voyage to the Levant could have taught our ancestors this means; since as yet no ALLEN or DILLENIUS had discovered how

1. Epist. Dedicatoria Hist. præfixa.
2. One by Mr. HAMNELL in 1709. De *Magia Hyperboreorum* who, on page 42, suggests that *Berserk frenzies were probably brought on by the devil*; the other by Mr. RAMELIUS, 1725, de *Furore Bersekio*, who, on page 24, regrets that *much as he would have liked to acquit the Berserks of the charge of dealing with the devil, he dare not take up their defence in the matter.*
3. Am. Exot. Fasc. 3. p. 649. s. Opii deglutiunt bolum, quo intentionis idea exasperatur, turbatur ratio, et infrænus redditur animus, adeo ut stricto pugione, instar tigridum rabidarum in publicum excurrant, obvios quosvis, sive amicos, sive inimicos trucidaturi.
4. de Med. Aeg. page 121.

to prepare opium from the poppy-buds of Europe, and there is still less reason to conclude that such attempts could have been thought of at that time in Sweden, although they subsequently succeeded in the case of LINDESTOLPE, one cannot reasonably trace Berserk frenzies to this intoxicating resin.

If *Atropa Bella Donna* were indigenous to our northern climes, the example[1] given by Mr. GMELIN the younger would throw considerable light on the question of Berserks. The same applies to several intoxicating substances found in India.[2]

A suitable preparation could also be made from the *Hemp* leaf, which would be sufficient to bring on this periodic bout of frenzy.[3] But since it is not yet certain whether the hemp of our climate corresponds to the southern variety which gives the Persians, Indians and Egyptians their *Bangve*, whose intoxicating properties have already been noted by GALENO, and of which the Turks still use a mixture to strengthen their tobacco, according to Dr. RUSSEL's observations,[4] like the ancient Scythians who, according to HERODOTUS, were made giddy in the head from the smoke of its seeds, thrown on hot stones, and as it is also certain that this East Indian plant was not known in the North at such an early period, there is no likelihood that it could have been used for this purpose.

Of those native plants of ours which might be considered in this connection, several are known to have some intoxicating properties, although not in sufficient degree to warrant attention, such as *Crambe maritima, Lolium tremulentum* and others of which either a large dose is required to obtain such a powerful effect or whose influence is more likely to have rendered the Berserks incapable of committing the excesses ascribed to them in as much as, having a more soporific effect, they bring on drowsiness and apathy. *Datura stramonium* should also merit special consideration as its properties, which are not unknown to our Physicians, have been the subject of new studies by the same Mr. GMELIN.[5]

Of all Swedish plants, however, I consider the Fly-Agaric, *Agaricus muscarius*, to be the one which really solves the mystery of the Berserks. Its use is so widespread in Northern Asia that there are hardly any nomadic tribes there that do not use it in order to deprive themselves of their feelings and senses that they may enjoy the animal pleasure of escaping the salutary bonds of

1. Russian Journey. Vol. 3, p. 361. 15 grains in wine made a Persian soldier light-headed.
2. Diss. Linn. Inebriantia. S. 3.
3. Alpin. *l.c.* p. 121. Kemph. *l.c.* p. 645.
4. Nat. Hist. of Aleppo. page 83.
5. GMELIN. *l.c.* Tom. 1. p. 43. A man who gathered the seeds of Datura at Voronezh was asked what he used them for, and replied that one put them in beer to make the intoxication greater.

reason. The Ostyaks, the Samoyed, the Yukagir and others use it daily, and the *Chukchi* whose rigorous icy climate does not produce this mushroom, obtain it by bartering their reindeer, which are their most valuable possession. The dose of this poisonous substance is from 1 to 4 mushrooms, according to size. The Ostyaks can only tolerate one, or use a decoction of 3.[1] The Kamchadal drink it with a decoction of *Epilobium*.[2] Those who use this mushroom first become merry, so that they sing, shout, *etc.*, then it attacks the functions of the brain and they have the sensation of becoming very big and strong; the frenzy increases and is accompanied by unusual energy and convulsive movements. *The sober persons in their company often have to watch them to see that they do no violence to themselves or others.* The raving lasts 12 hours, more or less. Then lassitude sets in, culminating in complete exhaustion and sleep. STELLER[3] reports the curious fact that the urine of persons under the influence of this mushroom possesses the same intoxicating properties. And that the Tungus Shamans, in the ceremonial use of their so-called magic drum, are accustomed to swallow a goodly draught of this urine so as to be able to fall into the epilepsy or ecstatic trance proper to this ceremony has been attested by Mr. GEORGI[1], with several pertinent comments, in his description of the peoples under the Russian Imperial Government. T. II, pages 329, 336.

What seems to me to point particularly to the Fly-Agaric in this case is the fact that its use is a custom from the part of Asia whence ODIN, with his Aesir, made his celebrated migration to our North. For although the distiller's art subsequently devised a short cut to this ignominious abuse, and the use of the mushroom accordingly ceased in the region of the Danube, it none the less spread from that point with the northward-migrating Hordes, which still used it. And that the history of the Berserks in our North begins with the arrival of ODIN, I find not only accepted by those who have produced all possible evidence from the dark annals of antiquity, but also consistent with the designs of a warlord who with a dozen raving men could make himself so feared and secure among alien tribes. The honour other warriors won for themselves by slaying a Berserk, whom they looked upon as a malefactor, seems to prove that the custom was not a native one. And as this mushroom, like similar substances, prematurely emaciates the human body and makes it insensitive and clumsy, it may have had the effect of inculcating ODIN's principle of going to Valhalla in good time, through a voluntary death, over the precipice, so that the honour of the proven hero might not be dimmed,

1. *Vide* Georgi [6] – RGW.
2. *Vide* Krasheninnikov [4] – RGW.
3. *Vide* Steller [5] – RGW.

especially as statecraft demanded that a Berserk, as the foremost militia-man, should be regarded as invincible.

[In Ödman's account I have printed in italic the crucial sentence, the source of much misunderstanding. (Mörner [45] relies on it, quoting it in full.) Here it is:

> The sober persons in their company [*i.e.*, those who have taken the fly-agaric – RGW] often have to watch them to see that they do no injury to themselves or others.

This statement is unjustified. A careful reading of our many Exhibits fails to disclose a single case where a Siberian tribesman under the influence of the fly-agaric threatened either himself or others with injury. On the contrary, the effect is to calm the subject, to put him into a benign mood. Nor is there any reason to believe that this is a question of dosage. Members of the lower orders of the *Russian* community, discovering in the 18th century a surprising new inebriant and jumping to the conclusion that it must be similar to vodka but even more extraordinary, threaten themselves with injury when under its influence (*Vide* [4], p. 236; [10], p. 249), but the attentive reader will note with a weary smile that, like the tiresome exhibitionists whom we have all known, in every case the inebriated man always takes care to have a friend at hand to arrest his hand. The one case in the *Russian* community of a suicide [4] is only hearsay, as Krasheninnikov takes pains to make clear. We may all suspect what that means in the context of the people Krasheninnikov was dealing with. He says the natives take the fly-agaric before going out to kill someone but he gives no support for his statement. He did not know the native language and he relied largely on Russian informants. His own colleague Steller does not repeat this, and the many later observers who knew the country from long experience, some of them professional linguists speaking the local languages, fail to confirm it. Krasheninnikov was an astute observer trying to arrive at the facts through a screen of questionable informants. He should be admired as a worthy period piece, certainly not quoted as an authority on the fly-agaric in 1968. After all, we no longer consult Benjamin Franklin or Lomonosov on physics, great and revered figures though they be. . . . As for Ödman, why should anyone cite him today?]

[44]

SCHÜBELER, Fredrik Christian. Viridarium Norvegicum I. Norway. 1886. pp. 224-226.

[After identifying *Amanita muscaria* by the various vernacular names current in the chief countries of western Europe, our author, the Norwegian botanist, continues: – RGW]

In old Norwegian historical writings there are many references to the fact that in olden times there was a particular type of warrior known as Berserkers, *i.e.*, men who at times were seized by a wild fury which temporarily doubled their strength and made them oblivious to bodily pain, but at the same time numbed all humanity and reason and made them resemble wild beasts. This rage, which was called Berserksgang, occurred not only in the heat of battle but also in the course of strenuous work, so that those who were possessed by it accomplished things that otherwise would appear beyond human strength.[1] This condition is said to have begun with trembling, chattering of the teeth and a sensation of cold, after which the face swelled up and changed colour. Along with this went a fierce passion, growing into a positive fury, during which they howled like wild beasts, bit the edges of their shields, and hewed down everything that came in their way without distinction of friend and foe. When the attack had passed it was followed by an intense apathy and lack of energy, which might last for a day or more.[2]

Even while I was a medical student, and later as a practising doctor, I had a special predilection for toxicology, and particularly the poisons that come from the vegetable world. I have a lively recollection that all that long time ago by comparing all the symptoms that appeared during the so-called Berserksgang, I came to the conclusion that this paroxysm could scarcely be anything else than a kind of intoxication, the symptoms of which closely resembled the effects of *Amanita muscaria.* When one compares with this the descriptions by Krascheninnikov, Steller, and Erman of the symptom seen in the state of intoxication reached by the Kamchadals and other peoples of north-east Asia after partaking of fly amanita, then the correspondence is so unmistakable that my previous hypothesis has all the appearance of fact. Some time after I had reached this stage, I discovered by serendipity that the Swedish Professor Samuel Ödman expressed the same opinion a hundred

1. Eyrbyggja Saga 28. Landnamabók III. 20. P. A. Munch. The History of the Norwegian People. Christiania. 1852-63. Part I, Vol. 2, pp. 172-3.
2. R. Keyser. Collected Works. Christiania, 1868, p. 355 ff.

years ago, though without offering any special evidence for the rightness of the theory.[1]

The reports given by the above-mentioned Russian authors may be summarised thus:[2] since the Kamchadals came into closer contact with the Russians they have begun to drink brandy, and to leave the use of fly amanita to their nomadic neighbours the Koryaks, for whom they now collect the fungi, which can profitably be exchanged for reindeer. – The first symptoms to appear in one who has drunk a 'liqueur' made from fly amanita and the juice of *Epilobium angustifolium* are: a characteristic trembling in every limb; half an hour later he begins to rage, and according to his temperament he then becomes either merry or morose. – The drunkenness has, in fact, a certain resemblance to that caused by wine or brandy, in that the drunken person, being more or less unconscious, almost invariably grows jolly and only rarely morose. The face becomes red and swollen, as if from extravasated blood, and the drunken men now begin to say and do many involuntary things. – Those who are only mildly affected feel themselves very light of foot and appear inclined to all kinds of bodily activity. The slightest irritation has an unduly strong effect on nerves rendered abnormally sensitive. Some display a muscular strength which in normal conditions would be impossible. Eye-witnesses have reported that a person who ordinarily would have great difficulty in even lifting such a load, has carried a sack of flour weighing 120 pounds a distance of 15 versts (10 miles). – 'The Kamchadals also confirmed what I had already been told in Tigilisk (Ust Tigil) about the intoxicating property of fly amanita, and assured me further that it is not eaten in Sedanka, but only collected for the Koryaks, who in winter will often give a reindeer for a single specimen.'[3] – 'The fly amanita (Muchamor) is very rare in the northern part of Kamchatka, and the Koryaks first learned its properties from the fact that the flesh of a reindeer that had eaten this fungus had just as intoxicating an effect as the plant itself. From the experience thus gained, instinct gradually taught them to enjoy the delights of this precious ware in the most economical way, by keeping the urine of a person intoxicated with fly amanita and later using it as a very effective intoxicant.' – 'As an example

1. Kongl. Vetenskaps Academines nya Handlingar (Royal Scientific Academy's Recent Proceedings). Vol. 5, 1874, p. 240 ff.

2. Cf. Emile Boudier. *Die Pilze in œkonomischer, chemischer und toxilogischer Hinsicht.* Uebertragen und mit Anmerkungen versehen von (The economic, chemical and toxicological aspects of fungi. Translated and annotated by) Dr. Th. Husemann. Berlin 1867, p. 118 ff.

3. George Kennan says that in spite of the fact that the Russian Government has forbidden the sale of fly amanita to the Koryaks, this trade still flourishes. He has, in fact, himself witnessed that a single specimen has been exchanged for furs to a value of 20 dollars. (George Kennan. *Tent Life in Siberia*. London, 1871, p. 139).

of the striking increase in muscular power, a man related that when he had swallowed a portion of fly amanita in the morning he could work without difficulty during the hay-making season from morning to night, and accomplish as much as would otherwise require three men.'

Our celebrated historian P. A. Munch is of the opinion that the Berserksgang was nothing but 'a periodically returning madness'.[1] I cannot share his view, but insist that the berserksgang was induced by some stimulant or other, and specifically by the fly amanita. The probability that this is so is greatly strengthened by the fact that the symptoms described above, both of Berserksgang in Norway and of the effects of fly amanita in Kamchatka, are presented with an accuracy as congruent as if they were written by a keen-witted doctor familiar with all kinds of sickness. Moreover the most peculiar symptoms that accompanied Berserksgang were always repeated in the same form, and the whole thing ceased some time after the introduction of Christianity (about A. D. 1000) had brought a purer ethical standard. It is true that Berserksgang was regarded, both by those who suffered from it and by onlookers, at times at least, as so great an infliction that even the Icelander Thorstein Ingemundsøn made a vow to the Gods in order to free his brother Thorer from this misfortune (Utími).[2] This may, however, also be explained by saying that Thorer in his better moments may have recognised his deplorable condition, but that he found it just as difficult to break free from it as those who now drug themselves with opium or hashish, or even those who have succumbed to an excessive consumption of spirits.

Here may be mentioned also what is related (in the Droplaugar sona-saga p. 3) of the Icelander Thrymketil;[3] but according to what Professor Konrad Maurer of Munich has told me personally, this account is nothing but an addition which an Icelander in the latter half of the last century wove into the Saga.

And if, as appears to me to be established beyond doubt, use was made of some stimulant or other, this cannot have been either beer, mead, wine or brandy, as the effect of all these is very different from what is described here. Moreover, brandy was not known in Norway before 1531, when it was called 'Aqua vitæ'.[4] And, of course, there can be no question of opium or hashish in this country so long ago. A motive for the constant use of fly

1. P. A. Munch. *The History of the Norwegian People.* Part 1, Vol. 1, p. 790.

2. Vatnsdöla Saga. Ch. 30. 37. – P. A. Munch, *The History of the Norwegian People.* Part 1, Vol. 1, pp. 790-1.

3. R. Keyser. *Collected Works.* Christiania, 1868, p. 356.

4. Collected Examples for the Language and History of the Norwegian People, 2, pp. 48-49. The English name 'whisky' is a corruption of the Celtic word 'usquebaugh', *i.e.*, 'water of life'. (F. W. Pavy. A Treatise on Food and Dietetics. 2nd Ed. London, 1875, p. 240)

amanita as an intoxicant can also be found in the fact that it produces the same phantasms and visions as hashish or opium.[1]

Although the old historical records do not say so, it seems reasonable to suppose that the means that was used to induce the state in question must have been kept a secret. The Berserkers were feared by all, and could, in a sense, enforce whatever they wished; it is therefore in the nature of things that they did all they could to retain this extraordinary respect from the common people. So the knowledge of the stimulant was probably passed on as a secret from man to man. Even in our own time, even in our own capital, it is well known that there are certain 'secret' recipes that are bequeathed from parents to children.

The less enlightened part of the population naturally saw the Berserksgang as a supernatural phenomenon, to be ranked with enchantment and suchlike wonders; it may even have been believed that the Berserkers were possessed by demons; but this cannot have been the case with well-informed people. That even as early as the beginning of the 11th century Berserksgang was regarded as a condition for which the sufferers were themselves responsible is clearly indicated by the following: before Erik Jarl left Norway, he called together in 1015 the noblemen and the most powerful peasants to consult them about the laws and government of the kingdom. At this meeting single combat was abolished, and Berserkers and robbers were outlawed.[2] In Thorlak's and Ketil's Icelandic Church Law[3] which was adopted in 1123 as the law of Iceland, there appears the following passage: 'If any man goes Berserk, he is to be punished with three years' banishment (Fjörbaugsgar), and the same applies to those men who are present if they do not bind him, but if they bind him no one shall be punished. Each repetition of the offence will he be punished.' Any comment on this would appear to be superfluous.

[45]

MÖRNER, Carl Th. Några erfarenhetsrön om de högre svamparna: Kritisk öfversikt. (Some observations on the higher fungi) Published in Upsala Läkareförenings Förhandlingar, Vol. XXIV, 1-2. Upsala. 1919.

[On November 1, 1918, Carl Mörner read a paper on the higher fungi before the Royal Scientific Society in Upsala. Mörner was a well known professor of physiology and an excellent amateur mycologist. In the discussion

1. Emile Boudier *l.c.*, p. 124 (Husemann's annotation).
2. Grettis Saga. Ch. 19.
3. Grágás, Finsen's Edition. Ch. 7, pp. 22-23.

that followed the reading of his paper Professor H. Hildebrandsson, who held a chair in meteorology, told of an episode that had happened in 1814 when Sweden and Norway were engaged in a short war with each other. When Mörner published his book on mushrooms in 1919 he inserted a paragraph giving the substance of what Hildebrandsson had said. This is in the second of the two paragraphs that we quote from Mörner's book: – RGW]

Described in popular terms, a mild or medium-severe attack of poisoning resembles a state of alcoholic intoxication in a person who cannot hold his liquor. 'The Kamchadales,'[1] to quote E. Fries, 'make an intoxicating drink [from the fly-agaric], and it has been suggested that the fly-agaric was used in our country in olden times to bring on the so-called berserk furies'. This last-mentioned piece of information is obviously based on S. Ödman's 'Attempt to Explain the Berserk Frenzies of the Old Norse Warriors on the Basis of Natural History' (1784), an interesting work of which a brief outline may be given at this point. After dismissing opium, belladonna, hemp and the like as out of the question for the North, particularly at that time, Ödman expresses the following opinion: 'Of all Swedish plants, however, I consider the Fly-agaric, *Agaricus muscarius*, to be the one which really solves the mystery of the Berserks'. Explaining his reasons for this belief, he gives the following classic description of the notorious amanita intoxication: 'Its use is so wide-spread in Northern Asia that there are hardly any nomadic tribes there that do not use it in order to deprive themselves of their feelings and senses that they may enjoy the animal pleasure of escaping the salutary bonds of reason.... The dose of this poisonous substance is from 1 to 4 mushrooms, according to size.... Those who use this mushroom first become merry, so that they sing, shout, *etc.*, then it attacks the functions of the brain and they have the sensation of becoming very big and strong; the frenzy increases and is accompanied by unusual energy and convulsive movements. *The sober persons in their company often have to watch them to see that they do no violence to themselves or others.* The raving lasts 12 hours, more or less. Then lassitude sets in, culminating in complete exhaustion and sleep'.

According to the observations of an officer of the Värmland regiment, fly-agaric was also used at a much later date in order to induce a good fighting spirit. In the course of an advance during the 1814 war, he noticed that some of the men were seized with frenzy and foaming at the mouth. This was traced, on investigation, to the cause just mentioned.[2]

1. Other authors also mention the Koryaks, Ostiaks, Samoyedes, and Chukchi as using fly-agaric for purposes of intoxication.

2. This was reported by Professor H. Hildebrandsson in connexion with a lecture given by the author to the Royal Scientific Society on 1 Nov. 1918.

[It is unfortunate that Mörner did not get from Hildebrandsson the source of his account; it would have been easy at the time to do this. Recent efforts by Bo Holmstedt, Professor of Toxicology at the Karolinska Institutet, Stockholm, to find confirmation for the episode have been unavailing: he can discover no local tradition in Värmland relating to it, and he is an extraordinarily resourceful man. Hildebrandsson's reputation as a meteorologist was high; his scholarship above reproach. But in this case he was not in his own field and he was volunteering a remark in a discussion following a lecture. Should not a careful person ask himself how much weight to give to such an utterance, even from a man of Hildebrandsson's standing? – RGW]

[46]

NORDHAGEN, Rolf. Fluesopp of Berserkergang. (Amanita and Berserk Frenzies) Published in the Norwegian newspaper *Aftenposten*, Jan. 11, 1930.

[Professor Nordhagen held the chair in botany at Oslo University. – RGW]

In an article, Dr. F. Grøn has launched an attack on the late Professor Schübeler's theory that Amanita were the cause of berserk frenzies. Since my own name is also mentioned in the article, on account of a popular piece I wrote for *A-Magasinet*, in which I took a sympathetic view of the theory, I should now like to be allowed to make a few remarks on the subject. For Doctor Grøn's article suffers from one grave defect – several very important elements are passed over in silence, although they are of crucial significance for anyone wishing to adopt a stand on the theory.

It would never have occurred to me to resurrect Professor Schübeler's theory had I not come across a thoroughly modern work on mushroom-poisoning which strongly supports the theory. I am thinking of the book by the eminent Swedish scholar, Professor Carl T. Mörner, entitled: 'On the Higher Mushrooms' (Uppsala, 1919). Since Mörner is a physician and professor of physiological chemistry, considerable importance must be attached to his statements.

In the chapter on Amanita, Mörner deals first with the two varieties which can be considered relevant in this context, that is, *Amanita muscaria* and *Amanita pantherina*. He quotes at length the Swedish professor S. Ödman, who in 1784, that is, long before Schübeler's time, published a famous dissertation entitled 'Attempt to Explain the Berserk Frenzies of the Old Scandinavian Warriors, on the Basis of Natural History', in which he maintains that the berserks know how to use Amanita in order to put themselves in a suitably martial mood. In this connection Mörner produces some extraordinarily interesting information which militates strongly in favour of the theory.

In the first place he mentions a detail which Professor H. Hildebrandsson reported in 1918 after a lecture given by Mörner at the Royal Scientific Society in Stockholm. During the advance of the Värmland regiment in 1814, a Swedish officer noticed that some of the men were 'raving' and foaming at the mouth. *An investigation revealed that these soldiers had eaten Amanita in order to be in proper fighting fettle!*

This piece of information strikes me as being extremely important and I am surprised that Dr. Grøn does not say a single word about either Mörner's book or this particular episode. The soldiers of the Värmland regiment cannot possibly have hit on the notion of eating Amanita by themselves. There must undoubtedly be a popular tradition behind the whole story which may very well have survived from Viking times. In his important articles on Norwegian folk medicine, Reichborn-Kjennerud has demonstrated the antiquity of many of the household remedies which are sometimes applied to both humans and animals even today in our rural areas.

Secondly, Mörner reports a very important case of mushroom-poisoning (no doubt from *Amanita pantherina*) which occurred at Malmö in 1908. A German workman had consumed a plateful of poisonous fungi. After a bare one and a half hours he became light-headed and had to be taken to hospital. Violent fits of mania soon ensued; the patient sang, shouted, laughed and ground his teeth as he jumped about on the bed and hit out with his arms. On account of this he had to be transferred to the psychiatric department and, in the course of this, he put up a furious resistance, manhandling the attendants. The spectacle continued until 4 a. m., when he fell asleep, and the next morning he was 'perfectly fit' again.

Speaking of this case, Mörner says that it may be described as a 'complete berserk frenzy'.

It would take far too long to discuss all the theories put forward on the subject during the past hundred years. But as Mörner is a specialist in toxicology (in his book he mentions no less than 170 different works that are basically toxicological in content!), it will not do to contend, as Dr. Grøn does, that modern toxicological research knocks the bottom out of the Ödman-Schübeler theory.

I also think that he passes rather lightly over all the accounts which exist of the use by primitive Siberian peoples of Amanita as an intoxicant. We cannot escape the fact that the old explorers' descriptions of the course taken by the 'intoxication' tallies very closely with the berserk symptoms.

That the intoxication very often takes a far quieter form is established beyond all doubt. But this by no means disproves the theory. Here as in the case of other poisons it depends on the size of the dose. It may be noted that Amanita, like tobacco, are highly prized by Northern Siberian tribes, and that they therefore undoubtedly economize with the precious substance. My colleague Professor H. U. Sverdrup has told me a hair-raising tale about the exorbitant price demanded for a single specimen of Amanita in the region of the Bering Strait. But that is another story, as Kipling says. Experience has obviously taught these people to arrive at the most appropriate dose. Besides, in the last few generations Amanita has steadily been supplanted by hard liquor all over Siberia, so conditions are now not nearly so favourable as they were for the study of Amanita intoxication.

Bergen, January 1930.

[47]

KUYLENSTIERNA-ANDRASSY, Stella. Pustan Brinner. Stockholm. 1948. pp. 92-93.

[The author is a Swedish lady married to an Hungarian nobleman, Count Imre Andrassy. The book is autobiographical. The title means, 'Burning Plain'. The chapter from which we quote is entitled, 'Svarta mässan i Fehérvár', or The Black Mass in Fehérvár. It tells of an experience that the author went through at the end of the second World War, in 1945, when the Russian forces occupied the town of Fehérvár momentarily. Her account follows: – RGW]

All private cars are to be instantly present at Apponyitér, the radio announced one day in November.

Imre [her husband] came and collected me in our car and then we drove to the market place where the cars were to assemble. A doctor with dressings was placed in our car. The marching column was formed and the caravan, numbering about 150 cars, snaked its way out of town.

As usual, we drove with secret orders. Only the chief of the column knew where we were headed. We drove southwards. It had been said in the news that the town of Fehérvár had been liberated from the Russians the same day. We guessed that we were on our way there to bring help to the liberated people.

Our guesses came true. The goal was the old Hungarian coronation town, a beautiful country town famous for its cathedral in baroque style. We arrived there in the afternoon. The town was small where before the war there lived only 40,000 people. But the town had great strategic value. Through it runs the auto route down to the Balkans and the main road to Vienna. Through it passed the big oil line from Lispe. The Germans had determined to hold it at any price, but one day when the Panzers became stuck in the bottomless clay the Russians had taken it by surprise. Now it had been recaptured by a counterattack. A Hungarian officer whom we met at the market place in front of the cathedral related to us how this had happened.

'The Russians came stealthily in the November mist immediately before dawn. They came through the maize fields which rustled where the harvest had not been brought in because of lack of people. They succeeded in passing over the Danube without being seen at the big marshes near Baja, where one thought it would be impossible for human beings to advance. The spearhead consisted of 'convict battalions', recruited from among the worst criminals and loose people in the Red Army. In their nostrils they had pieces of cotton soaked with a curious mushroom poison that transforms human beings into beasts. Some of them were captured by us. When the effects of the poison started to wear off, they shouted and cried as though obsessed. They raged and bit like dogs mad with rabies and thereafter fell to the ground in a deep sleep.'

[Under the tensions that must have existed at that horrible moment in Fehérvár, it was not to be expected that the details needed to satisfy scientists in peace-time would be forthcoming. Cotton wads soaked with 'mushroom poison' were in the Russian nostrils. How did they breathe? The inquirer naturally would like to know what kind of mushroom it was. There is no hint, in the Russian writings about war through the ages, of such a practice. The minor nationalities of Siberia, where the fly-agaric is used, certainly did not make up the contingent of Russian troops, since they are far too few in number and the Russian authorities frown on their consumption of the fly-agaric. The reports concerning the Altaïc peoples of Siberia indicate that mushrooms (except for *Fomes fomentarius*, used as tinder for making fire) play almost no rôle in their lives. – RGW]

CITATIONS FROM THE ṚGVEDA

CITATIONS FROM THE ṚGVEDA

The following verses from the ṚgVeda are cited in our text. Where the page numbers are printed in italic, the text of the verse is quoted in whole or in part.

INDEX

INDEX

Numbers in heavy brackets [ ] refer to the Exhibits.